# Advances in Neurobiology
Volume 38

**Series Editor**

Alexej Verkhratsky, University of Manchester
Manchester, UK

*Advances in Neurobiology* covers basic research in neurobiology and neurochemistry. It provides in-depth, book-length treatment of some of the most important topics in neuroscience including molecular and pharmacological aspects. The main audiences of the series are basic science researchers and graduate students as well as clinicians including neuroscientists (neurobiologists and neurochemists) and neurologists. *Advances in Neurobiology* is indexed in PubMed, Google Scholar, and the Thompson Reuters Book Citation Index.

Johannes Gräff • Steve Ramirez
Editors

# Engrams

## A Window into the Memory Trace

Springer

*Editors*
Johannes Gräff
Brain Mind Institute
School of Life Sciences, EPFL
Lausanne, Switzerland

Steve Ramirez
Department of Psychological & Brain Sciences
Boston University
Boston, MA, USA

ISSN 2190-5215 ISSN 2190-5223 (electronic)
Advances in Neurobiology
ISBN 978-3-031-62982-2 ISBN 978-3-031-62983-9 (eBook)
https://doi.org/10.1007/978-3-031-62983-9



This Springer imprint is published by the registered company Springer Nature Switzerland AG
The registered company address is: Gewerbestrasse 11, 6330 Cham, Switzerland

# Preface

More than a hundred years ago, the German Biologist Richard Semon (1859–1918) coined the term "engram" to delineate the neural foundation accountable for storing and recollecting memories, essentially evoking for the first time the possibility that there might be a physical substrate of a memory trace. Although multiple renowned neuroscientists and psychologists alike kept looking for the engram, its concept remained largely dormant until the late 2000s. In the last one and a half decade, however, the advent of numerous potent tools for investigating and influencing memory function at the cellular, neuronal circuitry and molecular level has ignited a resurge in enthusiasm for exploring the engram. As such, we believe the time is now ripe for the first textbook on this topic, which covers the concept of engram cells from various angles, a concept that is surely likely to evolve itself in the years to come.

In the first, introductory chapter, the book's editors Johannes Gräff and Steve Ramirez engage in a conversation about the past, present and future of memory and engram research, which includes a personal account of Steve's important contributions to the groundbreaking re-discovery of engram cells with his late friend Xu Liu in the early 2010s.

In the second part of the book, we learn how engram research has evolved from a single brain area of investigation to multiple ones (Chapters by Ramirez and Roy), and what new insights have been gained from engrams in two of the most heavily studied brain areas in memory research—the hippocampus and the prefrontal cortex (Chapters by Maren and DeNardo).

In part three of the book, we then dive into the synaptic and molecular underpinnings of engram functioning (Chapters by Kaang and Barco), and learn that engrams can be studied with different immediate early genes (IEGs), each with their own advantages and disadvantages (Chapter by Lin).

In the fourth part, we then move beyond the traditional studies of engram cells in the context of associative learning. We find that engram cells can also be deployed to better understand memory attenuation (Chapter by Gräff), and what role they play in memory generalization (Chapter by Hen), spatial memory (Chapter by Tanaka) as well as observational fear memory (Chapter by Kitamura).

Lastly, we reach beyond the restricted realm of the neurobiology of engram cells to a computational framework thereof (Chapter by Gerstner), and end with two chapters putting contemporary engram research into both a sociological and philosophical perspective (Chapters by Chiapperino and Ryan).

We hope that this book will stimulate, reactivate or even create new engrams in your, the reader's, brain, and warmly thank all the contributors for their timely and important contributions.

Brain Mind Institute
School of Life Sciences, EPFL
Lausanne, Switzerland
Johannes Gräff

Department of Psychological & Brain Sciences
Boston University
Boston, MA, USA
Steve Ramirez

# Acknowledgements

We would like to thank synapsstudios.ch for the matching illustrations in this book, as well as Stéphanie Debayle for her help in compiling all the chapters.

# Contents

# Contributors

**Lorena Andreoli** Okinawa Institute of Science and Technology Graduate University (OIST), Memory Research Unit, Okinawa, Japan

**Ángel Barco** Instituto de Neurociencias, Universidad Miguel Hernández – Consejo Superior de Investigaciones Científicas, Alicante, Spain

**Wei-li Chang** Departments of Psychiatry and Neuroscience, Columbia University, New York, NY, USA

Division of Systems Neuroscience, New York State Psychiatric Institute, New York, NY, USA

**Luca Chiapperino** STS Lab, Institute of Social Sciences, Faculty of Social and Political Sciences, University of Lausanne, Lausanne, Switzerland

**Laura DeNardo** Department of Physiology, University of California Los Angeles, Los Angeles, CA, USA

**Kaitlyn E. Dorst** Department of Psychological and Brain Sciences, Boston University, Boston, MA, USA

Department of Neuroscience, Tufts University School of Medicine, Boston, MA, USA

Graduate Program for Neuroscience, Boston University, Boston, MA, USA

**Joslyn Doupe** Neuroscience Graduate Program, SUNY Upstate Medical University, Syracuse, NY, USA

**Miguel Fuentes-Ramos** Instituto de Neurociencias, Universidad Miguel Hernández – Consejo Superior de Investigaciones Científicas, Alicante, Spain

**Chiara Gastaldi** Brain Mind Institute – School of Computer and Communication Sciences – School of Life Sciences, EPFL, Lausanne, Switzerland

**Wulfram Gerstner** Brain Mind Institute – School of Computer and Communication Sciences – School of Life Sciences, EPFL, Lausanne, Switzerland

**Johannes Gräff** Brain Mind Institute, School of Life Sciences, EPFL, Lausanne, Switzerland

**Daehee Han** Center for Cognition and Sociality, Life Science Institute, Institute for Basic Science (IBS), Daejeon, South Korea

Interdisciplinary Program in Neuroscience, College of Natural Sciences, Seoul National University, Seoul, South Korea

**James E. Hassell Jr.** Department of Psychological and Brain Sciences and Institute for Neuroscience, Texas A&M University, College Station, TX, USA

**Rene Hen** Departments of Psychiatry and Neuroscience, Columbia University, New York, NY, USA

Division of Systems Neuroscience, New York State Psychiatric Institute, New York, NY, USA

**Hyunsu Jung** Center for Cognition and Sociality, Life Science Institute, Institute for Basic Science (IBS), Daejeon, South Korea

Interdisciplinary Program in Neuroscience, College of Natural Sciences, Seoul National University, Seoul, South Korea

**Bong-Kiun Kaang** Center for Cognition and Sociality, Life Science Institute, Institute for Basic Science (IBS), Daejeon, South Korea

Interdisciplinary Program in Neuroscience, College of Natural Sciences, Seoul National University, Seoul, South Korea

Department of Biological Sciences, College of Natural Sciences, Seoul National University, Seoul, South Korea

**Takashi Kitamura** Department of Psychiatry, University of Texas Southwestern Medical Center, Dallas, TX, USA

Department of Neuroscience, University of Texas Southwestern Medical Center, Dallas, TX, USA

Peter O'Donnell Jr. Brain Institute, University of Texas Southwestern Medical Center, Dallas, TX, USA

**Chaery Lee** Center for Cognition and Sociality, Life Science Institute, Institute for Basic Science (IBS), Daejeon, South Korea

Department of Biological Sciences, College of Natural Sciences, Seoul National University, Seoul, South Korea

**Yingxi Lin** Department of Psychiatry, O'Donnell Brain Institute, University of Texas Southwestern Medical Center, Dallas, TX, USA

**Stephen Maren** Department of Psychological and Brain Sciences and Institute for Neuroscience, Texas A&M University, College Station, TX, USA

**Meizhen Meng** Department of Psychiatry, O'Donnell Brain Institute, University of Texas Southwestern Medical Center, Dallas, TX, USA

Neuroscience Graduate Program, University of Texas Southwestern Medical Center, Dallas, TX, USA

**Fionn M. O'Sullivan** School of Biochemistry and Immunology, Trinity College Dublin, Dublin, Ireland

Trinity College Institute of Neuroscience, Trinity College Dublin, Dublin, Ireland

**Francesco Panese** STS Lab, Institute of Social Sciences, Faculty of Social and Political Sciences, University of Lausanne, Lausanne, Switzerland

**Kritika Ramesh** Department of Psychiatry, University of Texas Southwestern Medical Center, Dallas, TX, USA

**Steve Ramirez** Department of Psychological & Brain Sciences, Boston University, Boston, MA, USA

**Dheeraj S. Roy** Department of Physiology and Biophysics, Jacobs School of Medicine and Biomedical Sciences, State University of New York at Buffalo, Buffalo, NY, USA

**Tomás J. Ryan** School of Biochemistry and Immunology, Trinity College Dublin, Dublin, Ireland

Trinity College Institute of Neuroscience, Trinity College Dublin, Dublin, Ireland

Florey Institute of Neuroscience and Mental Health, Melbourne Brain Centre, University of Melbourne, Melbourne, VIC, Australia

Child & Brain Development Program, Canadian Institute for Advanced Research (CIFAR), Toronto, ON, Canada

**Kazumasa Z. Tanaka** Okinawa Institute of Science and Technology Graduate University (OIST), Memory Research Unit, Okinawa, Japan

**Joseph I. Terranova** Department of Anatomy, Midwestern University, Downers Grove, IL, USA

**Krithika Vasudevan** Department of Psychological and Brain Sciences and Institute for Neuroscience, Texas A&M University, College Station, TX, USA

**Tushar D. Yelhekar** Department of Psychiatry, O'Donnell Brain Institute, University of Texas Southwestern Medical Center, Dallas, TX, USA

**Zachary Zeidler** Department of Physiology, University of California Los Angeles, Los Angeles, CA, USA

**Ying Zhang** Center for Life Sciences & IDG/McGovern Institute for Brain Research, School of Life Sciences, Tsinghua University, Haidian District, Beijing, China

# About the Editors

**Johannes Gräff** is Associate Professor and Director of the Laboratory of Neuroepigenetics at the Brain Mind Institute of the Ecole Polytechnique Federale de Lausanne (EPFL) in Lausanne, Switzerland. Trained as a biologist at the University of Lausanne, Johannes Gräff obtained his PhD in Neuroscience at ETH Zurich before pursuing his postdoctoral studies at the Picower Institute of Learning and Memory at the Massachusetts Institute of Technology in Boston, USA. His laboratory studies the molecular underpinnings of memory formation, storage and change, with a particular emphasis on epigenetic mechanisms.

**Steve Ramirez** is Associate Professor of Psychological and Brain Sciences at Boston University and a former Junior Fellow of Harvard University. He received his BA in Neuroscience from Boston University and went on to receive his PhD in Neuroscience at MIT. His lab focuses on imaging and manipulating memories throughout the mammalian brain, with a particular emphasis on artificially modulating memories to alleviate symptoms associated with pathologies of the brain.

# Part I
# Introduction

# Engrams to Remember: A Conversation Between Johannes Gräff and Steve Ramirez

**Johannes Gräff and Steve Ramirez**

**Abstract** Johannes Gräff (JG): Steve, in preparation for this conversation, I pulled out the book "In search of memory" by Eric Kandel from my bookshelf. Obviously one big question is, given that this book was written more than 20 years ago: Are we there yet? Have we found memory?

**Keywords** Engrams · Heisenberg · Memory trace · IEG · Cellular

Steve Ramirez (SR): Oh, I love it! I don't think we found it, but I think that it's not completely invisible anymore. I think that we've found the tip of the iceberg. In particular, in the last 15 years of engram research we've gotten some pretty clear snapshots of what memory looks like when passing through the brain. But we certainly haven't gotten the full picture of it, let alone of what the physiological signatures of memory looks like in the brain. This is exciting because it gives us so many more unknowns to discover!

**JG: That tip of the iceberg is obviously growing with more and more people becoming interested and convinced about engram cells having at least something to do with memory. So, now, 20 years after Kandel's book and a good decade after the discovery of engram cell, how would you define what a memory is?**

SR: The only thing that we can really confidently say is that when experience enters the brain, it modifies the brain—and, somehow, those modifications persist to let recall happen. In other words, there is an enduring change in the brain as a result of experience. And I really like that definition because the term "enduring change"

J. Gräff (✉)
Brain Mind Institute, School of Life Sciences, EPFL, Lausanne, Switzerland
e-mail: johannes.graeff@epfl.ch

S. Ramirez (✉)
Department of Psychological & Brain Sciences, Boston University, Boston, MA, USA
e-mail: dvsteve@bu.edu



J. Gräff, S. Ramirez (eds.), *Engrams*, Advances in Neurobiology 38,
https://doi.org/10.1007/978-3-031-62983-9_1

doesn't restrict it to only the genetics, the synapses, the ensemble or the circuit level, or even to a brain region, rather that this change—as undefined as it may be—happens somehow somewhere enabling it to be revisited through recollection. It doesn't necessarily zoom in or out to any level of analysis because it most likely involves all those levels of analysis.

**And all these levels of analysis are very much needed to bring engram research, or memory research, more generally speaking, to the next level.**

Exactly. I also think the way that we define an engram can often be constrained or dictated by the tools that we use to study it. For instance, if, in the 90s, the field had discovered Npas4 before cFos as the first immediate early gene related to memory, then all the engram studies might have started from an Npas4-centric perspective. However, it happened to be cFos that was initially discovered that we then really dove into. Had it been some calcium sensor or another immediate early gene, or if our current miniscopes had been developed in the 90s, then our definition of an engram would have evolved in a manner that was constrained by the technique that was used to discover it. This is why I think that nowadays, all of these techniques are important and can give us a different insight into the levels of understanding of what an engram actually is, but this is why I also try to be unrestrained by the techniques that we're using to analyze the engram. If anything, understanding that every technique has its own restrictions is already an important step forward, because it entails that by only using one technique we cannot capture the full picture of what a memory is.

**So, if I hear you right, then it's almost like there are two concepts of the engram. There's an engram in the sense of Richard Semon, the engraving of an experience, or the approximation of the memory trace. And then there is what we usually refer to, namely an engram tool, such as optogenetics. And it goes without saying that we are only as good as the tools we have. And oftentimes we get asked, for one particular engram study, is this the whole memory? To which the answer has to be an obvious no. Yet, the engram as a tool provides an unprecedented resolution by which memories can be queried, which then begs the question: Do you think that all our memories are stored as an engram in Semon's sense, or otherwise asked, is there such a thing as one engram per given memory?**

This is a good question. Although it's not the most scientific approach, I sometimes like being my own test subject and force myself to recall something and then ask myself, what is it like when I recall that memory? And I find that it's really hard to recall only one memory without thinking of others associated with it. For example, if I think of a memory the last time I saw my parents, it's immediately flooded with other times when I saw my parents and we were doing different things and they're wearing different things in each one of them and we're saying different things. Therefore, it's almost like we can't actually fully recall the exact same memory twice, so to speak. Or, in other words, like the saying that you can't step in the same river twice. So, in that sense, it's hard for me to believe that we have one specific engram per memory, but rather that these engrams might actually be way more interlinked and flexible and constantly changing.

**It almost seems to me that there is a Heisenberg-like uncertainty principle in memory cells. We cannot ask a memory without actually altering it because**

**inevitably recalling the memory renders it labile and prone to change. Your engram, as you recall it, might be influenced by your current environment, through your other memories, your mood and so on.**

**Coming back to the dichotomy of engram cells, which, on the one hand, are the physical approximation of the memory trace, and, on the other hand, our tool to approximate it. Currently, the field is heavily based on cFos as a marker and that has been working great. There are also the Arc and Npas4 camps that need to be mentioned. Yet, all of them capture essentially the activity of excitatory cells. Do you think that other cell types in the brain would also have such a thing as an engram?**

I would assume that if we were to have some kind of tool by which we could look at all cells indiscriminately and simultaneously, for example, in a fear conditioning task in a rodent's brain, we would probably see activity in virtually every cell type. There would certainly be astrocytes and glia and microglia, as well as oxytocin, dopamine and serotonin and all sorts of different cell types coordinating their activity, perhaps along different time scales and perhaps doing very different computations that enable different parts of that fear memory to be formed and stored. Similar to an orchestra playing a symphony, the brass section would be doing its part and the percussion sections would do its part. Who knows what the conductor is, but the conductor would be doing their part too. And that is also why I think that we have all these highly interesting research papers coming out recently that look at astrocytes' contributions to fear learning or at an oxytonergic engram or at a dopaminergic engram. We're just starting to really appreciate that all of these cell types are involved in different parts of learning and memory. So, I would be surprised if there was a cell type that was not involved in learning and memory. I think memory is just what the brain does.

**This implies that there must be engrams we don't know yet. And that we haven't invented the tools to actually study them. Related to this, the field of engram research is generally speaking limited to the mouse, because mice are amenable for genetic manipulations, virus-based infections and so on. So, the question arises: Do you think we will ever be able to say that we have a similar engram concept in the human brain?**

I sure hope so! There's no law of physics that requires breaking in order to figure out how memory or how an engram works in the human brain. It's just a really complicated problem. If we flashback a hundred years ago, the idea of having something in your back pocket that could act as a calculator, a computer, calling someone halfway across the world, being able to zoom across the world and enable communication, was pure science fiction. That almost didn't make sense in the world as it existed 100 years ago. So, all we have to do now is a thought experiment 100 years forward and ask what the equivalent of a smartphone would be. If we apply that to neuroscience, our understanding of memory might have completely different concepts 100 years from now, which might sound completely foreign to us today. But I think that's a good thing because similar to telecommunication, the field of neuroscience is tremendously advanced. So, I would say that as long as humanity is around, it's an inevitability that we'll figure out how memory works. It's just a very tough problem, that we are only now starting to dive into with the tools and techniques that we currently have.

**You actually already answered another question that I had and which was, where do you see engram research in the future? I wasn't thinking about a hundred years from now, but more like 10 years from now.**

Take Richard Semon: For him, memory was the irritable substance. But what was the irritable substance? In the meantime we've been able to update this concept insofar as we now know that the irritable substance involves neurons and glia and dopaminergic cells as well as different IEGs and calcium transients. Thus, the irritable substance has now been redefined to brain activity and patterns of brain activity across cell types and time scales. And now, a similar thing might be happening with the concept of engram cells, which might become our modern-day irritable substance that in ten or a hundred years from now will be defined in another direction that gives us a better understanding of what it is that we're working with.

**Coming to the discovery of engram cells. Do you remember a Eureka moment during your time in the Tonegawa lab where you were said to yourself, this discovery could indeed be quite something?**

100%. It's one of the few very vivid episodic memories I have of the past. So, I remember that when I joined the Tonegawa lab, Xu Liu and a couple of other postdocs were working on developing these activity-dependent viruses and there were like 66 of them that they had tried and failed and tried or that were leaky, not dox-sensitive, or simply didn't do the things that we thought that they would do. That was the hard part because they tried different IEG promoters like the SARE promoters, synthetic promoters, different copies of the cFos promoter and so on. And when I joined the lab, I remember thinking there was a bunch of projects that I could have worked on. Xu and I got along really well, and we developed the idea of activating a memory, which was exciting, but we still thought that it was kind of far off. However, there was one virus that seemed promising and we made an outline of, like, let's do a dox on control, a dox off control and see if it's activity dependent or not. One day, Susumu caught us in the hallway talking about the experimental design, and he suggested that instead of being perfectionists, why not just try the real experiment? See if light-activated freezing would actually work. And if it doesn't, you can figure out if it was because of the virus, if it was because of leakiness or whatever, but just try the million-dollar experiment.

At first, Xu and I were a little bit annoyed because we anticipated that we would be just going down another path of failure. But in the end, we reasoned that Susumu does have a point. "Let's just try it." And we didn't really have much to lose. We decided to aim for CA1 because all of the literature was very CA1-focused. We injected five animals to target CA1 with the TRE-dependent optogenetic constructs. For four out of the five animals, we shine the light. Nothing happens, and we thought, "That's fine. It's just that science doesn't work most of the time." But then in the fifth animal, when we turned the light on, the animal does freeze, and we were cautiously excited. In fact, it's the video that I show all the time during my talks. But, of course, this begged the question of why didn't the other four freeze? To find out, we did histology on that animal, and we realized that we had completely messed up the surgery. In fact, we had gone about 200 microns deeper in the brain and hit the dentate gyrus. Whereas in the other four animals, we had successfully targeted CA1, but nothing happened. So, we thought, "Let's try the dentate with the next round of animals." And sure enough, the majority of the animals in that experiment—which we did blinded and properly controlled—did show light-induced freezing, which we were particularly excited about.

But, to be honest, the real aha! moment came as we were trying to wrap up that paper. At that time, Tomas Ryan had joined the lab and started his project on amnesia and engrams. To do so, he first needed to replicate the light-induced freezing in the lab. And he did! Meaning that it wasn't a spurious result on our end or something confounded by, say, the deodorant we were wearing or that we were sneezing in the behavior room or that we were aggressive with handling the animals. In other words, we now had the confirmation that this finding was not just specific to the secret sauce that Xu and I had, and that's when things really took off for us in the lab.

**Thanks for sharing. There are so many things about science in your narrative here. There is trial and error, failure, serendipity. What if you hadn't misinjected that animal? The importance of controls, the importance of independent replications... And ultimately also the encouragement of your PI to just try it. I remember, when you two started to show these data at the institute's retreat, everybody sensed that this was kind of something big. But who would have known back then of how big the field became over the years?**

**You mentioned your collaborator Xu... What is the engram that you keep of him? He sadly passed away some years ago. So, is the engram a positive one, or does the sadness outweigh?**

Oh, man, it's a difficult question. But in a sense, it gives the engram purpose. Probably the first word that comes to mind when thinking about Xu is just genuine gratefulness for not just having learned so much science with him but also because we teamed up so well and we were so comfortable sharing success with each other. Like at the institute's retreat that you mentioned, that was one of the first times that we decided to do our presentation together because we were co-first authors and we could champion the idea of a team-oriented spirit in science. This was a lifelong value we both shared.

And the sadness has, in the meantime, grown into more of an appreciation for the fragility of engrams and the fragility of life, that that's a brutal reality that we have to wrestle with. So, the purpose that it's given for me, is that it's the only way to outlive our own skin. It's through the engrams that we impart onto others. And I appreciate that fact so much more now after his passing, even though, it is still very sad that that person cannot be conscious there to experience those memories. I think that I went through the roller coaster of excitement and then grief and loss and then growing past the grief. And I think that you never really grow beyond the grief. You just grow your life around it. And then I think it makes you appreciate the enduring change that person left in your brain; it really lets me appreciate what that change is and what it means because it lets you really feel the power of an engram from within.

**Related to this, Takashi Kitamura is working on social engrams and how an episode witnessed by an observer can instill an engram onto their brain. So, your own recollection really speaks to that. And I think what you shared with us also speaks to another point, namely, as with all people who are dear to our hearts who pass away first there's grief and sadness and then over time the engram that was negative for a while can become a positive one again. Put into perspective, this also speaks for the malleability of the engram. So, we owe a lot to Xu and to you for the engram field. And it's fair to say, in the engram field nowadays you are considered a hero of the engram. May I ask you, who are your heroes?**

I think there are both historical and contemporary heroes, and then a few that meet both aspects. Brenda Milner and Sue Corkin, for instance. We owe so much to what we know about the hippocampus in memory to Brenda Milner and Sue Corkin and everyone else implicated in the original HM studies. How bold and risky it must have been to start with a patient and cataloging the different kinds of amnesia that may have resulted from the damage of the brain surgery to some random brain area that we didn't know much about? And then the whole hippocampus learning and memory field really took off. In that sense, those two are both historical and contemporary heroes, in my opinion. On the historical side, there is definitely also Richard Semon and Endel Tulving, about whom I really appreciate how much foresight they had to say that memory and cognition are something that the brain does. Because, in essence, this was not really obvious before. In the days of the ancient Greeks, "feeling" was something that the heart did, with black bile and other bodily fluids holding our different temperaments and emotions. Then, more contemporary, there is the work in the mid-2000s that Sheena Josselyn did that put the "let's look at memories" topic on the map, in terms of capturing and localizing memories to specific sets of cells. When her work coincided with Mark Mayford's work, multiple strikes of lightning seem to have been captured in the same bottle around the same time, with both of them trying to identify a stable correlate of memory. And then there was also Sheena's work of trying to find or bias a correlative memory and subsequently kill those cells to erase a memory. It just made the idea of targeting memories at the cellular level in the brain a real thing. Something that the field could now play with. In that sense, Sheena's conceptual and Mark's transgenic technological advances were both instrumental, without which the field wouldn't be where it is today. In addition, the field of engrams owes a lot to some of John Gazowski's work with catFISH and the notion that the same set of cells will be activated again by the same memory. The mere idea that cells may be reactivated again when recollecting an experience was highly influential for us. I think that this was another one of these rare moments when the field really conceptually advanced. Together with the technological advances such as opto- and chemogenetics, this then really enabled the search for the engram to take off.

**Indeed, the catFISH method you mentioned is instrumental in that it actually allows to assess one of Semon's criteria of the engram, namely that the original cells become reactivated by recall and have a content storage inherent to them. So how do you picture the following: You remember the discovery of**

**the engram cells in the Tonegawa lab more than 10 years ago, as if it was yesterday. And we all have our childhood memories that we remember. So, are engrams stable for that long? How do you imagine this decade-old memories to be stored in the brain? And related to that, why do we keep some memories and not others?**

The fact that we can recall things that are decades old suggests that there's some trace of that memory that exists in the brain for that amount of time. Now, whether or not it's the same trace is a different question. It could also be that the cells that were involved in holding on to that memory of what I did 10 years ago might not be reactivated at all today. It might also be a different set of cells recruiting different circuits in the brain, suggesting that the original memory has morphed over time. Yet there's still a consistency to it. And I think this is the paradox of memory research: Memory feels stable, even though the underlying biology appears to be unstable. But I'm not sure how to reconcile it yet, other than it might be only a subset of those cells, a very small percentage of those cells or a subset of those synapses that are truly the stable correlate of that memory that happens to be still wired appropriately in the brain. At least that's the way I imagine it.

**I agree. For me, there almost has to be a cellular correlate that remains because otherwise, how would it be possible? It would actually be quite a big risk to transfer the storage of a very important, highly salient memory onto a completely different subset of cells. Which then begs the next question. We all have these decade-old memories that we didn't think about for long, but then all of a sudden that memory is there again, triggered, for example, by a specific odor, such as a perfume or cheese. How is this brought about? Because clearly the cells were not doing anything other than just being around. But then there's this trigger and the whole memory is back, or at least we think the whole memory is back. How do you picture that?**

Yeah, I had that moment recently where I went to a Mexican restaurant and they had this dish, something with pickled vegetables, that is a typical side for a Salvadorian dish. And I had it, this ratatouille moment where I was just like, oh my God, I remember the last time I had this with my family, like years ago! And it was instantaneous. It wasn't even that I was vaguely remembering it; it was totally a full-blown episodic experience that came back to life over just the right combination of taste buds that were tickled in the right way. I think that some memories take the low road when it comes to being accessed, whether it's through odor or through taste or... They seem to have direct access to the episodic memory systems of the brain that bring them back to life. And maybe there are indeed some memories that are particularly privileged in that respect, as I don't think it's all of our memories, but there might be some that were formed probably during childhood that really were important experiences for one reason or another that we had in our lives, that our biology considered that experience important or memorable or salient enough to really make that memory last.

**Perhaps one of the underexplored areas in engram research thus far is the role of sleep. And that of dreams, come to think of it, because obviously when**

**we dream, we mix reality with our emotions. So, projecting ourselves a little bit into the future, how do you reckon that sleep affects engram stability?**

We spend a third of our life sleeping, unconscious, and yet during that unconscious process... the brain is very much awake! For example, there is certainly hippocampal activity going on, such as memory replay, and these processes are indispensable for memory consolidation. One theory that I'm very partial to is that we have memories not so much as to recall the past, but as a basis for predicting the future, to stay one step ahead of the present. And if we can use and recombine memories to stay one step ahead of the present, that gives us some kind of biological advantage. Thus, in sleeping and, in particular, dreaming, it might be our brain's attempt of practicing being ahead of the present: Let's recombine all of our past experiences into some Picasso-like experience where it's nonsensical and it doesn't seem to have any order to it other than it's the different things that we've experienced in the past that our brain is recombining into practice, like fictitious events of the future. So, you're using the past to just imagine all of the different universes that your body might have to go through so that you're able to go through one of them tomorrow. And if it's anything what you've mentally rehearsed in sleep, you're a little bit more prepared or a little bit more ready for it. In that sense, dreaming might be a way of using the past and recombining it to imagine all the different crazy futures that might be possible.

**I really like this notion that our memories are there to prepare ourselves for what's next. The more pronounced the memory capacity, the higher the preparedness for what is yet to come. On that thought, we could come to an end of this interview and conclude by saying that we hope that this book is not just a memory of the past, but a preparation for what's next.**

Yeah, I love it because it's saying that we hope that the textbook gives accounts of past research that already has happened as well as of contemporary research that currently is happening in the engram world so that we can then all begin to imagine and make the next years of research possible.

**A canvas on which future research can be drawn!**

# Part II
# Engram Cells and Their Mnemonic Role Across Different Brain Structures

# Engrams: From Behavior to Brain-Wide Networks

**Kaitlyn E. Dorst and Steve Ramirez**

**Abstract** Animals utilize a repertoire of behavioral responses during everyday experiences. During a potentially dangerous encounter, defensive actions such as "fight, flight, or freeze" are selected for survival. The successful use of behavior is determined by a series of real-time computations combining an animal's internal (i.e., body) and external (i.e., environment) state. Brain-wide neural pathways are engaged throughout this process to detect stimuli, integrate information, and command behavioral output. The hippocampus, in particular, plays a role in the encoding and storing of the episodic information surrounding these encounters as putative "engram" or experience-modified cellular ensembles. Recalling a negative experience then reactivates a dedicated engram ensemble and elicits a behavioral response. How hippocampus-based engrams modulate brain-wide states and an animal's internal/external milieu to influence behavior is an exciting area of investigation for contemporary neuroscience. In this chapter, we provide an overview of recent technological advancements that allow researchers to tag, manipulate, and visualize putative engram ensembles, with an overarching goal of casually connecting their brain-wide underpinnings to behavior. We then discuss how hippocampal fear engrams alter behavior in a manner that is contingent on an environment's physical features as well as how they influence brain-wide patterns of cellular activity. Overall, we propose here that studies on memory engrams offer an exciting avenue for contemporary neuroscience to casually link the activity of cells to cognition and behavior while also offering testable theoretical and experimental frameworks for how the brain organizes experience.

---

K. E. Dorst
Department of Psychological and Brain Sciences, Boston University, Boston, MA, USA

Department of Neuroscience, Tufts University School of Medicine, Boston, MA, USA

Graduate Program for Neuroscience, Boston University, Boston, MA, USA

S. Ramirez (✉)
Department of Psychological & Brain Sciences, Boston University, Boston, MA, USA
e-mail: dvsteve@bu.edu

 
J. Gräff, S. Ramirez (eds.), *Engrams*, Advances in Neurobiology 38,
https://doi.org/10.1007/978-3-031-62983-9_2

**Keywords** Network · Activity pattern · Brain · Hippocampus

## Episodic Memory and the Hippocampus

Episodic memory is the brain's representation of personally experienced events. It is often defined along three dimensions: informational ("what"), temporal ("when"), and spatial ("where") components (Tulving 1983; Manns and Eichenbaum 2006). Episodic memory involves at least four mechanistic stages, namely encoding, storing, retrieving, and updating, all of which involve coordination of populations of cells across brain structures. Encoding involves making associations of the sights, sounds, smells, and other contextual cues of one's lived experiences with their environment. For example, I am actively encoding the experience of writing this chapter during a power outage: Candles are lit throughout the apartment, many of which smell of clean linen, the cat is hiding under my feet, and I am feeling a sense of irritation at the leak in my ceiling. Storing episodic memories, termed consolidation, involves stabilizing those features into the neural connections active during the encoding stage. As I rest in the darkness of this power outage, the events of this experience will be imprinted in my brain. Retrieval re-engages those underlying neural connections to produce the recollection of those sights, sounds, smells, and cues. Telling this story to a friend later in time will put my brain into a partially similar state as during encoding. This "mental time travel" during retrieval is a unique feature of episodic memory that allows us to tell these stories as if we are still experiencing them (Tulving 1983, 2002; Eichenbaum and Cohen 2001). And finally, memory updating is perhaps a ubiquitous process during which the recollection of a memory renders it malleable and susceptible to modification, including strengthening or weakening the memory, or introducing new information altogether. As I recall the memory of writing this opening, its episodic details presumably become molten–the candle, the cat, the ceiling, when the power was restored–all are all bits of mnemonic information that the process of recollection is capable of modifying.

This concept of episodic memory is evolutionarily conserved across species and relies on a specialized region of the brain called the hippocampus. The most famous recorded clinical evidence of the hippocampus being crucial for episodic memory processing was the case of Henry Molaison (1926–2008), previously known as patient H.M. (Scoviille and Milner 1957; Penfield and Milner 1958). As a result of a childhood injury, H.M. developed intractable epilepsy, and his quality of life greatly diminished due to the severity of his untreatable seizures. Later in his young adult life, H.M. opted for a surgical procedure to locate the seizure-producing areas of his brain and subsequently remove them. A sizable portion of H.M.'s brain was bilaterally removed, and the hippocampus was one of the main structures that was eliminated. The seizures stopped but at the cost of some of H.M.'s cognitive abilities. He developed anterograde amnesia, which is an impairment for generating new episodic memories. H.M. was unable to remember the hospital staff or what he had for breakfast each morning. Yet he was able to remember old childhood memories and learn new motor tasks, which suggested that the hippocampus was not essential for long-term episodic memory and that the brain contains separate memory systems (e.g., episodic, motor, emotional) dedicated to processing their specific features (Gabrieli 1998). This case was foundational for a plethora of research surrounding facets of episodic memory and contributions of the hippocampus.

Scientists can either observe or perturb the brain to test for the sufficiency or necessity of the hippocampus in mediating episodic memory. Around the time of H.M., Wilder Penfield (1891–1976), a Canadian neurosurgeon, developed the Montreal Procedure, which stimulated various portions of the brain to determine their respective functional roles (Ladino et al. 2018). Remarkably, a handful of participants who received stimulation in areas neighboring the hippocampus recalled vivid episodic memories. Modern technology such as functional magnetic resonance imaging can examine metabolic activity within the hippocampus as a participant is describing details surrounding an episodic memory. Relatedly, rodent models have been used to study the hippocampus and episodic memory at a finer spatial resolution. Specialized surgical procedures were developed to target the hippocampus directly. Like the Montreal Procedure, electrodes can be lowered into the rodent hippocampus for either recording or electrical stimulation. This type of procedure allows researchers to observe and manipulate neural activity in vivo. Electrode recordings of the rodent hippocampus generated substantial evidence for populations of cells that code for space and time and other multi-dimensional features of the episodic event of traversing an arena (O'Keefe and Dostrovsky 1971; Wood et al. 2000; Fyhn et al. 2004; Moser et al. 2008; Eichenbaum 2014). Hippocampal populations also generate sharp-wave ripples, which occur when many cells are synchronously active, as a form of memory consolidation (Joo and Frank 2018). Additionally, electrical stimulation of cells within the rodent hippocampus induces long-term potentiation (LTP), strengthening the synaptic connections between cells and offering a putative mechanism for memory storage and/or retrieval.

Behaviorally, to test for a rodent's ability to encode, store, and retrieve episodic memories, researchers often use a contextual fear conditioning paradigm (CFC). Rodents are placed into a chamber with grid flooring that can pass electrical current and are given a handful of minutes to freely explore. Then, a series of mild shocks are passed into the flooring, and the rodent elicits an unconditioned response (Maren 2008; Pavlov 1927). Throughout the conditioning session, there is an increase in freezing behavior, which is defined as the cessation of all movement with the exception of breathing and is an established rodent behavioral metric indicative of fear (Grossen and Kelley 1972; Bouton and Bolles 1980; Fanselow 1980; Blanchard and Blanchard 1989). Once the experiment is over, the rodents are placed back into their homecage for the fear memory to consolidate. Over subsequent days, the rodents can be placed back into the conditioning chamber, without shock delivery, to test for their capacity to retrieve the fear memory. The defensive freezing persists during this retrieval session, indicating that the rodents recall the original episodic memory of getting shocked in that particular environment. Electrical stimulation that induces LTP in the rodent hippocampus enhances fear learning during CFC (Lynch 2004;

Bliss et al. 2018). Conversely, lesions or inactivation of the hippocampus severely impair the defensive behavior during this retrieval test (Kim and Fanselow 1992; Kim et al. 1993; Maren et al. 1997; Good et al. 2007). Without an intact hippocampus, animals often fail to retain the recent episodic memory formed during CFC, thus adding to the body of evidence for the hippocampus in regulating episodic memory.

## A Brief History of Memory Engrams

The idea that there is a physical imprint of an episodic memory in our brains fascinates both the layperson and the scientist. This notion, termed a memory "engram," was originally theorized by Richard Semon (1859–1918) as a lasting physical modification within the brain's state and structure to support memory processing (Semon 1921; Schacter et al. 1978; Josselyn et al. 2017). Subsequent work led by Karl Lashley aimed to identify the precise locus of the engram through a series of lesion experiments in rodents. Although he failed to find the site of a putative engram, his main conclusion was that there were concerted efforts to mediate a memory trace throughout many areas of the brain (Tonegawa et al. 2015; Josselyn et al. 2017). Donald Hebb (1904–1985), who was a student of Lashley and worked briefly with Penfield, took the mantle of elucidating engram mechanisms and proposed the idea of LTP—that groups of neurons must be coactive during the formation of a memory and their synapses would subsequently be strengthened to hold on to this newly-encoded mnemonic information (Josselyn et al. 2017). Methods in traditional electrophysiology across species observed correlated single-unit activity of neurons during both the formation and retrieval of mnemonic events (Olds et al. 1972; Miyashita 1988), thus accumulating evidence that memory footprints exist across populations of coactive cells, akin to memory engram ensemble. However, the technology at the time limited the extent to which these scientists could probe a specific memory engram within the rodent's brain.

An explosion in technological advancements in the twenty-first century revolutionized neuroscience–especially in the field of memory engrams. Activity-dependent "tagging" strategies were developed to identify putative engram ensembles that leveraged the nature of a class of genes called immediate early genes (IEGs). These IEGs are transiently active due to neural activity, making them ideal candidates for identifying engram ensembles across learning (Abraham et al. 1991; Dragunow 1996; Guzowski et al. 2005). This tagging system must be temporally regulated using an inducible system. For example, the "TetTag" strategy utilizes the IEG cFos to drive the expression of a tetracycline transactivator (tTA) gene (Reijmers et al. 2007). The tTA gene product, in turn, can bind to its respective tetracycline response element (TRE), which will ultimately induce the expression of the proximal gene, which can be a simple fluorescent reporter or an effector gene (as further outlined below). By using doxycycline, a tetracycline derivative, tTA binding to TRE can be blocked, thus preventing this "TetTag" system from labeling randomly activated cells. Removing doxycycline from the animal's system can, in contrast, open a window during which active cells can become susceptible to labeling. Researchers now widely use tagging strategies to test for both the necessity and sufficiency of engram ensembles in episodic memory.

The first set of experiments that targeted discrete memory-bearing engram cells tested for the necessity of populations in the amygdala, a region known for assigning emotional valence to memories, in producing fear (Han et al. 2007, 2009). These pioneering experiments selectively labeled and ablated an auditory fear conditioning engram in the lateral amygdala (LA). By first biasing a population of LA cells to process an auditory fear memory engram ensemble, the researchers then expressed diphtheria toxin receptors to prime cells for pharmacological ablation. Systemic injection of diphtheria toxin before a memory retrieval session caused specific cell death within the engram ensemble that accompanied a deficit in defensive freezing when presented with the training tone. Importantly, ablation of a random set of LA cells did not alter the behavioral retention of the fear memory during the test phase, demonstrating that ablation of the specific engram was necessary for memory disruption.

Later, the popularity of optogenetics allowed researchers to label engram ensembles with light-sensitive proteins, called opsins, to manipulate their activity. During the onset of light, opsins undergo a conformational change to allow ions to flow freely down their concentration gradients. Inhibitory opsins such as halorhodopsin or archaerhodopsin hyperpolarize cells, while excitatory opsins such as channelrhodopsin (ChR2) excite cells during the presentation of light (Boyden 2011). Importantly, when light is turned off, cells will stabilize their activity towards their baseline. Using optogenetics to test for the necessity of engram ensembles for episodic memory, researchers leveraged tagging strategies to insert inhibitory opsins to cells active during CFC. Silencing an engram ensemble during a retrieval test effectively silenced that cell population and prevented fear memory expression, as measured by a lack of freezing behavior when the mice were placed back in the original training context (Denny et al. 2014; Tanaka et al. 2014). Interestingly, mice increased their freezing behavior after the cessation of light during the experimental session, demonstrating that the engram was re-engaged and memory expression was still retained. Optogenetic inactivation of a hippocampal CFC engram also produced a large-scale reduction of endogenous IEG expression in the hippocampus and other regions implicated in learning and memory such as the prefrontal cortex (Kitamura et al. 2017), the amygdala (Tanaka et al. 2014), the lateral entorhinal cortex, and perirhinal cortex (Tanaka et al. 2014), suggesting that memory-bearing engrams are necessary to functionally recruit circuitry to memory expression. These works are modernized complements to Lashley's original search for an engram.

Simultaneous efforts around the world used tagging and optogenetics to test the sufficiency of engram ensembles for mediating memory. Tagging hippocampus cells processing a CFC engram with ChR2 permitted researchers to demonstrate that these small populations of hippocampus cells were sufficient to drive the behavioral expression of a memory (Liu et al. 2012). In this study, mice freely explored a novel arena during a baseline period and optogenetic stimulation of the original

CFC engram produced robust freezing responses strictly during the presentation of light (i.e., there was no freezing behavior when the light was off), suggesting a behavioral output indicative of fear memory retrieval. Importantly, this alteration in freezing behavior was not seen in two control conditions in which there was (1) no ChR2 in the tagged CFC engram or (2) there was no shock presentation during the tagging session; ChR2 activation of this "context-only" engram was not sufficient to produce freezing. This work demonstrated that hippocampus cells active during the formation of a fear memory were sufficient to drive the behavioral expression of the memory.

## Expanding the Engram's Behavioral Repertoire

Since these seminal works, there were many related studies that manipulated memory-bearing engram ensembles across stages of learning, health and disease, and environmental contingencies. First, hippocampal-mediated engrams have been shown to be labile upon reactivation; artificial manipulation of the engram ensemble can alter both contextual representation and emotional valence (Ramirez et al. 2013; Redondo et al. 2014; Ohkawa et al. 2015). This method of artificial memory induction used optogenetics to induce a retrieval-like state of the engram while encoding new salient information into memory, thus overriding the original information encoded in the engram ensemble. Additionally, engram reactivation work began testing its efficacy in producing changes in behavior under amnesic states such as Alzheimer's Disease (AD) or anisomycin (ANI) administration (Ryan et al. 2015; Roy et al. 2016). Remarkably, optogenetic reactivation of the tagged hippocampal CFC engram ensemble in both the AD and ANI mouse model produced a behavioral expression of fear memory retrieval (i.e., freezing), suggesting that there was a deficit in engram reactivation under naturalistic conditions.

In concurrent work, tagging and reactivating hippocampal engrams of positive emotional valence (i.e., male-female social interactions, food or drug-based rewards) rescued depression-like behavioral phenotypes across a battery of tests (Ramirez et al. 2015). In this study, male mice that were chronically stressed via restraint stress underwent a tail suspension test (TST), sucrose preference test (SPT), open field test (OFT), and elevated plus maze test (EPMT). Optogenetic activation of the positive hippocampal engram increased struggle time in the TST, sucrose consumption in the SPT, time in the center of the OFT, and time in the open arms of the EPMT strictly during the presentation of light, which were abolished in the absence of light. This was among the first studies to show within-animal modulation of behavior across test contingencies, suggesting that hippocampal engram ensembles were not fixed to produce a singular behavioral output.

This body of work suggests that hippocampal CFC engram ensembles are flexible in mediating different behavioral phenotypes upon reactivation contingent on test conditions. Freezing is an ethological readout of fear in rodent models and is modulated by CFC engram reactivation. Yet there are a variety of defensive behaviors animals can choose from when presented with danger or other potentially harmful stimuli (Bouton and Bolles 1980; Fanselow 1980, 2022). Earlier rodent ethological work suggests that defensive behaviors exist in a hierarchical repertoire and that they are selected based on environmental contingencies such as the possibility to escape or hide in proximity of danger, as shown, for example, with experiments that exposed rodents to innately fearful stimuli such as predator odor trimethylthiazoline or a looming disk (Rosen et al. 2008; Salay et al. 2018; Shang et al. 2018). These general findings support the notion that when animals are in spatially restricted environments, they default to more passive defensive behaviors such as freezing and more active responses such as avoidance or hiding in large arenas or spaces with shelters at fixed locations. Fittingly, a recent report demonstrated an avoidance-like behavioral phenotype when a CFC engram was reactivated in a large arena without the presence of any overt freezing responses even though they demonstrated freezing behavior during the initial fear memory encoding during engram tagging (Chen et al. 2019). One key possibility that could contribute to this difference in defensive behavioral outputs during hippocampal CFC engram reactivation is that there are different environmental features that bias whether or not animals adopt an active or passive defensive strategy. Indeed, hippocampal-based systems are crucial for integrating sensory information to generate spatial representations of the environment for navigation, and this information is continuously updated as the animal explores this new context. In any "predefense" scenario where there is no fear-eliciting threat per se, exploratory behaviors such as stretch approach posture and rearing are selected and presumably adaptive (Blanchard and Blanchard 1989; Molewijk 1995; Sturman et al. 2018; Fanselow 2022). This new contextual information is integrated into memory and can influence how the animal behaves during subsequent exposure to fear-eliciting stimuli.

To test the idea that hippocampal CFC engrams are capable of flexibly modulating defensive behaviors, we used a within-animal design where all rodents underwent reactivation across simple environments of various sizes (Fig. 1; Dorst et al. 2024). As expected, freezing increased during the tagging period, where the fear memory was encoded into the dedicated engram ensemble. Moreover, when animals were placed in the smallest arena, optogenetic activation of a hippocampal engram produced robust light-induced freezing. Interestingly, however, there was no observable light-induced freezing in those same animals when they were in the largest arena. It is important to note that mice increased their freezing behavior during the tagging session, yet light-induced freezing was modulated across environments. However, this observed trend in the freezing behavior could be due to a phenomenon called "encoding specificity," which necessitates contextual similarity across encoding and retrieval conditions to produce significant memory retrieval (Tulving 1983; Frankland et al. 2019; Jung et al. 2023). As the small arena was a similar size to the conditioning arena, follow-up experiments sought to determine if conditioning size influences subsequent light-induced freezing across the three environments (Dorst et al. 2024). Animals conditioned in a larger arena still increased their freezing behavior, yet hippocampal CFC engram reactivation did not

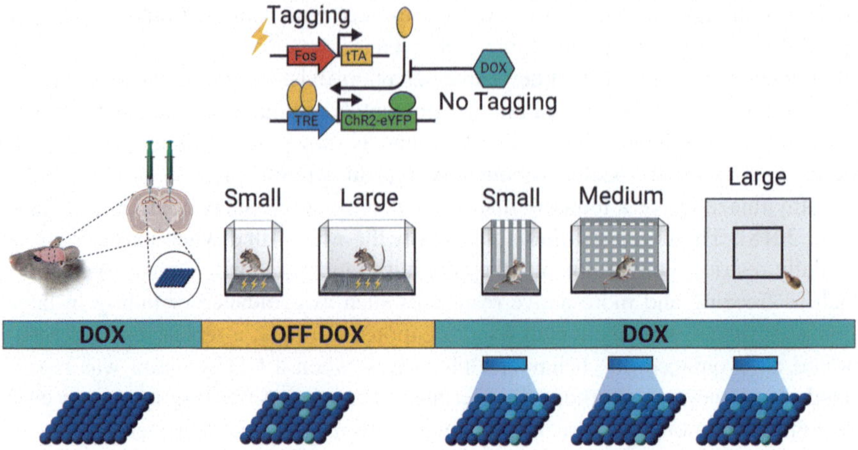


**Fig. 1** Tagging strategy and behavioral schedule for activity-dependent labeling of a hippocampal Contextual Fear Conditioning (CFC) engram and reactivation across environments. Mice are taken off of doxycycline (DOX) two days prior to hippocampal CFC engram tagging. During the CFC session, mice explore an environment (i.e., small or large chamber) and are administered unconditioned stimuli (e.g., mild foot shocks) to tag active cells. Mice are then placed back on the DOX diet after CFC to prevent further tagging of cells. Hippocampus cells expressing an effector of interest (e.g., channelrhodopsin-2) are then reactivated in each environment of different sizes (small, medium, large) while measuring behavioral outputs

produce any discernible trends in light-induced freezing. This indicates that the cells that were recruited into the dedicated CFC engram ensemble are not "freezing" cells that are solely fixed to produce freezing behavior. Instead, animals adapt to environmental contingencies to alter their behavioral responses, even when the fear memory engram is reactivated. Future studies may examine the effects of hippocampal CFC engram reactivation and light-induced freezing in more complex environmental conditions such as introducing an exit or shelter that warrants the rodent to engage in more navigation-based escape strategies. For example, during an initial baseline period, rodents would be able to integrate new sensory representations of this complex environment into memory to create a cognitive map for proper spatial navigation. Instead of defaulting to freezing, we predict that animals would engage in more ambulatory responses and that the latency for seeking either an exit or shelter would decrease at the onset of hippocampal CFC engram reactivation when compared to control conditions. This work underscores the dynamic nature of memory-guided behavior.

## Hippocampal Engrams Engage Brain-Wide Networks

There are a variety of interleaving brain-wide pathways that mediate behavior. These areas serve functions such as sensory detection, information integration, and commanding behavioral outputs (Fendt and Fanselow 1999; Gross and Canteras 2012; Tovote et al. 2015; Silva et al. 2016). During the initial exposure to fear-eliciting stimuli, contextual information is relayed to cortical areas such as sensory cortices, the hippocampus and the amygdala via the thalamus. This information must be rapidly processed by the brain to dictate appropriate defensive responses. These integration steps also occur within these regions of the brain and other thalamic and hypothalamic subnuclei, and behavioral output is mediated by midbrain regions such as the periaqueductal gray. The extensive characterization of the contribution of these regions to behavior utilized observational or perturbational studies to demonstrate both necessity and sufficiency. These experiments would examine the individual region or its participation in anatomical circuits or systems that usually involve two or three regions. Yet, there are enduring brain-wide activation patterns during salient experiences, with certain regions acting as the major sources of activity.

Much engram work has also shown that reactivating hippocampal engram ensembles generated alterations in cFos activation in regions that are important for mediating learning and affective states, such as the amygdala, nucleus accumbens, lateral septum, and various hypothalamic nuclei (Ramirez et al. 2013, 2015; Redondo et al. 2014; Ryan et al. 2015). A few of these aforementioned regions have direct anatomical connections to the hippocampus. Yet, these interacting systems are functionally connected, which is defined as the "temporal coincidence of spatially distant neurophysiological events" (Friston 1994) to produce a behavior. This term is often used in human neuroimaging, where blood oxygenation is used as a measurement of functional connectivity while the subject performs a cognitive task in a scanner (Hillman 2014). This approach is too costly for use in rodent models, so researchers examine endogenous brain-wide co-expression of IEGs such as cFos. Technological advancements in tissue clearing, immunolabeling, optical microscopy, and machine learning allow researchers to take unbiased approaches to quantify cFos throughout the intact rodent brain. These rich datasets are then used to compare cFos expression within and across experimental conditions.

A branch of mathematics called graph theory is used to study the relationships across objects of large-scale systems such as the brain (Bullmore and Sporns 2009; Bassett and Mattar 2017). It is often used to map out airport systems, electric grids, and suggested friends on social sites such as Facebook, among other things. Objects, defined as nodes in a network, can be airports, power sources, people, or brain regions. A significant pairwise relationship or connection is defined as an edge that connects two nodes, a friendship or an airline route. Network neuroscientists can calculate the pairwise correlation of endogenous cFos expression from all subjects in the experimental condition across two regions (and iterate through all possible pairs of regions), which acts as the defining edge between the two brain regions (Terstege and Epp 2022). These edges can connect multiple nodes, showing many significant pairwise relationships, with the resulting visualization (i.e., a graph) looking more like a spider web of information about the relationships of these objects. Additional analyses can cluster these nodes to organize the graph based on similar correlation patterns as well as find nodes of high importance, termed a "hub node." Using these analyses, researchers can elucidate a functional connectivity profile of cFos expression as a result of hippocampal engram reactivation.

To directly test the idea that reactivating a hippocampal engram produces brain-wide alterations in cFos functional connectivity, Dorst, Senne, and colleagues devised an experiment in which rodents underwent CFC in a small arena and the tagged engram ensemble was optogenetically reactivated in either a novel small arena or a large arena, which promoted and discouraged light-induced freezing, respectively (Dorst et al. 2024, Fig. 1). An additional group of animals that underwent naturalistic fear memory retrieval (i.e., no CFC engram manipulation) was added to this set of experiments as a positive control to compare and contrast network features (Fig. 2) across all forms of memory manipulation (i.e., naturalistic vs. artificial) (Dorst et al. 2024). There were similar amounts of freezing in animals that underwent CFC engram reactivation in the small arena and those that underwent natural retrieval, demonstrating that fear memory manifests as defensive behavior. Interestingly, all networks had unique topological features, such as the number of edges, clusters, and the size and composition of each cluster. Although freezing levels were similar across the group that underwent CFC engram reactivation in the small arena and the group that underwent natural retrieval, their network structures were different (Dorst et al. 2024). This is an example of biological degeneracy, where underlying functional dynamics can yield similar outputs (i.e., behavior) (Marder and Taylor 2011). Conversely, the network structures of the two groups of CFC engram reactivation in the small and large arenas were more similar than any other network even though their behavioral readouts were unique (i.e., freezing or no freezing). All three experimental networks converged onto shared hub nodes–the basolateral amygdala and the infralimbic area. These two regions act in concert with the hippocampus over a broad fear memory system, which suggests that both artificial and naturalistic memory are engaging similar pathways to mediate memory expression regardless of behavioral outcomes. Other hubs were unique to each network, which opened up exciting avenues to further test the contributions of these areas on memory expression under various conditions. Taken together, these works show that memory engrams are capable of producing flexible brain-wide network states.

## Conclusions

Since the advent of the traditional "tag and manipulate" engram work, there have been countless experiments that have activated cellular ensembles to modulate memory. Yet, there is more to look forward to in the near future. Fear conditioning has been a particularly potent protocol to induce a negative affective state such as fear, and memory retrieval is sufficient to produce defensive behaviors such as freezing or avoidance. However, there are a plethora of experiences that are classified as negative or aversive and are also encoded as memory engrams. For example, a predator exposure test offers a multi-modal experience that induces risk assessment, avoidance behavior, and freezing (Yang et al. 2004). Tagging and reactivating a "predator exposure engram" may produce similar behavioral phenotypes across

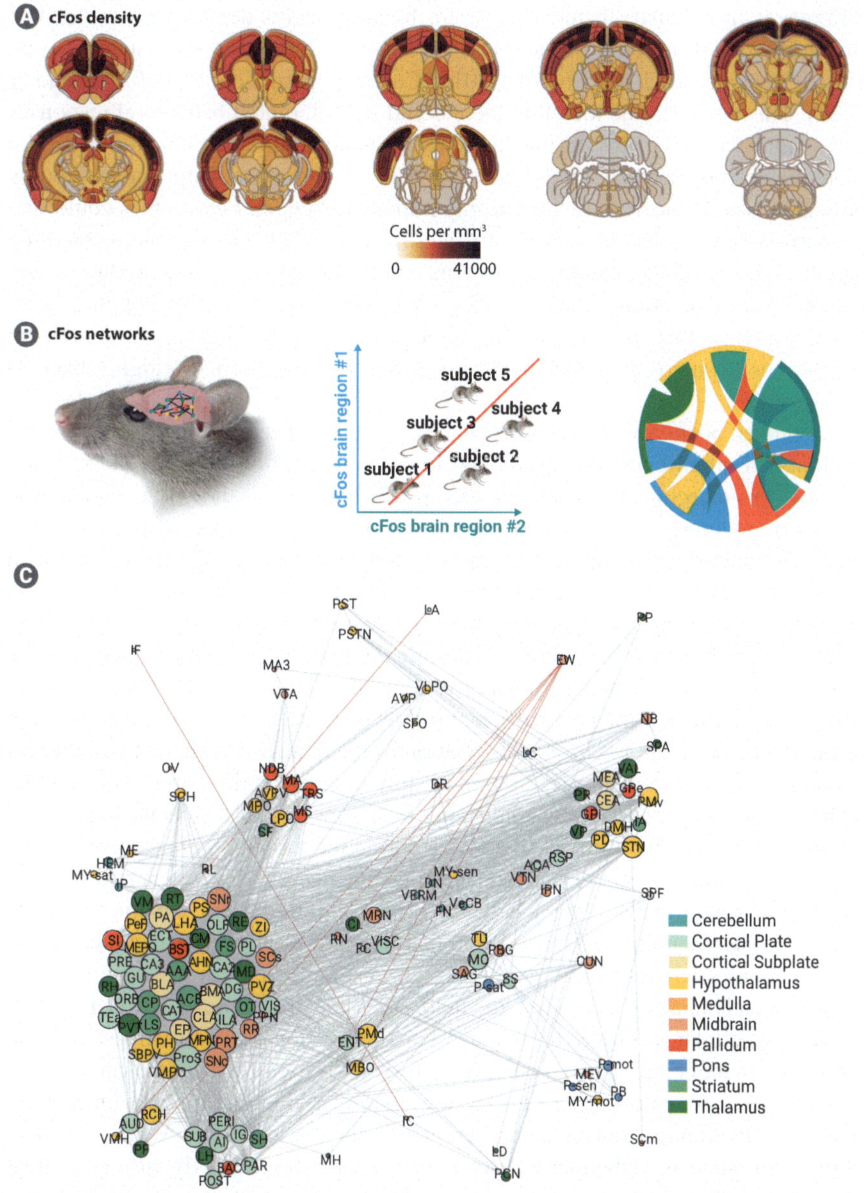


**Fig. 2** (**a**) Example heatmap of rodent brain-wide cFos density to measure functional activity corresponding to a given behavior or circuit manipulation. (**b**) This brain-wide activity can then be correlated across regions and across animals to construct unbiased networks (**c**) for further analyses (e.g., topological features, hubs, nodes, and edges)

varying environmental conditions. Importantly, many of these aforementioned studies have solely examined the behavioral consequences of fear engram manipulation in male rodents. However, there has been emerging evidence showing sex differences in defensive behaviors, where female rodents demonstrate a blend of freezing and darting responses and less vocalization during threat (Blanchard et al. 1991; Gruene et al. 2015; Shansky 2018; Colom-Lapetina et al. 2019; Ma and Jr 2022). This points to the idea that there could be differential behavioral responses across males and females during fear engram manipulation, which would yield a more nuanced perspective on the memory-induced behavioral outputs mediated by biological sex. Concurrently, hippocampal-mediated engram ensembles could disproportionately engage the neural substrates mediating behavior depending on other variables that mediate external and internal states. Reactivation may produce a corresponding set of brain-wide networks, each with unique and overlapping functional patterns that provide useful large-scale data for future computational approaches to interrogate and use as predictive frameworks for bridging back to biology.

Moreover, positive engrams are an exciting avenue of research, as the behavioral readouts are often operant/instrumental in nature and offer ethologically relevant metrics. An emerging and promising therapeutic approach is the active recollection of positive memories, which attenuate brain- and body-wide stress responses in real time. The underlying cellular mechanisms supporting these effects, as well as a putative therapeutic link between activated memories and degeneration, for instance, are virtually unknown. Surprisingly, there is a dearth of studies that investigate how memories are linked to degeneration, despite the fact that memory is precisely the aspect of cognition affected. Promisingly, as discussed in this chapter, hippocampus cells active during learning are sufficient for the neuronal and behavioral expression of memory; repeated or "chronic" activation of these cells is sufficient to enhance, suppress, and restore memories both in healthy mice and animal models of PTSD, addiction, and depression, thus confirming their generalizable capacity to alleviate pathological states. Excitingly, cells active during the formation of negative memories show unique genetic enrichment profiles associated with degeneration and a subset of psychiatric disorders, whereas cells active during positive memory formation upregulate neuroprotective markers that themselves are putative clinically relevant targets; each respective engram offers a unique strategy for manipulating cognition and behavior. Therefore, the field's ongoing discoveries raise the possibility that the cellular substrates for memory can bi-directionally control disease. Thus, a primary goal of engram research moving forward is to understand the causal basis of memory in the brain and to probe how its underlying circuits and brain-wide structure-function relationships are disrupted in the diseased state. While disorders of the brain, such as Alzheimer's Disease, have been extraordinarily difficult to treat because most treatments rely on drug-based approaches that were developed over half a century ago with minimal improvement, memory is a core symptom of dementia that is often studied as an effect of the degenerated brain rather than a cause. Patients, for instance, do not have stress or reward-related circuitry artificially perturbed over a lifetime; rather, endogenous biological processes, including memory, have direct access to this circuitry and therefore form a causal, albeit understudied, connection between cognition and its influence on biological outputs. We speculate, therefore, that a promising area of future research involves extracting physiological/genetic signatures unique to memory engrams that can then be sculpted back onto a diseased brain as putative treatments and that artificially manipulating memories can itself be viewed as a tailored 21st century means by which to promote health in an organism.

## References

Abraham WC, Dragunow M, Tate WP (1991) The role of immediate early genes in the stabilization of long-term potentiation. Mol Neurobiol 5:297–314

Bassett D, Mattar M (2017) A network neuroscience of human learning: potential to inform quantitative theories of brain and behavior I Elsevier Enhanced Reader

Blanchard RJ, Blanchard DC (1989) Attack and defense in rodents as ethoexperimental models for the study of emotion. Prog Neuro-Psychopharmacol Biol Psychiatry 13:S3–S14

Blanchard DC, Shepherd JK, Carobrez ADP, Blanchard RJ (1991) Sex effects in defensive behavior: baseline differences and drug interactions. Neurosci Biobehav Rev 15:461–468

Bliss TVP, Collingridge GL, Morris RGM, Reymann KG (2018) Long-term potentiation in the hippocampus: discovery, mechanisms and function. e-Neuroforum 24:A103–A120

Bouton ME, Bolles RC (1980) Conditioned fear assessed by freezing and by the suppression of three different baselines. Anim Learn Behav 8:429–434

Boyden ES (2011) A history of optogenetics: the development of tools for controlling brain circuits with light. F1000 Biol Rep 3:11

Bullmore E, Sporns O (2009) Complex brain networks: graph theoretical analysis of structural and functional systems. Nat Rev Neurosci 10:186–198

Chen BK, Murawski NJ, Cincotta C, McKissick O, Finkelstein A, Hamidi AB, Merfeld E, Doucette E, Grella SL, Shpokayte M, Zaki Y, Fortin A, Ramirez S (2019) Artificially enhancing and suppressing hippocampus-mediated memories. Curr Biol 29:1885–1894.e4

Colom-Lapetina J, Li AJ, Pelegrina-Perez TC, Shansky RM (2019) Behavioral diversity across classic rodent models is sex-dependent. Front Behav Neurosci 13:45

Denny CA, Kheirbek MA, Alba EL, Tanaka KF, Brachman RA, Laughman KB, Tomm NK, Turi GF, Losonczy A, Hen R (2014) Hippocampal memory traces are differentially modulated by experience, time, and adult neurogenesis. Neuron 83:189–201

Dorst KE, Senne RA, Diep AH, De Boer AR, Suthard RL, Leblanc H, Ruesch EA, Pyo AY, Skelton S, Carstensen LC, Malmberg S, McKissick OP, Bladon JH, Ramirez S (2024) Hippocampal engrams generate variable behavioral responses and brain-wide network states. J Neurosci 10:e0340232023

Dragunow M (1996) A role for immediate-early transcription factors in learning and memory. Behav Genet 26:293–299

Eichenbaum H (2014) Time cells in the hippocampus: a new dimension for mapping memories. Nat Rev Neurosci 15:732–744

Eichenbaum H, Cohen NJ (2001) From conditioning to conscious recollection: memory systems of the brain. Oxford University Press, New York

Fanselow MS (1980) Conditional and unconditional components of post-shock freezing. Pavlov J Biol Sci 15:177–182

Fanselow MS (2022) Negative valence systems: sustained threat and the predatory imminence continuum. Emerg Top Life Sci 6:467–477

Fendt M, Fanselow MS (1999) The neuroanatomical and neurochemical basis of conditioned fear. Neurosci Biobehav Rev 23:743–760

Frankland PW, Josselyn SA, Köhler S (2019) The neurobiological foundation of memory retrieval. Nat Neurosci 22:1576–1585

Friston KJ (1994) Functional and effective connectivity in neuroimaging: a synthesis. Hum Brain Mapp 2:56–78
Fyhn M, Molden S, Witter MP, Moser EI, Moser M-B (2004) Spatial representation in the entorhinal cortex. Science 305:1258–1264
Gabrieli JDE (1998) COGNITIVE NEUROSCIENCE OF HUMAN MEMORY. Annu Rev Psychol 49:87–115
Good MA, Barnes P, Staal V, McGregor A, Honey RC (2007) Context- but not familiarity-dependent forms of object recognition are impaired following excitotoxic hippocampal lesions in rats. Behav Neurosci 121:218–223
Gross CT, Canteras NS (2012) The many paths to fear. Nat Rev Neurosci 13:651–658
Grossen NE, Kelley MJ (1972) Species-specific behavior and acquisition of avoidance behavior in rats. J Comp Physiol Psychol 81:307–310
Gruene TM, Flick K, Stefano A, Shea SD, Shansky RM (2015) Sexually divergent expression of active and passive conditioned fear responses in rats. elife 4:e11352
Guzowski JF, Timlin JA, Roysam B, McNaughton BL, Worley PF, Barnes CA (2005) Mapping behaviorally relevant neural circuits with immediate-early gene expression. Curr Opin Neurobiol 15:599–606
Han J-H, Kushner SA, Yiu AP, Cole CJ, Matynia A, Brown RA, Neve RL, Guzowski JF, Silva AJ, Josselyn SA (2007) Neuronal competition and selection during memory formation. Science 316:457–460
Han J-H, Kushner SA, Yiu AP, Hsiang H-L, (Liz), Buch T, Waisman A, Bontempi B, Neve RL, Frankland PW, Josselyn SA (2009) Selective erasure of a fear memory. Science 323:1492–1496
Hillman EMC (2014) Coupling mechanism and significance of the BOLD signal: a status report. Annu Rev Neurosci 37:161–181
Joo HR, Frank LM (2018) The hippocampal sharp wave–ripple in memory retrieval for immediate use and consolidation. Nat Rev Neurosci 19:744–757
Josselyn SA, Köhler S, Frankland PW (2017) Heroes of the engram. J Neurosci 37:4647–4657
Jung JH, Wang Y, Mocle AJ, Zhang T, Köhler S, Frankland PW, Josselyn SA (2023) Examining the engram encoding specificity hypothesis in mice. Neuron. Available at: https://www.sciencedirect.com/science/article/pii/S0896627323002039. Accessed April 7 2023
Kim JJ, Fanselow MS (1992) Modality-specific retrograde amnesia of fear. Science 256:675–677
Kim JJ, Rison RA, Fanselow MS (1993) Effects of amygdala, hippocampus, and periaqueductal gray lesions on short- and long-term contextual fear. Behav Neurosci 107:1093–1098
Kitamura T, Ogawa SK, Roy DS, Okuyama T, Morrissey MD, Smith LM, Redondo RL, Tonegawa S (2017) Engrams and circuits crucial for systems consolidation of a memory. Science 356:73–78
Ladino LD, Rizvi S, Téllez-Zenteno JF (2018) The Montreal procedure: the legacy of the great Wilder Penfield. Epilepsy Behav 83:151–161
Liu X, Ramirez S, Pang PT, Puryear CB, Govindarajan A, Deisseroth K, Tonegawa S (2012) Optogenetic stimulation of a hippocampal engram activates fear memory recall. Nature 484:381–385
Lynch MA (2004) Long-term potentiation and memory. Physiol Rev 84:87–136
Ma L, Jr M (2022) Sounding the alarm: sex differences in rat ultrasonic vocalizations during Pavlovian fear conditioning and extinction. eNeuro 9(6):ENEURO.0382–22.2022
Manns JR, Eichenbaum H (2006) Evolution of declarative memory. Hippocampus 16:795–808
Marder E, Taylor AL (2011) Multiple models to capture the variability in biological neurons and networks. Nat Neurosci 14:133–138
Maren S (2008) Pavlovian fear conditioning as a behavioral assay for hippocampus and amygdala function: cautions and caveats. Eur J Neurosci 28:1661–1666
Maren S, Aharonov G, Fanselow MS (1997) Neurotoxic lesions of the dorsal hippocampus and Pavlovian fear conditioning in rats. Behav Brain Res 88:261–274
Miyashita Y (1988) Neuronal correlate of visual associative long-term memory in the primate temporal cortex. Nature 335:817–820

Molewijk HE (1995) The ambivalent behaviour "stretched approach posture" in the rat as a paradigm to characterize anxiolytic drugs. Psychopharmacology 121(1):81
Moser EI, Kropff E, Moser M-B (2008) Place cells, grid cells, and the brain's spatial representation system. Annu Rev Neurosci 31:69–89
O'Keefe J, Dostrovsky J (1971) The hippocampus as a spatial map. Preliminary evidence from unit activity in the freely-moving rat. Brain Res 34:171–175
Ohkawa N, Saitoh Y, Suzuki A, Tsujimura S, Murayama E, Kosugi S, Nishizono H, Matsuo M, Takahashi Y, Nagase M, Sugimura YK, Watabe AM, Kato F, Inokuchi K (2015) Artificial association of pre-stored information to generate a qualitatively new memory. Cell Rep 11:261–269
Olds J, Disterhoft JF, Segal M, Kornblith CL, Hirsh R (1972) Learning centers of rat brain mapped by measuring latencies of conditioned unit responses. J Neurophysiol 35:202–219
Pavlov IP (1927) Conditioned reflexes: an investigation of the physiological activity of the cerebral cortex. Oxford University Press
Penfield W, Milner B (1958) Memory deficit produced by bilateral lesions in the hippocampal zone. AMA Arch Neurol Psychiatry 79:475–497
Ramirez S, Liu X, Lin P-A, Suh J, Pignatelli M, Redondo RL, Ryan TJ, Tonegawa S (2013) Creating a false memory in the hippocampus. Science 341:387–391
Ramirez S, Liu X, MacDonald CJ, Moffa A, Zhou J, Redondo RL, Tonegawa S (2015) Activating positive memory engrams suppresses depression-like behaviour. Nature 522:335–339
Redondo RL, Kim J, Arons AL, Ramirez S, Liu X, Tonegawa S (2014) Bidirectional switch of the valence associated with a hippocampal contextual memory engram. Nature 513:426–430
Reijmers LG, Perkins BL, Matsuo N, Mayford M (2007) Localization of a stable neural correlate of associative memory. Science 317:1230–1233
Rosen JB, Pagani JH, Rolla KLG, Davis C (2008) Analysis of behavioral constraints and the neuroanatomy of fear to the predator odor trimethylthiazoline: a model for animal phobias. Neurosci Biobehav Rev 32:1267–1276
Roy DS, Arons A, Mitchell TI, Pignatelli M, Ryan TJ, Tonegawa S (2016) Memory retrieval by activating engram cells in mouse models of early Alzheimer's disease. Nature 531:508–512
Ryan TJ, Roy DS, Pignatelli M, Arons A, Tonegawa S (2015) Engram cells retain memory under retrograde amnesia. Science 348:1007–1013
Salay LD, Ishiko N, Huberman AD (2018) A midline thalamic circuit determines reactions to visual threat. Nature 557:183–189
Schacter DL, Eich JE, Tulving E (1978) Richard Semon's theory of memory. J Verbal Learn Verbal Behav 17:721–743
Scoviille WB, Milner B (1957) Loss of recent memory after bilateral hippocampal lesions. J Neurol Neurosurg Psychiatry 20(1):11
Semon RW (1921) The mneme. G. Allen & Unwin ltd, London/New York
Shang C, Chen Z, Liu A, Li Y, Zhang J, Qu B, Yan F, Zhang Y, Liu W, Liu Z, Guo X, Li D, Wang Y, Cao P (2018) Divergent midbrain circuits orchestrate escape and freezing responses to looming stimuli in mice. Nat Commun 9:1232
Shansky RM (2018) Sex differences in behavioral strategies: avoiding interpretational pitfalls. Curr Opin Neurobiol 49:95–98
Silva BA, Gross CT, Gräff J (2016) The neural circuits of innate fear: detection, integration, action, and memorization. Learn Mem 23:544–555
Sturman O, Germain P-L, Bohacek J (2018) Exploratory rearing: a context- and stress-sensitive behavior recorded in the open-field test. Stress Amst Neth 21:443–452
Tanaka KZ, Pevzner A, Hamidi AB, Nakazawa Y, Graham J, Wiltgen BJ (2014) Cortical representations are reinstated by the hippocampus during memory retrieval. Neuron 84:347–354
Terstege DJ, Epp JR (2022) Network neuroscience untethered: brain-wide immediate early gene expression for the analysis of functional connectivity in freely behaving animals. Biology 12:34
Tonegawa S, Liu X, Ramirez S, Redondo R (2015) Memory engram cells have come of age. Neuron 87:918–931

Tovote P, Fadok JP, Lüthi A (2015) Neuronal circuits for fear and anxiety. Nat Rev Neurosci 16:317–331

Tulving E (1983) Elements of episodic memory. Oxford University Press

Tulving E (2002) Episodic memory: from mind to brain. Annu Rev Psychol 53:1–25

Wood ER, Dudchenko PA, Robitsek RJ, Eichenbaum H (2000) Hippocampal neurons encode information about different types of memory episodes occurring in the same location. Neuron 27:623–633

Yang M, Augustsson H, Markham CM, Hubbard DT, Webster D, Wall PM, Blanchard RJ, Blanchard DC (2004) The rat exposure test: a model of mouse defensive behaviors. Physiol Behav 81:465–473

# Memory Storage in Distributed Engram Cell Ensembles

**Ying Zhang and Dheeraj S. Roy**

**Abstract** One of the most fascinating aspects of the brain is its ability to acquire new information from experience and retain it over time as memory. The search for physical correlates of memory, the memory engram, has been a longstanding priority in modern neurobiology. Advanced genetic approaches have led to the localization of engram cells in a few brain regions, including the hippocampus and cortex. Additionally, engram cells exhibit learning-induced, persistent modifications and have at least two states, active and silent. However, it has been hypothesized that engrams for a specific memory are distributed among multiple brain regions that are functionally connected, referred to as a unified engram complex. Recent tissue-clearing techniques have permitted high-throughput analyses of intact brain samples, which have been used to obtain a map of the engram complex for a contextual fear memory. Careful examination of these engram complex maps has revealed a potentially underappreciated contribution of subcortical regions, specifically thalamic nuclei, to memory function. These more holistic studies support the unified engram complex hypothesis for memory storage and have important implications for understanding dysfunctional engrams in the context of human disease.

**Keywords** Active · Silent · Clearing · Engram complex · Thalamus

Revealing the material basis of memory remains a central goal of modern neuroscience (Dudai and Morris 2013). A memory engram is the enduring physical or chemical changes that occur in the brain upon learning, representing acquired memory

Y. Zhang
Center for Life Sciences & IDG/McGovern Institute for Brain Research,
School of Life Sciences, Tsinghua University, Haidian District, Beijing, China
e-mail: y_zhang@mail.tsinghua.edu.cn

D. S. Roy (✉)
Department of Physiology and Biophysics, Jacobs School of Medicine
and Biomedical Sciences, State University of New York at Buffalo, Buffalo, NY, USA
e-mail: dsroy@buffalo.edu

 
J. Gräff, S. Ramirez (eds.), *Engrams*, Advances in Neurobiology 38,
https://doi.org/10.1007/978-3-031-62983-9_3

information, and it is thought that recall is realized by the expression of these changes (Semon 1904). Neuronal ensembles that are activated by learning hold engrams, and a reactivation of these neurons gives rise to the recall of the specific memory (Josselyn and Tonegawa 2020).

Many influential attempts have been made to localize engrams in the brain. As fruit flies form an olfactory memory, some mushroom body neurons show long-term modifications in an axon branch-specific manner (Davis 2011). The cerebellum in both rodents and rabbits is considered a brain region in which motor memories are stored. It was reported that different cell types of the rabbit cerebellum encode different aspects of memories for eyeblink conditioning, providing one of the first demonstrations of multiple putative engram cell populations within a single brain region contributing to complementary memory functions (Thompson 2005). Initially reported in vertebrate model systems, the transcription factor CREB upregulates the expression of immediate early genes (IEGs) in response to recent neural activity (Sheng and Greenberg 1990). This signaling pathway resulted in two widely used approaches to identify putative engram cells: (1) Overexpressing CREB in a subset of neurons in a memory brain region (e.g., amygdala, hippocampus) prior to an upcoming behavioral task allows these higher CREB cells to store memory information (Han et al. 2009; Park et al. 2016; Silva et al. 2009); and (2) By visualizing IEG-positive cells following encoding and recall in the same animal, it became possible to identify the subset of encoding-activated cells that are reactivated during recall, which may contribute to the memory recall process (Vazdarjanova and Guzowski 2004; Deng et al. 2013). Building on these approaches, advances in genetics more recently made it possible to specifically tag encoding-activated neurons in the rodent brain during a user-defined time window together with their manipulation using chemogenetics/optogenetics-like techniques during post-encoding epochs. When applied to the hippocampus, it was demonstrated that artificial reactivation of encoding-activated neurons led to robust memory recall (Liu et al. 2012), suggesting that these ensembles were indeed engram cells. Further support for these engram cells came from the observation that these cells exhibited higher input synaptic strength post-encoding relative to neighboring non-engram cells (Ryan et al. 2015). This synaptic plasticity was maintained for several days after learning, correlated with an increase in structural plasticity (i.e., dendritic spines) restricted to engram cells, and these engram cells showed preferential connectivity with engram cells, but not non-engram cells, in downstream brain regions.

Cellular protein synthesis is necessary for long-term synaptic plasticity and memory consolidation. Interestingly, memory engrams are retained under protein synthesis inhibition-induced retrograde amnesia (Roy et al. 2017a). These engram cells can be activated by optogenetic stimulation for a full-fledged recall up to eight days after learning but not by stimulation using natural recall cues (thus, amnesia). These amnesic engram cells do not show persistent synaptic/structural plasticity; however, their preferential engram cell-to-engram cell connectivity is preserved. The authors referred to this state as “silent” engram cells, indicating that engrams may be maintained in at least two different states: silent or active, where the latter refers to engram cells that are effectively reactivated by natural cues to drive recall and exhibit correlates of long-lasting, learning-induced plasticity. The retention of memory information under amnesia suggested that the time-limited protein synthesis following learning is dispensable for memory storage but may be necessary for effective memory retrieval processes. Consistent with this idea, optogenetic memory recall under amnesia was shown to be stimulation strength-dependent, with low-power stimulation eliciting only partial recall. Finally, the authors showed that silent engram cells can be converted to active engram cells by the overexpression of α-p-21-activated kinase 1 (PAK1), which increased the spine density and hence connectivity of engram cells (Roy et al. 2017a). These findings indicate that memory information may be retained in the form of a silent engram under amnesic states and supports the hypothesis that memory is stored as the specific connectivity between engram cells. This hypothesis is also supported by another study in which optogenetically induced long-term depression (LTD) of amygdala cells impaired existing conditioned behavioral responses (Nabavi et al. 2014). However, subsequent optogenetically induced long-term potentiation (LTP) of the same cells restored natural cue-evoked memory recall. Therefore, the memory information must have persisted in the brain, possibly in the form of preferential engram connectivity, even after the amygdala synapses were depressed. Still, the lack of synaptic potentiation prevented successful memory retrieval.

While these studies have greatly enhanced our understanding of localized memory engrams in a few brain regions as well as the different states of engram cells, the engram theory suggests that a given memory is stored not in a single engram ensemble localized in a brain site but instead in a unified engram complex–a network of engram cell ensembles connected and distributed in dispersed brain regions (Semon 1904). Until recently, a thorough mapping of a unified engram complex for a specific memory was not achieved.

## Mapping Engram Cells Across the Brain

The most common strategy used to identify putative engram cells across brain regions has been to start by quantifying the number of IEG-positive cells per region immediately following a memory behavior test. Building on such initial datasets, functional neuronal manipulations in candidate regions were used to provide causal evidence for their contributions to memory function. In an early pioneering attempt (Wheeler et al. 2013), the authors characterized neurons expressing the IEG c-Fos in 84 brain regions, including cortical, hippocampal, thalamic, and midbrain nuclei, following contextual fear memory recall either 1- or 36-days post-training. While both timepoints are considered long-term memory, weeks after training (e.g., the 36-day timepoint) it is thought that memories become increasingly dependent on brain regions other than those in the medial temporal lobe system (including the hippocampus), such as cortical regions, which is a process referred to as systems consolidation of memory. They identified several brain regions co-activated by recall, among which 21 regions were targeted in subsequent loss-of-function experiments (Vetere et al. 2017). Specifically, a chemogenetic approach was used to inhibit the neuronal activity of these regions individually following fear training for 10 consecutive days, after which memory recall tests were performed. Four regions, namely hippocampal CA1, thalamic nucleus reuniens (RE), laterodorsal (LD) thalamus, and lateral septum (LS), were necessary for systems consolidation, supporting the hypothesis that these regions may serve as network hubs for remote memories. Although this work revealed a fear memory network, the use of standard histological techniques did not permit a complete examination of the entire mouse brain, and since the causal experiments were not limited to training-induced c-Fos-positive cells, direct evidence in support of the presence of engram cells was not obtained.

The development of tissue-clearing techniques, such as CLARITY (Chung et al. 2013) and iDISCO (Renier et al. 2014), combined with advanced light-sheet microscopy, has enabled high-throughput analyses of intact brain samples. Brain-wide mapping of behavior-activated neurons (i.e., based on IEG expression) has been applied to foot shocks and cocaine administration, pharmacologically induced brain states, whisker stimulation, parental behavior, and fear memory (Renier et al. 2016; Ye et al. 2016; DeNardo et al. 2019). To characterize neurons activated by either foot shocks or cocaine administration in a brain-wide manner, a transgenic mouse line was used to link the expression of the IEG Arc to a fluorescent reporter in a user-defined time window (Ye et al. 2016). CLARITY-processed brain samples led to the finding that foot shocks and cocaine recruit a core group of overlapping brain regions, including subregions of the amygdala (basolateral or BLA, central or CeA), prefrontal cortex (PFC), and nucleus accumbens (NAc). The authors focused on PFC ensembles and found that shock- and cocaine-activated neurons exhibit distinct projection targets. They also demonstrated that shock- and cocaine-activated neurons are sufficient to drive aversive and appetitive behavioral responses, respectively. Comparable experiments were performed using iDISCO to map brain-wide activity patterns in several conditions, including the administration of antipsychotic haloperidol, whisker-related exploration, and parenting female mice (Renier et al. 2016). Whole brain immunostaining for several IEGs, such as c-Fos, Arc, Egr-1, pS6, and Npas4, was performed following iDISCO-based tissue clearing. The authors identified critical brain regions for each behavioral experiment based on neuronal activity increases across the entire brain. For instance, haloperidol administration resulted in robust activation of globus pallidus, dorsal striatum, and NAc, whereas parenting female mice showed the highest density of activated neurons in the anterior region of the hypothalamus. While these findings suggested that some of these regions may contain engram cells, functional experiments in which the underlying cells are perturbed were not carried out to demonstrate this possibility. Another group examined auditory fear memory using a transgenic mouse line in which the expression of the IEG c-Fos was linked to a fluorescent reporter in a user-defined time window together with iDISCO (DeNardo et al. 2019). By comparing brain-wide activity patterns between fear training and memory recall at multiple time points (1-, 7-, 14-days post-training), the authors identified a critical role for PFC in memory recall at remote time points (i.e., 14-days post-training). In addition to providing causal evidence for PFC engram cells, they also noted that the brain regions that co-vary with PFC neuronal activity change over time; 1-day post-training PFC clustered with the amygdala, midline thalamic nuclei, and higher-order visual areas, whereas 14-days post-training PFC primarily clustered with cortical regions such as anterior cingulate (ACC), temporal association, and entorhinal (EC). Further support for these PFC engrams contributing to remote memory recall was provided by engram reactivation evidence showing a time-dependent increase in the reactivation of PFC engram cells, which is a key criterion of the engram theory. While these studies represent the first attempts to apply tissue-clearing approaches for brain-wide activity mapping, engram cells were demonstrated only in a limited brain region (i.e., PFC) rather than in a brain-wide manner.

More recently, engram cells were mapped across the mouse brain using an improved tissue-clearing approach known as SHIELD, which exhibits better fluorescent protein stabilization as compared to CLARITY/iDISCO (Park et al. 2018). During a contextual fear memory paradigm, brain-wide neural activity was visualized by linking the expression of the IEG c-Fos to tdTomato fluorescence (Fig. 1) (Roy et al. 2022a). SHIELD-processed brain samples were imaged using a high-speed selective plane illumination microscope (SPIM) followed by a neural

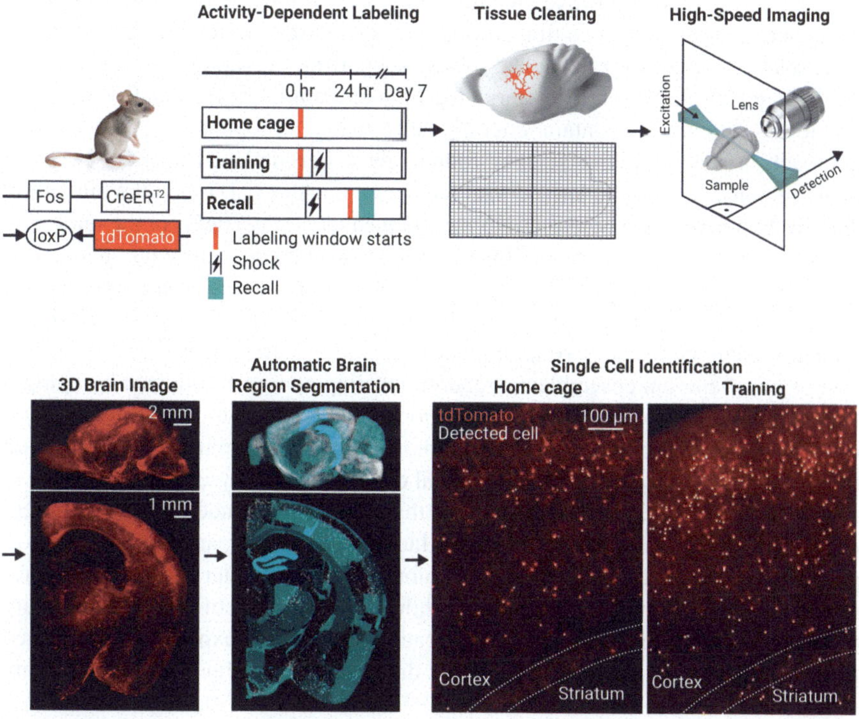


**Fig. 1** Brain-wide activity mapping. A transgenic line links the activity of the IEG c-Fos to fluorescent tdTomato expression through the mouse brain in a user-defined behavioral window. State-of-the-art tissue clearing, whole-brain sample microscopy, and computational approaches permit the automatic detection of behavior-activated neurons in each brain region at single-cell resolution

network-based automatic cell counting approach. Brain-wide activity maps for fear encoding and recall from 247 regions revealed the activation pattern in the cerebral cortex, amygdala, thalamus, striatum, hypothalamus, and hippocampus as well as their subdivisions. While recall activated fewer neurons than encoding, brain regions most strongly activated by recall were also strongly activated by encoding. Because conducting optogenetic or chemogenetic manipulations that permit an assertion of engrams in hundreds of brain regions that display activity-dependent labeling is impractical, the authors developed an analytical approach to narrow down the list of brain regions. To rank-order the list of putative engram-containing brain regions, they devised an engram index based on the concept that engrams are held by neuronal ensembles activated by learning and reactivated to support recall (Semon 1904). The latter suggests elevated neuronal activity in putative engram brain regions induced by recall. The numerator of the engram index equation computes the increase in neural activity for a given brain region during encoding as compared to baseline (i.e., home cage). In contrast, the denominator computes the difference between the activation levels of encoding and recall. In doing so, the latter decreases the engram index for brain regions with high activity during either encoding or recall, which also helps exclude regions primarily responsive to the foot shocks themselves (i.e., false positive candidate engram regions). Hippocampal subdivisions, amygdala, and association cortical brain structures were among the regions with relatively high engram index values, supporting the validity of this engram index approach. Importantly, by targeting high engram index regions that were not previously known to contain engram cells using an optogenetic reactivation approach in a neutral context, they found robust light-induced memory recall in many regions, such as the laterodorsal tegmental nucleus (LDTg) in the hindbrain, anterior hypothalamic nucleus (AHN), periaqueductal gray (PAG) in the midbrain, and ventrolateral orbital cortex (VLO). Although most optogenetic engram reactivation studies have employed 20 Hz protocols, this study suggested that frequencies closer to the in vivo firing rates of each brain region/cell type were most effective at inducing memory recall. Two additional observations strengthened the identification of novel engram ensembles; a random labeling approach followed by optogenetic reactivation could not induce memory recall, and optogenetic reactivation experiments applied to regions that did not show high engram index values did not result in memory recall. Further, functional connectivity analyses led the authors to suggest that there is a core set of regions that are similarly modulated by different engram cell manipulations. Core regions, including EC, PAG, and PFC, may represent hub regions for a contextual fear memory that exert greater influence on memory processes. Together, this study shows how the application of brain-wide activity mapping aided by an engram index and causal reactivation experiments permitted the discovery of engram cell ensembles distributed throughout the mouse brain (Fig. 2).

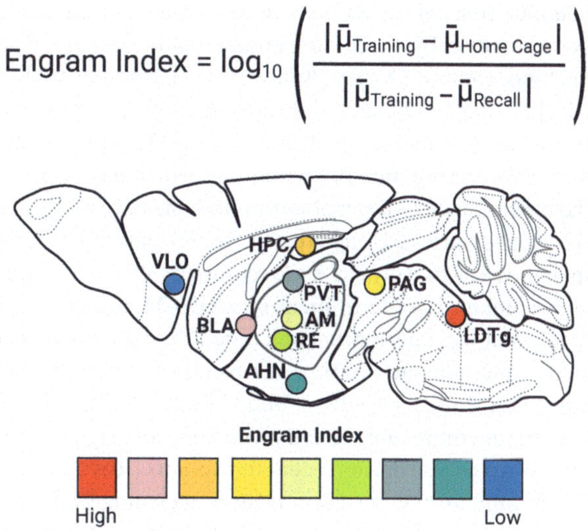


**Fig. 2** A partial list of constituent engram ensembles in the unified engram complex for a contextual fear memory. The engram index helps rank-order putative engram-containing brain regions based on a brain-wide activity map, which helps guide the powerful functional intervention experiments needed to confirm the existence of engrams. Hippocampus (HPC)

## Thalamic Contributions to Memory

Several research groups have identified memory engrams in cortical regions as well as subcortical structures, namely the hippocampus and amygdala (Josselyn and Tonegawa 2020). Other than these well-known learning and memory structures, nuclei within the thalamus have repeatedly been noted in brain-wide activity mapping studies (Vetere et al. 2017; Roy et al. 2022a). Evolutionary studies have suggested that the expanded neocortex in humans, as compared to other mammalian species, underlies our exceptional executive function capabilities. Decades of research on cortical processes have demonstrated that each cortical region performs one or more executive functions. These cortical functions critically depend on the integration of internal states (emotion) and external information (sensory inputs). The thalamus, being a crucial region for sensory processing and multi-sensory integration with both bottom-up and top-down connectivity, is in a powerful anatomical and functional position to guide cortical processes (Roy et al. 2022b). The thalamus contains ~50 subnuclei that together gives rise to the major excitatory input to the entire cortex, which in turn sends return projections to one or more thalamic nuclei (Sherman and Guillery 1996). This connectivity pattern has led to the idea that the thalamus also integrates information being processed across the cortex. While past thalamic research has focused mostly on sensory processing (i.e., their relay role), the importance of non-relay contributions for memory processing has recently been emphasized. Human subjects with localized thalamic lesions not only exhibit widespread cortical dysfunction but also show cognitive function deficits, supporting a hub-like role for the thalamus in regulating cortical functions (Shine et al. 2023).

Complementary studies using animal models have also pointed towards a critical role for the thalamus in learning and memory. For instance, optogenetic manipulations of the RE thalamus revealed contributions to memory specificity and generalization by processing contextual information from the PFC that is routed to the hippocampus (Xu and Sudhof 2013). Another study showed an active role of the RE thalamus in contextual discrimination, specifically for hippocampus-dependent memories (Ramanathan et al. 2018). PFC neurons show elevated neural activity during a working memory paradigm, which has been causally linked to inputs from the mediodorsal (MD) thalamus (Bolkan et al. 2017). It is well accepted that memories can be retrieved long after the initial learning experience. Paraventricular (PVT) thalamus acts as a crucial node recruited into cortico-amygdalar networks to retrieve and maintain long-term memories (Do-Monte et al. 2015). There is growing evidence that the anterior thalamic nuclei (ATN), which is composed of three subdivisions anterodorsal (AD), anteroventral (AV), and anteromedial (AM), support episodic memory (Aggleton and O'Mara 2022). It is thought that AD neurons participate in contextual memory encoding through their projections to the retrosplenial cortex (RSC) during recent time points (Roy et al. 2021), while being disengaged via an inhibitory connection from hippocampal CA3 during systems consolidation to support remote memory recall (Vetere et al. 2021). In contrast, AV neurons are dispensable for recent contextual memory encoding but are necessary for working memory processes (Roy et al. 2022c). AM neurons convey cortico-hypothalamic information to hippocampal circuitry for the generation of context-specific, high-valence memories, both at recent and remote time points (de Lima et al. 2017; Toader et al. 2023).

In a brain-wide engram mapping study (Roy et al. 2022a), the authors found high engram indices in AM, PVT, and RE thalamus, suggesting that these regions contain engrams for contextual fear memory. Optogenetic reactivation of putative engram cells in each of these three thalamic nuclei resulted in robust memory recall in a neutral context, representing the first direct demonstration of thalamic engrams (Fig. 2). Crucially, when MD thalamus was reactivated in a similar manner memory recall was not detected, consistent with the fact that MD showed a lower engram index as compared to AM/PVT/RE thalamus. These experiments support the idea that subcortical structures, in particular the thalamus, play an underappreciated role in memory.

## Reactivation of Engram Cells During Recall

To support the presence of engram cells in a brain region, at least a part of these cells must be reactivated during successful memory recall (Semon 1904). A strategy that has been widely used to study engram reactivation includes fluorescently tagging learning-activated ensembles followed by their reactivation, based on IEG (e.g., c-Fos) expression, induced by recall (Ryan et al. 2015). Applying this strategy to individual brain regions has revealed engram reactivation in hippocampal, cortical, and amygdala subregions (Josselyn and Tonegawa 2020). Because IEG staining has been challenging in whole brain preparations, engram reactivation studies have been mostly limited to one or a few brain regions. A novel staining technology known as eFLASH (electrophoretically driven fast labeling using affinity sweeping in hydrogel) enables deep penetration of antibodies in whole brain samples along with uniform labeling (Yun et al. 2019). By combining brain-wide tagging of learning-activated ensembles and brain-wide c-Fos staining after recall using eFLASH, the first brain-wide engram reactivation experiments were performed (Roy et al. 2022a). The authors observed that ~36% (i.e., 88/247) of the learning-activated brain regions analyzed had engram reactivation above chance levels following contextual fear memory recall. These experiments not only revealed engram reactivation in known hippocampal and amygdala subregions but also showed reactivation in many thalamic, cortical, midbrain, and brainstem structures. Interestingly, when they compared the candidate regions identified by their engram index approach with these engram reactivation results, over 60% of the regions were shared between analyses. Specifically, many probable engram-carrying regions with high engram index values (e.g., LDTg, cunciform nucleus, nucleus of the lateral olfactory tract, medial pretectal area, and tuberomammillary nucleus) also exhibited high engram reactivation. This observation supports the validity of employing an engram index approach to identify putative engram-containing brain regions with high probability, especially because brain-wide engram reactivation studies remain technically challenging.

Although optogenetic reactivation of an individual engram cell ensemble results in significant levels of memory recall, it does not reach the level of recall attained by natural recall cues (Liu et al. 2012). This observation, together with the fact that over 35% of all learning-activated brain regions are reactivated for memory recall (Roy et al. 2022a), suggests that the full level of recall may be achieved by reactivating multiple engram cell ensembles at the same time. This possibility was tested using a chemogenetic neuronal activation approach based on the excitatory hM3Dq DREADDs receptor (Roy et al. 2022a). A DREADD approach was preferred over optogenetics since the former allows for the simultaneous activation of several hM3Dq-expressing engram populations distributed throughout the brain. The authors reactivated a single engram cell ensemble (hippocampal CA1), two engram cell ensembles (CA1, BLA) tagged in the same animals, three engram cell ensembles (CA1, BLA, AM) tagged in the same animals, and four engram cell ensembles (CA1, BLA, AM, RE) tagged in the same animals. While CA1 engram cell reactivation resulted in memory recall, the reactivation of two and three engram cell ensembles conferred enhanced memory recall strength (Fig. 3). This finding indicated that the simultaneous reactivation of multiple engram cell ensembles results in greater memory recall than their subsets. Interestingly, they did not observe further enhancement of memory recall by the simultaneous reactivation of four engram cell ensembles, suggesting that there may be an upper limit in the capacity of memory engram cell reactivation to modulate behavioral outputs. Nevertheless, since the simultaneous reactivation of three- and four-engram cell ensembles resulted in memory recall strength comparable to natural memory recall, this may reflect a mechanism by which the brain adjusts the strength of memory recall depending on the importance of the memory for current and future decisions. Complementing these observations

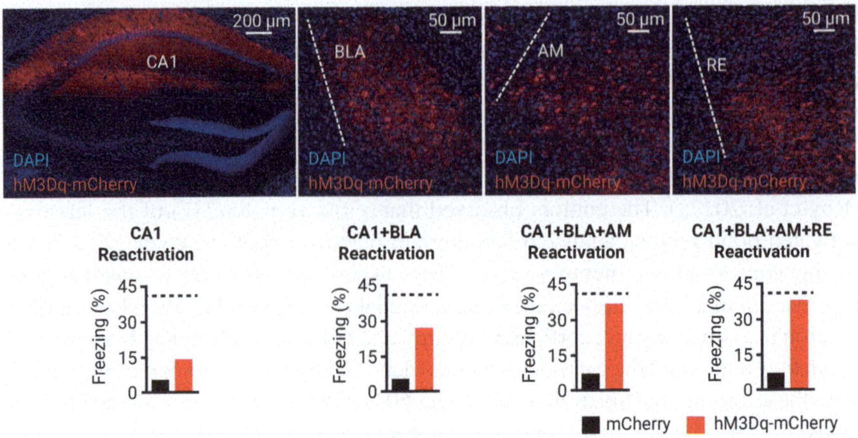


**Fig. 3** Simultaneous reactivation of multiple engram cell ensembles. Reactivation of three or four engram cell ensembles distributed throughout the brain in the same animals confers enhanced memory recall strength that is comparable to recall induced by natural recall cues (dashed line in graphs indicate natural recall level)

is the finding of comparable neural activity patterns across brain regions between natural memory recall and recall induced by engram cell reactivation, suggesting that the reactivation of engram cells recreates at least part of the activity pattern corresponding to natural memory recall (Roy et al. 2022a).

In line with the observation that over a third of all learning-activated brain regions exhibit significant engram reactivation following memory recall (Roy et al. 2022a), one interpretation is that engrams are stabilized because of cellular consolidation (i.e., stable engrams). However, due to the relatively low overlap between learning-activated and recall-activated neurons (~12–35%) (Roy et al. 2022a), another hypothesis is that engrams may change due to consolidation with neurons dropping out of or into the engram (i.e., dynamic engrams). Using a spiking neural network model, the evolution of engram cells and their selectivity with consolidation was probed (Tome et al. 2022). This model predicted that many learning-activated neurons would be removed from the engram as consolidation progressed, while other neurons that were not activated by learning were recruited into the engram, supporting the dynamic engram idea. Interestingly, the evolution of engram cells correlated with engrams being initially encoded in an unselective state that becomes selective with consolidation. The authors proposed that learning-activated neurons may be forced out of the engram due to strong inhibition and therefore only neurons that are highly responsive to the training cues will remain in the engram after consolidation. In support of these modeling observations, using a contextual fear conditioning task in mice, it was demonstrated that memory selectivity emerged after a delay period of 12 hours following training. Also, the level of recall-induced reactivation of learning-activated hippocampal dentate gyrus (DG) neurons declined with consolidation, a phenomenon that required intact DG cholecystokinin-expressing (CCK) interneurons. Together, this study showed that memories could be encoded by dynamic engrams that become selective with cellular consolidation, which requires local inhibitory activity during recall.

Are there engram-containing brain regions that exhibit heightened neural activity during recall as compared to encoding? Although the engram theory may not directly predict such recall-enhanced regions (Semon 1904), three brain regions showed higher recall activation as compared to memory encoding in a brain-wide engram mapping experiment (Roy et al. 2022a). These regions are the olfactory cortex (Ol), lateral hypothalamus (LH), and AV thalamus. The authors suggested that the activation of engram cells by recall cues in these regions is enhanced to support specialized recall processes. A previously reported example of a brain region in the literature that is crucial for memory recall and shows enhanced neuronal activity (i.e., c-Fos) during recall is the dorsal subiculum (dSub) (Roy et al. 2017b). While the mechanism underlying such recall-enhancing processes remains unknown, one possibility is the presence of local recurrent circuitry in such brain regions. If the recurrent circuitry is strengthened during the post-encoding consolidation time window, subsequent reactivation of engram cells by recall cues may result in the firing of connected cells in the region. In this scenario, these brain regions will show greater numbers of c-Fos activated neurons by recall as compared to memory encoding. These observations suggest that while engrams are reactivated to support memory recall, the recall process itself offers an opportunity to modify features of engrams (e.g., synaptic plasticity, connectivity, etc.).

## Perspectives for Future Engram Research

The concept that a memory is stored not just in a single engram cell ensemble but in learning-induced enduring changes in multiple functionally connected neuronal ensembles was suggested by Richard Semon ("unified engram complex") (Semon 1904) and Donald Hebb ("neurons that fire together wire together") (Hebb 1949). The experimental evidence for this concept came from an observation that gene expression is altered by experience in widespread, behaviorally defined neural circuits (Helmstetter et al. 2008), multiunit recording experiments that identified distributed brain regions involved in memory formation (Gabriel 1993), and an analysis of engram cells from multiple hippocampal subfields and the amygdala (Ryan et al. 2015; Choi et al. 2018). Further support for the unified engram complex has since been provided by brain-wide activity mapping studies, which have revealed the hitherto most comprehensive set of brain regions where the component engrams are localized for a single engram complex (Vetere et al. 2017; DeNardo et al. 2019; Roy et al. 2022a). The distributed nature of engram cell ensembles for a specific memory has suggested that the memory engram within an individual brain region may contribute a subset of the overall memory information (Eichenbaum 2016). For example, hippocampal engrams are thought to primarily contribute contextual information by acting as an index for cortical memories of various sensory modalities (Teyler and DiScenna 1986; Teyler and Rudy 2007), whereas amygdala engrams hold valence information for a given experience (Morrison and Salzman 2010; Tovote et al. 2015). In addition, cortical engrams such as those in the retrosplenial and prefrontal cortices may support spatial navigation (Vann et al. 2009) and top-down control of memory retrieval (Tomita et al. 1999), respectively. Further, engrams in auditory (Fritz et al. 2005) and olfactory cortices (Shakhawat et al. 2014) may support auditory recognition memory and odor-induced learned behaviors, respectively. Although brain-wide activity mapping has been a powerful approach to further our understanding of memory engrams, current methodologies have at least three limitations that could be addressed in the future. One limitation is that current approaches cannot be used to identify silent engram cells, which are formed in certain brain regions during encoding but are not reactivated by natural recall cues for recall, although they can be optogenetically reactivated (Roy et al. 2017a). A second limitation is that current approaches have been restricted to brain regions where IEGs are expressed, which is the case for most of the brain, but not all regions express IEGs. Finally, the slow dynamics of IEGs (i.e., tens of minutes to a few hours) have not allowed researchers to distinguish between encoding-activated and post-encoding-activated ensembles.

An exciting question about engrams is whether it is possible to identify transcriptional signatures that could distinguish engram cells from their non-engram neighbors. In one study, hippocampal engram cells labeled by the IEG Arc following contextual fear conditioning were profiled 24 hours later (i.e., the long-term memory time point) (Rao-Ruiz et al. 2019). Only engram cells showed CREB-dependent transcriptional changes linked to the co-expression of Arc, Atf3, Penk, and Kcnq3, which are genes known to play a role in synaptic and structural plasticity. The authors showed that overexpressing CREB in learning-activated neurons disrupts the co-expression network of Arc, Atf3, Penk, and Kcnq3, resulting in memory deficits 72 hours after encoding. This behavioral observation supports the hypothesis that CREB-induced changes in this four-gene network specific to engram cells are required for their long-lasting maturation and stability. Another study characterized chromatin accessibility, 3D genomic architecture, and gene expression of hippocampal engram cells labeled by the IEG Arc (Marco et al. 2020). They found that while memory encoding significantly modifies the chromatin accessibility state of engram cells, only minor changes in accessibility are seen after cellular consolidation and recall. Unexpectedly, it was noted that memory encoding results in an initial increase in enhancer accessibility without the expected transcriptional changes. Further, engram reactivation uses a subset of the encoding-primed enhancers to bind to their respective promoters to upregulate genes involved in mRNA transport and local protein translation (including Kif5, Eif, Rpl, and Rps), providing strong evidence that memory recall modulates the state of engram cells. Transcriptional profiling has also been applied to remote memory engram cells in the cortex (Chen et al. 2020). The authors found ~100 differentially expressed genes specifically in engram cells; these included genes for transcriptional regulation, vesicle exocytosis, transmembrane transport, dendritic spine organization, and long-range intracellular transport. They also identified remote memory-specific transcriptional changes in non-neuronal cortical cells (i.e., astrocytes and microglia), which included ligand-receptor genes known to support communication between such non-neuronal populations and neighboring excitatory neurons (i.e., engram cells). Based on these observations, it is likely that non-neuronal cells contribute to the maintenance of remote memory engrams.

As our mechanistic insights into the recruitment, stabilization, and reactivation of memory engrams mature, studying how engrams are altered in disease states will become a priority. For instance, many of the upregulated genes in engram cells across remote memory are linked to memory-affecting neuronal disorders (Chen et al. 2020). Multiple groups have examined hippocampal engram cells in mouse models of Alzheimer's disease (AD) (Roy et al. 2016; Perusini et al. 2017). In one study, although early AD mice were amnesic in long-term memory tests, direct optogenetic activation of hippocampal engram cells labeled by c-Fos resulted in robust memory recall (Roy et al. 2016). Amnesia is correlated with decreased dendritic spine density of engram cells. Optogenetic LTP applied to incoming perforant path synapses of engram cells restored both spine density and long-term memory recall induced by natural recall cues. Critically, the long-term memory rescue required intact hippocampal engram cells with restored spine density. Another study showed that direct optogenetic activation of hippocampal engram cells labeled by Arc is sufficient to improve recall in AD mice (Perusini et al. 2017). These observations suggest that strategies applied to engram cells and their circuits can support long-lasting improvements in cognitive functions, which may provide insights and therapeutic value for future approaches that rescue memory in AD as well as other cognitive disorders.

**Acknowledgments** This work was supported by the School of Life Sciences and the IDG/McGovern Institute for Brain Research (Y.Z.) and the Warren Alpert Distinguished Scholar Award and NIH 1K99NS125121-01 (D.R.).

## References

Aggleton JP, O'Mara SM (2022) The anterior thalamic nuclei: core components of a tripartite episodic memory system. Nat Rev Neurosci 23:505–516

Bolkan SS et al (2017) Thalamic projections sustain prefrontal activity during working memory maintenance. Nat Neurosci 20:987–996

Chen MB, Jiang X, Quake SR, Sudhof TC (2020) Persistent transcriptional programmes are associated with remote memory. Nature 587:437–442

Choi JH et al (2018) Interregional synaptic maps among engram cells underlie memory formation. Science 360:430–435

Chung K et al (2013) Structural and molecular interrogation of intact biological systems. Nature 497:332–337

Davis RL (2011) Traces of Drosophila memory. Neuron 70:8–19

de Lima MA, Baldo MV, Canteras NS (2017) A role for the anteromedial thalamic nucleus in the acquisition of contextual fear memory to predatory threats. Brain Struct Funct 222:113–129

DeNardo LA et al (2019) Temporal evolution of cortical ensembles promoting remote memory retrieval. Nat Neurosci 22:460–469

Deng W, Mayford M, Gage FH (2013) Selection of distinct populations of dentate granule cells in response to inputs as a mechanism for pattern separation in mice. elife 2:e00312
Do-Monte FH, Quinones-Laracuente K, Quirk GJ (2015) A temporal shift in the circuits mediating retrieval of fear memory. Nature 519:460–463
Dudai Y, Morris RG (2013) Memorable trends. Neuron 80:742–750
Eichenbaum H (2016) Still searching for the engram. Learn Behav 44:209–222
Fritz J, Mishkin M, Saunders RC (2005) In search of an auditory engram. Proc Natl Acad Sci USA 102:9359–9364
Gabriel M (1993) In: Vogt BA, Gabriel M (eds) Neurobiology of Cingulate Cortex and Limbic Thalamus. Birkhauser
Han JH et al (2009) Selective erasure of a fear memory. Science 323:1492–1496
Hebb DO (1949) The organization of behavior: a neuropsychological theory. Wiley
Helmstetter FJ, Parsons RG, Gafford GM (2008) Macromolecular synthesis, distributed synaptic plasticity, and fear conditioning. Neurobiol Learn Mem 89:324–337
Josselyn SA, Tonegawa S (2020) Memory engrams: recalling the past and imagining the future. Science 367:eaaw4325
Liu X et al (2012) Optogenetic stimulation of a hippocampal engram activates fear memory recall. Nature 484:381–385
Marco A et al (2020) Mapping the epigenomic and transcriptomic interplay during memory formation and recall in the hippocampal engram ensemble. Nat Neurosci 23:1606–1617
Morrison SE, Salzman CD (2010) Re-valuing the amygdala. Curr Opin Neurobiol 20:221–230
Nabavi S et al (2014) Engineering a memory with LTD and LTP. Nature 511:348–352
Park S et al (2016) Neuronal allocation to a hippocampal engram. Neuropsychopharmacology 41:2987–2993
Park YG et al (2018) Protection of tissue physicochemical properties using polyfunctional crosslinkers. Nat Biotechnol 37:73–83
Perusini JN et al (2017) Optogenetic stimulation of dentate gyrus engrams restores memory in Alzheimer's disease mice. Hippocampus 27:1110–1122
Ramanathan KR, Ressler RL, Jin J, Maren S (2018) Nucleus reuniens is required for encoding and retrieving precise, hippocampal-dependent contextual fear memories in rats. J Neurosci 38:9925–9933
Rao-Ruiz P et al (2019) Engram-specific transcriptome profiling of contextual memory consolidation. Nat Commun 10:2232
Renier N et al (2014) iDISCO: a simple, rapid method to immunolabel large tissue samples for volume imaging. Cell 159:896–910
Renier N et al (2016) Mapping of brain activity by automated volume analysis of immediate early genes. Cell 165:1789–1802
Roy DS et al (2016) Memory retrieval by activating engram cells in mouse models of early Alzheimer's disease. Nature 531:508–512
Roy DS, Muralidhar S, Smith LM, Tonegawa S (2017a) Silent memory engrams as the basis for retrograde amnesia. Proc Natl Acad Sci USA 114:E9972–E9979
Roy DS et al (2017b) Distinct neural circuits for the formation and retrieval of episodic memories. Cell 170:1000–1012
Roy DS et al (2021) Anterior thalamic dysfunction underlies cognitive deficits in a subset of neuropsychiatric disease models. Neuron 109:2590–2603
Roy DS et al (2022a) Brain-wide mapping reveals that engrams for a single memory are distributed across multiple brain regions. Nat Commun 13:1799
Roy DS, Zhang Y, Halassa MM, Feng G (2022b) Thalamic subnetworks as units of function. Nat Neurosci 25:140–153
Roy DS et al (2022c) Anterior thalamic circuits crucial for working memory. Proc Natl Acad Sci USA 119:e2118712119
Ryan TJ, Roy DS, Pignatelli M, Arons A, Tonegawa S (2015) Engram cells retain memory under retrograde amnesia. Science 348:1007–1013

Semon R (1904) Die Mneme als erhaltendes Prinzip im Wechsel des organischen Geschehens. Wilhelm Engelmann

Shakhawat AM et al (2014) Visualizing the engram: learning stabilizes odor representations in the olfactory network. J Neurosci 34:15394–15401

Sheng M, Greenberg ME (1990) The regulation and function of c-fos and other immediate early genes in the nervous system. Neuron 4:477–485

Sherman SM, Guillery RW (1996) Functional organization of thalamocortical relays. J Neurophysiol 76:1367–1395

Shine JM, Lewis LD, Garrett DD, Hwang K (2023) The impact of the human thalamus on brain-wide information processing. Nat Rev Neurosci 24:416–430

Silva AJ, Zhou Y, Rogerson T, Shobe J, Balaji J (2009) Molecular and cellular approaches to memory allocation in neural circuits. Science 326:391–395

Teyler TJ, DiScenna P (1986) The hippocampal memory indexing theory. Behav Neurosci 100:147–154

Teyler TJ, Rudy JW (2007) The hippocampal indexing theory and episodic memory: updating the index. Hippocampus 17:1158–1169

Thompson RF (2005) In search of memory traces. Annu Rev Psychol 56:1–23

Toader AC et al (2023) Anteromedial thalamus gates the selection and stabilization of long-term memories. Cell 186:1369–1381

Tome DF et al (2022) Dynamic and selective engrams emerge with memory consolidation. Preprint at bioRxiv. https://doi.org/10.1101/2022.03.13.484167

Tomita H, Ohbayashi M, Nakahara K, Hasegawa I, Miyashita Y (1999) Top-down signal from prefrontal cortex in executive control of memory retrieval. Nature 401:699–703

Tovote P, Fadok JP, Luthi A (2015) Neuronal circuits for fear and anxiety. Nat Rev Neurosci 16:317–331

Vann SD, Aggleton JP, Maguire EA (2009) What does the retrosplenial cortex do? Nat Rev Neurosci 10:792–802

Vazdarjanova A, Guzowski JF (2004) Differences in hippocampal neuronal population responses to modifications of an environmental context: evidence for distinct, yet complementary, functions of CA3 and CA1 ensembles. J Neurosci 24:6489–6496

Vetere G et al (2017) Chemogenetic interrogation of a brain-wide fear memory network in mice. Neuron 94:363–374

Vetere G et al (2021) An inhibitory hippocampal-thalamic pathway modulates remote memory retrieval. Nat Neurosci 24:685–693

Wheeler AL et al (2013) Identification of a functional connectome for long-term fear memory in mice. PLoS Comput Biol 9:e1002853

Xu W, Sudhof TC (2013) A neural circuit for memory specificity and generalization. Science 339:1290–1295

Ye L et al (2016) Wiring and molecular features of prefrontal ensembles representing distinct experiences. Cell 165:1776–1788

Yun DH et al (2019) Ultrafast immunostaining of organ-scale tissues for scalable proteomic phenotyping. Preprint at bioRxiv. https://doi.org/10.1101/660373

# Hippocampal Engrams and Contextual Memory

**Krithika Vasudevan, James E. Hassell Jr., and Stephen Maren**

**Abstract** Memories are not formed in a vacuum and often include rich details about the time and place in which events occur. Contextual stimuli promote the retrieval of events that have previously occurred in the encoding context and limit the retrieval of context-inappropriate information. Contexts that are associated with traumatic or harmful events both directly elicit fear and serve as reminders of aversive events associated with trauma. It has long been appreciated that the hippocampus is involved in contextual learning and memory and is central to contextual fear conditioning. However, little is known about the underlying neuronal mechanisms underlying the encoding and retrieval of contextual fear memories. Recent advancements in neuronal labeling methods, including activity-dependent tagging of cellular ensembles encoding memory ("engrams"), provide unique insight into the neural substrates of memory in the hippocampus. Moreover, these methods allow for the selective manipulation of memory ensembles. Attenuating or erasing fear memories may have considerable therapeutic value for patients with post-traumatic stress disorder or other trauma- or stressor-related conditions. In this chapter, we review the role of the hippocampus in contextual fear conditioning in rodents and explore recent work implicating hippocampal ensembles in the encoding and retrieval of aversive memories.

**Keywords** Contextual learning · Aversive memories · Hippocampus · DREADDs

K. Vasudevan · J. E. Hassell Jr. · S. Maren (✉)
Department of Psychological and Brain Sciences and Institute for Neuroscience, Texas A&M University, College Station, TX, USA
e-mail: maren@tamu.edu



J. Gräff, S. Ramirez (eds.), *Engrams*, Advances in Neurobiology 38,
https://doi.org/10.1007/978-3-031-62983-9_4

## Introduction

The capacity to learn and remember from experience is a core function of the nervous system that allows organisms to adaptively respond to their environment. Any given experience has the capacity to yield multiple memories, some of which are transient and others that might persist for the lifetime of the animal. This information is encoded by multiple memory systems in the brain and includes both procedural (implicit) and declarative (explicit) content. Consider the following example. Imagine that you are stung behind the ear by a bee while standing in front of the chocolate fountain during an outdoor wedding reception. Moments later, you feel your lips begin to swell, your throat close, and breathing becomes a struggle. You fall to the ground, gasping for air in the throes of a potentially deadly anaphylactic reaction. In this scenario, you would undoubtedly encode multiple memories of the life-threatening event. Indeed, you would associate the sound of buzzing insects with both the painful sting and the consequent severe allergic reaction. Buzzing sounds would retrieve this implicit memory and elicit a conditioned fear response that might yield sweating, elevated heart rate, agitation and a subjective state of fear. You would also encode an explicit memory of the event that would include the circumstances surrounding the event, including the time (late afternoon), place (neighbor's backyard), people (guests of honor), objects or items (chocolate fountain and buzzing bees), and internal states (fear and panic) associated with the event. This information, which collectively describes the "context" of the experience, is central to the episodic memory of the event. Unlike implicit memories, an explicit episodic memory permits "mental time travel" and allows you to recount the details of the near-fatal experience (Tulving and Thomson 1973).

As this example illustrates, context is an integral part of episodic memory—recollection evokes not only the memory of what occurred but also memory of where and when it occurred (Spear 1973). The critical role of context in memory retrieval is illustrated by "encoding specificity," the observation that content is better retrieved in the context in which it was encoded (Tulving and Thomson 1973). Contextual information is often central to recollection; without contextual details, one if often left with only a vague sense of familiarity. Consider the "the butcher on the bus" scenario (Mandler 1980), in which you encounter a man on a bus who seems familiar, but you cannot immediately collect how you know them. With time and effort, you might begin to retrieve specific details about where you last saw this man; you may recall that he wore an apron or sold you a cut of meat. As you piece these details together, it becomes clear that this is the person who works behind the meat counter at your grocery store. Encountering the butcher out of context complicated recollection of who they were—a process that would have been much easier had you encountered him in the grocery store, the context in which you had come to know him.

The role of context as a memory retrieval cue is particularly pronounced in the cases of fearful, traumatic, or otherwise negative events, the memories that can be particularly strong (Maren et al. 2013). Patients with post-traumatic stress disorder

(PTSD) or other trauma- and stressor-related disorders will often avoid contexts that remind them of the aversive event. Although the symptomatology of PTSD and other trauma- and stressor-related disorders is multifaceted and complex, animal models of these symptoms have provided considerable insight into learning and memory processes central to the pathogenesis of the disorders. Of particular interest to this chapter is the role of contextual learning and memory in maladaptive fear disorders. A common and robust behavioral paradigm used to explore this phenomenon (and the one that will be the focus of this chapter) is Pavlovian fear conditioning, a procedure in which animals learn to associate a cue or a context to an aversive event, often a mild foot shock. After fear conditioning, animals will have behavioral and autonomic responses, such as freezing and increased blood pressure, that can be used as indices of conditioned fear (LeDoux 2000; Maren 2001; Morgan et al. 1995)

Contexts are multimodal and include both exteroceptive (external) and interoceptive (internal) stimuli (Bouton et al. 2021; Maren et al. 2013). The representation of context in the brain is dynamic and flexible, allowing neuronal ensembles to respond to multiple stimuli across contexts (Bakker et al. 2008; Rolls 2016). This dynamic representation is particularly clear in the hippocampus, which is implicated in a variety of memory tasks including declarative (Eichenbaum 1999, 2004; Scoville and Milner 1957; Tulving and Markowitsch 1998) and working memory (Churchwell and Kesner 2011; Jin and Maren 2015; Lee and Kesner 2002; Sakurai 1994) and is a key structure in the encoding of context information (Corcoran et al. 2005; Hirsh 1974; Holland and Bouton 1999; Kennedy and Shapiro 2004; Maren et al. 1997). The hippocampus communicates with other brain structures involved in cognition and learning to integrate context information into memories (Fanselow 1990; Matus-Amat et al. 2004). These contextual representations are required for animals to learn about events occurring within the contexts they experience (Fanselow 1990; Matus-Amat et al. 2004).

Decades of work have shown that the hippocampus has an important role in encoding spatial information when animals are introduced to novel contexts (reference of reviews on HPC and space). For example, single-unit recordings in the hippocampus reveal the existence of place cells (Fyhn et al. 2004; Moser et al. 1995, 2015; O'Keefe et al. 1998), which are most active when an animal is in a certain location (also called the cell's "place field"); spatial firing properties have also been observed in several brain structures interconnected with the hippocampus. Of course, hippocampal neurons also encode non-spatial information, and contemporary accounts suggest that the hippocampus forms integrated contextual representations that include both spatial and non-spatial information (Eichenbaum 1996; Smith and Mizumori 2006; Wood et al. 1999). Indeed, the hippocampus contributes to contextual fear learning by not only representing the conditioning context but also stimulus-stimulus relationships within that context (Maren 2016, 2000). The hippocampus, therefore, can alternately be thought of as a cognitive map (O'Keefe and Nadel 1978) that includes both spatial and non-spatial information (Shapiro and Eichenbaum 1999; Smith and Mizumori 2006) and can use that information to index memories permanently stored in cortical regions (Goode et al. 2020; Teyler and DiScenna 1986).

Due to the central role of the hippocampus in contextual learning and memory, there has been considerable interest in understanding the neuronal mechanisms underlying these memories. Much of what we have learned concerning hippocampal function has been derived from permanent lesions, which are the consequence of clinical interventions (Scoville and Milner 1957), disease (Press et al. 1989) or experimental manipulations in the laboratory (O'Keefe and Nadel 1978; Jarrard 1993; Fanselow 2000; Bannerman et al. 2004). Global loss of function in the brain can be very informative; however large-scale and nonspecific manipulations do not inform the cellular and circuit organization of memory. Discerning the individual functions of these structures requires more precise manipulations. Recent work has focused on isolating the neuronal ensembles or "engrams" (Josselyn and Tonegawa 2020) within the hippocampus that support episodic memory. This work has been enabled by developments in activity-dependent tagging technologies that have enabled the isolation of specific engrams in the hippocampus and other structures of interest during temporally precise windows of cell activity (Liu et al. 2014; Mayford 2014; Roy and Tonegawa 2017; Tonegawa et al. 2015). These technologies have also allowed for manipulating these isolated ensembles, which holds promise as an avenue for future clinical intervention. This chapter will discuss the role of the hippocampus in contextual learning and memory, the role that hippocampal ensembles play in these processes, as well as behavioral and technological strategies for identifying, isolating, and manipulating these ensembles.

## Context Processing and the Hippocampus

Decades of work have revealed a critical role for the hippocampus in learning and memory processes, particularly in forms of memory that depend on context (such as episodic memory; (Holland and Bouton 1999; Ji and Maren 2007; Maren and Holt 2000; Maren et al. 2013; Smith and Mizumori 2006). Contexts may include stimuli as varied as the physical place, time, and internal state of the animal, setting a specific stage for memory formation and recollection (Bouton 2002). However, the hippocampus is not monolithic, and there are functional and anatomical specializations across hippocampal cell layers, as well as along its longitudinal axes, that bear a role on its memory. For example, the dorsal hippocampus (DH) has a critical function in contextual and spatial learning and memory (Fanselow and Dong 2010; Moser and Moser 1998; Moser et al. 1995), whereas the ventral hippocampus (VH) has a more prominent role in regulating stress (Herman et al. 2005), anxiety (Jimenez et al. 2018; Parfitt et al. 2017), and emotional behavior (Fanselow and Dong 2010; Xu et al. 2016). Although the DH and VH contribute differently to a variety of psychological processes, they are not entirely functionally distinct and exhibit considerable functional overlap. For instance, the VH exhibits context-dependent activity and coordinates the contextual retrieval of emotional memories (Hobin et al. 2003; Jin and Maren 2015) and the DH contains ensembles that, when active, induce retrieval of emotional memories (Bernier et al. 2017; Chen et al. 2019; Garner et al.

2012; Josselyn and Tonegawa 2020; Ressler et al. 2021). Within the hippocampus, different cell layers of the hippocampus are implicated in different aspects of memory, with CA1 and dentate gyrus (DG) most heavily implicated in contextual and spatial learning (Igarashi et al. 2014; Tsien et al. 1996) and CA2 playing a prominent role in social learning (Caruana et al. 2012; Tzakis and Holahan 2019; Watarai et al. 2021). Kheirbek and colleagues (2013) utilized a transgenic mouse strain that expressed the yellow light-activated chloride pump halorhodopsin in DG granule cells and found that inhibition of the entire dDG via light during encoding, but not during retrieval, impaired retrieval of conditioned fear (Kheirbek et al. 2013). Furthermore, they did not observe an effect of dDG inhibition on the retrieval of cued fear conditioning (Kheirbek et al. 2013). They conclude that the dDG plays a role in encoding contextual fear but not in its retrieval. A similar experiment by Bernier et al. (2017) yielded similar results. Intriguingly, this outcome is not obtained when only DG cells active during fear conditioning are inhibited. Instead, this experimental design suggests that the DG is involved in retrieving context fear (Denny et al. 2014). Computational models of the hippocampus such as the "Bayesian Context Fear Algorithm" model (BACON; Krasne et al. 2015) are able to mimic both results when run with their respective populations of inhibited cells (Bernier et al. 2017). These models suggest that nonspecific suppression (or reactivation) of the DG encompasses various context representations and that global modulation of DG activity either reactivates multiple context representations or suppresses all of them equally, effectively normalizing the strength of the context representations being retrieved and masking any effect on retrieval.

During learning, the hippocampus has been theorized to serve a role in indexing memories (Teyler and DiScenna 1986; Goode et al. 2020) and forming spatial and contextual memories of events (O'Keefe and Nadel 1978). Indexing theory suggests that the hippocampus is critical for establishing associative links between cortical networks that store the details of an episode. Memory retrieval involves recruiting these cortical networks to enable recollection. The best evidence for this model comes from partial engram reactivation—the hippocampus may perform pattern completion for "incomplete" engram reactivation, resulting in the full behavioral effect despite a lacking "overall" engram activity (Nalloor et al. 2012; Tanaka et al. 2018). Of course, extensive data also indicate that the hippocampus encodes spatial and contextual information during learning (Jarrard 1995; Ross and Slotnick 2008; Mizumori et al. 1999; Smith and Bulkin 2014; Chun and Phelps 1999). For example, physiological recordings of CA1 neurons have revealed their critical role in encoding place information as animals learn about novel environments (Leutgeb et al. 2004; Epsztein et al. 2011; Cohen et al. 2017). Notably, indexing and cognitive mapping functions are not mutually exclusive, and it is possible that hippocampal cells behave in accordance with both theories. Place cells that contribute raw spatial information may also serve as index cells that communicate with downstream neocortical regions to modulate retrieval (Moita et al. 2004). In addition to integrating context information into memories, the hippocampus also helps distinguish between separate contexts. In rodents, tactile and olfactory stimuli are particularly strong features of context that regulate memory retrieval (Huckleberry et al. 2016). The

DG, in particular, aids in context discrimination via sparse activity across many granule cells (Jung and McNaughton 1993; Leutgeb et al. 2007). Furthermore, suppressing long-term potentiation in the DG by ablation of NMDA receptors impairs pattern separation and completion (McHugh et al. 1996, 2007).

During contextual fear conditioning, animals make an association between the unconditioned stimulus (US) and the context in which it occurs. Interestingly, animals must first form a context representation before learning any context-US association. This feature of contextual conditioning was revealed by studies that observed that animals do not learn fear in a context in which footshock is delivered immediately after they are placed in a chamber (Blanchard et al. 1976; Fanselow 1990). This "immediate shock deficit" can be overcome if animals are either exposed to the testing context before shock or if significant time (several minutes) elapses between chamber entry and shock (Blanchard et al. 1976; Fanselow 1986; Landeira-Fernandez et al. 1995, 2006). This phenomenon is reflected in the differential expression of the immediate early gene Arc in the hippocampal subfields: placement in a novel context rapidly induces Arc in CA3, whereas Arc activity in CA1 requires several minutes in the novel context (Pevzner et al. 2012). This is consistent with studies showing that CA3 is implicated in rapidly acquiring new information (Kesner 2007), whereas CA1 plays a significant role in consolidating contextual information once acquired (Daumas et al. 2005). These data reveal that an animal requires time to form a context representation to remember that a biologically significant event has occurred within that context (Fanselow 1990; Matus-Amat et al. 2004).

Consistent with this, decades of work have shown that hippocampal lesions or inactivation impair contextual learning and memory. However, it is now apparent that the timing of these manipulations has a considerable influence over the nature of the impairment. For example, neurotoxic lesions made in the hippocampus within days of fear conditioning produce dense retrograde amnesia for contextual fear (Kim and Fanselow 1992; Maren et al. 1997). However, the same lesions made prior to conditioning failed to affect the acquisition of contextual fear (Maren et al. 1997; Frankland et al. 1998). This has led to the proposal that at least two processes might underlie the acquisition of contextual fear memories (Maren 2001). By this account, context learning might involve either (1) discrete cues in the environment (odor, light level, ambient sound, et cetera) that individually come into association with shock (i.e., an "elemental" strategy) or (2) a process by which the animal forms an integrated, unified representation of the entire context (i.e., a "conjunctive" or "configural" strategy) which then comes into association with shock (Maren et al. 1997; Rudy and O'Reilly 2001). Considerable evidence indicates that the hippocampus contributes to the formation of configural representations of context that are essential for context-US associations (Duncan et al. 2018; Holland and Bouton 1999; Maren et al. 1997; Matus-Amat et al. 2004). For example, the hippocampus is necessary for the context preexposure facilitation effect (CPFE), which describes the attenuation of the immediate shock deficit by context preexposure (Matus-Amat et al. 2004). Inactivating the hippocampus during context preexposure eliminates the CPFE (Chang et al. 2007; Matus-Amat et al. 2004). However, hippocampal damage or inactivation doesn't impede learning entirely; other structures such as the prefrontal cortex (Dixsaut and Gräff 2021; Beeman et al. 2013; Marek et al. 2013; Zelikowsky et al. 2013) and the basolateral amygdala (Maren 2001, 2003; Maren and Quirk 2004; Marek et al. 2013) and circuits between them provide compensatory mechanisms for learning that preserve the ability to learn when the hippocampus is offline (Maren 2001; Maren et al. 1997; Wiltgen et al. 2006). Rats conditioned after hippocampal damage show intact contextual fear conditioning, but this memory appears to result from elemental associations (Maren et al. 1997; Matus-Amat et al. 2004).

Although early work suggested that hippocampal damage spares auditory-cue fear conditioning (Kim and Fanselow 1992; Phillips and LeDoux 1992; Anagnostaras et al. 1999), there are now several studies revealing a role for the HPC in freezing to discrete CSs. For example, trace and long-delay fear conditioning require the hippocampus, and the hippocampus may be required for the generalization of fear to auditory CSs when training and test conditions differ (Maren et al. 1997; Quinn et al. 2008; Ji and Maren 2007; Maren 2008; Moyer et al. 1990). In addition, the VH has been shown to be involved in both the acquisition (Maren and Holt 2000; Zhang et al. 2001) and retrieval (Richmond et al. 1999). The involvement of the hippocampus in cued fear may be graded—the DH has more influence over fear retrieval when auditory CS-US associations are weaker (Quinn et al. 2008), suggesting a sliding scale for hippocampal involvement in auditory fear conditioning that is likely not based on conditioning modality.

## Hippocampal Engrams as the Substrate of Context Memory

Within the hippocampus, context learning is mediated by a variety of cellular processes, but ultimately, memory retrieval is believed to depend on the reactivation of cellular ensembles—so-called "engrams"—that were active during memory acquisition (Alvarez and Squire 1994; Guzowski et al. 1999; Treves and Rolls 1994). Cells are recruited to the engram during learning and by synaptic modifications between engram cells (Govindarajan et al. 2006; Minatohara et al. 2015; Ryan and Frankland 2022; Shapiro and Eichenbaum 1999). Engram cells are reactivated during presentations of stimuli associated with learning, leading to memory retrieval (Semon 1921).

A neuron can be involved in multiple engrams at once—memories that are formed closer in time are more likely to exhibit overlap among cells contributing to their respective engrams (Cai et al. 2016). Moreover, the engram for memory need not be reactivated completely to induce retrieval of that memory (Nalloor et al. 2012; Tanaka and McHugh 2018). Nalloor and colleagues (2012) found that animals exposed to a context and reintroduced to the same context had approximately 44% overlap in the neurons activated within the dorsal hippocampus. In contrast, they also demonstrated that animals exposed to a context and underwent contextual fear conditioning had reduced overlap in the neurons activated within the dorsal hippocampus, suggesting a "partial remapping" of the context (Nalloor et al. 2012). The pattern of cellular activity during memory acquisition can also impact the nature of the engram. For example, Garner and colleagues (Garner et al. 2012) found that artificial excitation of an engram representing a particular context (context A) during contextual fear conditioning in another context (context B) resulted in a "hybrid" contextual association with shock. Both contexts A and B elicited freezing behavior after this procedure. They also found that the activity of dCA1 and dCA3 was positively correlated with artificially driven memory retrieval and that the artificially excited cells were not a necessary component of the memory trace (Garner et al. 2012). Notably, reactivating these "neutral" contextual engrams in animals that have undergone extinction does not induce fear renewal when reactivated in a novel context (Liu et al. 2012). These findings suggest that the reactivation of "extra-engraphic" cells is most impactful on the engram during its formation, during which they become part of the engram.

## Methods for Engram Capture and Manipulations

In the last decade, tools have emerged that allow for the selective tagging and manipulation of neuronal ensembles engaged during behavioral experience (Garner et al. 2012; Josslyn and Tonegawa 2020). When isolating an engram, it is important to not only label the cells involved in memory but also to limit the temporal window in which these cells are tagged. This is especially true in the case of fear conditioning because the conditioning procedure is typically quite short in duration. Many engram-tagging technologies address this problem by utilizing immediate early genes (IEGs), which have low expression in quiescent cells, are expressed transiently when cells become active, and quickly decay after expression (Sheng and Greenberg 1990). IEGs serve regulatory functions in the nervous system (Lanahan and Worley 1998; Sheng and Greenberg 1990) but are also induced during the acquisition and expression of memory (Tischmeyer and Grimm 1999; Bozon et al. 2003; Davis et al. 2003; Feldman et al. 2010); their transience and activity-dependent expression makes them useful tools for isolating active cells. However, although IEGs report active neurons during specific experiences, they do not differentiate unique engrams in the brain. Indeed, inhibition of DG "fear engrams" prior to fear retrieval impairs freezing to the conditioning context, but inhibition of the entire DG does not (Bernier et al. 2017; Denny et al. 2014). The specificity of the ensemble modulating fear responding, in this case, conveys the importance of isolating distinct engrams associated with a behavioral experience.

Many (if not most) current technologies for labeling engrams take advantage of IEG expression to define a temporal window in which a particular ensemble of neurons is active. Genetic approaches using viral vectors in both wild-type and transgenic mice can be used to express genes of interest (typically effector and reporter proteins) in cells that are active during a particular behavioral experience. This strategy permits both the visualization and manipulation of active neuronal ensembles during memory encoding or retrieval, for example. There are many approaches to accomplish activity-dependent tagging of engram cells. For example, the "TetTag" approach utilizes a c-Fos IEG promoter in combination with a Tet-Off tetracycline-transactivator system. In this system, activity-dependent labeling is normally prevented in the presence of tetracycline (or a tetracycline analog, typically doxycycline), which is delivered in the animal's diet; removal of tetracycline is required to open the labeling window (typically 48 hours before the targeted behavior). This system provides broad temporal control over activity-dependent labeling in the brain (Redondo et al. 2014; Reijmers et al. 2007; Zhang et al. 2019; Ressler et al. 2021) and can be deployed with either transgenic mice or with viral vectors (Reijmers et al. 2007; Garner et al. 2012; Chen et al. 2019; Ressler et al. 2021; Zaki et al. 2022). Other engram labeling techniques take advantage of FosTRAP and ArcTRAP approaches in transgenic mice (Guenthner et al. 2013), which express gene targets in a temporally restricted tamoxifen-dependent manner. In these mouse lines, the labeling window is opened with systemic injection of tamoxifen (or its metabolite 4-hydroxytamoxifen). This labeling window is relatively short-lived (4–6 hours), making these techniques well-suited to tag neuronal ensembles with time-limited activity. Arc-CreER transgenic mice use a similar logic, expressing Cre-recombinase in Arc-positive cells in a tamoxifen- or 4-OHT-dependent manner (Denny et al. 2014; Lacagnina et al. 2019).

In some cases, more precise measurements of cell activity (including the temporal resolution of engrams) are required. Calcium activity is a useful predictor of cell activity (Dani et al. 1992; Ross 1989; Svoboda et al. 1997; Yuste et al. 1992) and is commonly used as a visual indicator for activity (Akerboom et al. 2012; Girven and Sparta 2017; Grienberger and Konnerth 2012; Wang et al. 2021), making it a good candidate for engram-isolation technologies. Indeed, multiple calcium- and activity-dependent imaging techniques such as CANE [capturing activated neuronal ensembles; (Sakurai et al. 2016)] and FLARE [fast light- and activity-regulated expression; (Wang et al. 2017)] have been developed in recent years.

## Capturing Hippocampal Engrams During Contextual Learning

Early engram labeling work focused on identifying and manipulating fear-conditioning memories. In the first paper of its kind, Han et al. (2009) showed that they could genetically recruit amygdala neurons into an engram and then tag and selectively eliminate those cells (Han et al. 2009). Selective deletion of these amygdala engram neurons resulted in impaired memory of the aversive memory in mice. In subsequent work, Garner and colleagues (2012) used a transgenic mouse model to introduce new cells into an engram by reactivating a neutral ensemble during learning and making those new cells necessary for memory retrieval. In complementary work, Liu and colleagues (2012) used TetTag transgenic mice to express

ChR2 to tag active cells during fear conditioning. Returning mice to a familiar context resulted in increased freezing during light-on epochs where ChR2 was exciting engram cells and low freezing during light-off epochs, comparable to freezing levels during context habituation (Liu et al. 2012). They also found that light activation of cells tagged during neutral-context exposure did not impact freezing during test, suggesting that the observed increases in freezing are due to reactivation of a fear memory instead of a nonspecific light-induced effect.

Reactivation of fear engrams is associated with the behavioral expression of conditioned freezing (Denny et al. 2014; Li et al. 2020; Tayler et al. 2013; Yoshii et al. 2017; Zaki et al. 2022). In a study that pioneered engram labeling methods, Reijmers and colleagues (2007) found that cells tagged during fear conditioning in mice were preferentially reactivated during retrieval in mice that successfully retrieved the fear memory; the degree of reactivation positively correlated with the degree of conditioned freezing. Other studies have reported that inhibiting fear engrams results in reduced conditional freezing, which implies a failure to retrieve the fear memory (Denny et al. 2014; Li et al. 2020; Tayler et al. 2013; Yoshii et al. 2017; Zaki et al. 2022). Interestingly, artificial reactivation of fear engram cells can lead to freezing behavior, even after behavioral performance has been disrupted by protein synthesis inhibitors (Ryan et al. 2015). This indicates that manipulations that appear to eliminate memory may spare neuronal engram ensembles that can drive performance when reactivated.

An interesting question is whether hippocampal engrams can be modified. Redondo and colleagues found that DG context engrams, representing emotional valence, can be updated when reactivated with an opposite-value unconditional stimulus (Redondo et al. 2014). Fear memories can also be linked to other memories via concurrent activation; indeed, some recent studies have found that context fear can be artificially induced in mice by optogenetically stimulating a DG context engram during fear conditioning in another context (Liu et al. 2014; Ramirez et al. 2013). Although these experiments do not isolate the conditioning memory itself, the specificity of the induced context fear memory suggests that fear memories can be integrated into existing memory by using the existing memory as a CS for further conditioning.

## Capturing Hippocampal Engrams During Memory Retrieval

The engram formed during fear conditioning is capable of inducing fear when reactivated in neutral contexts; however, there is no complete overlap of active cells and engram cells during fear retrieval (Denny et al. 2014; Bernier et al. 2017). The fear retrieval engram, while typically overlapping with the fear engram, does not contain the full set of information encoded by the fear engram. Fear retrieval necessarily occurs in a different temporal context (time passes after learning) and often in a different spatial context, as well. Would reactivating a hippocampal fear engram captured during a memory retrieval test (rather than conditioning) lead to the behavioral expression of fear?

Few studies have examined the impact of reactivating the retrieval engram on later retrieval. Khalaf et al. (2018) showed that greater reactivation of a remote fear memory ensemble in the DG was associated with more successful fear suppression during fear extinction. They also found that inhibition of the retrieval engram impairs fear extinction (Khalaf et al. 2018). These findings suggest that reactivation of the remote retrieval ensemble is necessary to successfully suppress fear; however, they do not address whether or not the reactivation of the retrieval ensemble increases fear the way reactivation of a conditioning ensemble does. Additionally, the retrieval ensemble was tagged directly by placing the animal in the conditioning context. Although this is a standard retrieval procedure in the laboratory, it is difficult in clinical practice because of how distressing the fearful context can be for patients; there are also logistical and practical issues with in vivo exposure. Returning a patient to a fearful context is not always possible or safe. A combat veteran with PTSD, for example, may not be able to return to the physical site of combat and, even if they could, may vehemently refuse to do so. Indirect or illusory retrieval of the conditioning context becomes the most effective way to prompt fear retrieval (Runyan and Dash 2005; Soeter and Kindt 2015). Consequently, clinicians primarily rely on imaginal exposure to reactivate memory of trauma. We therefore sought to develop a procedure in rats that would more closely mimic the imaginal retrieval procedures used in the clinic and reasoned that if fear memories can be retrieved without physical exposure to the fearful context, these memory representations can be isolated and disrupted well after the traumatic event.

To this end, we developed a "covert capture" procedure that relies on a retrieval cue to indirectly reactivate a memory of the context in which fear conditioning occurred (Ressler et al. 2021). The covert capture of hippocampal fear memories was enabled by conditioning rats with a "backward conditioning" procedure. In typical classical conditioning paradigms, the CS precedes the US—so-called "forward" conditioning. Forward conditioning results in strong excitatory associative conditioning to the CS, which during conditioning is directly predictive of US onset (Heth and Rescorla 1973). The predictive power of the CS translates to other contexts and persists even when rats are extinguished to the conditioning context (Bouton and Peck 1989; Bouton and King 1983; Hermann et al. 2016). This model of a US following a CS by a salient time difference is, however, not always representative of the ways in which CS-US associations are formed in maladaptive fear disorders such as PTSD. In backward conditioning, the US precedes the CS. Although the backward CS is experienced in proximity to shock USs, it is not predictive when shock occurs (if anything, it predicts when shock will not occur). Despite this, rats will come to show conditioned fear of a backward CS (Chang et al. 2003; Heth and Rescorla 1973). Work by Chang and colleagues (2003) demonstrated that this is not because the backward CS predicts the US, but because it is associated with the conditioning context, which itself is a strong predictor of the US. Accordingly, we have shown that freezing to a backward CS is attenuated by extinguishing the conditioning context (Ressler et al. 2021). Thus, the backward CS evokes conditional fear by retrieving a memory of the conditioning context.

After establishing that backward CSs could indirectly and covertly retrieve contextual fear memories, we next explored whether the presentation of the backward CS activated hippocampal neurons. Rats underwent either forward or backward conditioning, with a subset of each group designated as "NoTest" controls. After habituating the rats to the eventual test context, rats received either forward or backward conditioning in a novel context. The following day they were returned to the habituation context for a retrieval test (while the NoTest rats were left in their home cages); all rats were sacrificed 90-min after retrieval testing to assess hippocampal c-Fos expression. Presentation of the forward or backward CS-induced c-Fos throughout the hippocampus. Although Fos expression was similar in area CA1 in both forward- and backward-conditioned rats, c-Fos expression in the DG was significantly higher in the backward-conditioned group. This confirms that the hippocampus (and particularly DG) is engaged during the covert retrieval of contextual memories in the presence of a backward CS.

Does the presentation of the backward CS reactivate hippocampal engrams formed during fear conditioning? To examine this question, we utilized the TetTag approach to label hippocampal ensembles during backward fear conditioning. This approach utilizes DREADDs (designer receptors exclusively activated by designer drugs), which are G protein-coupled receptors expressed in cells but only active in the presence of clozapine-N-oxide, an otherwise inactive clozapine analog. Rats were injected with a viral cocktail containing AAV9-TRE-hM3Dq-mCherry-rBG (which expresses the excitatory DREADD hM3Dq) and AAV9-cFos-tTA-bGH and placed on a DOX diet. Forty-eight hours before conditioning, they were removed from the DOX diet. After labeling hippocampal ensembles associated with backward conditioning, the rats were returned to a DOX diet and underwent a retrieval test, after which they were sacrificed for Fos immunohistochemistry. Neurons that expressed both the mCherry reporter and c-Fos represent hippocampal engram cells that were active during conditioning and reactivated during the test. Relative to controls, the presentation of the backward CS in conditioned rats led to a significantly higher number of co-labeled cells.

Given that the backward CS reactivates hippocampal engram cells, we next sought to determine whether we could covertly capture the contextual fear engram when retrieved by the backward CS. Animals received the fos-tTA and TRE-hM3Dq viruses in the hippocampus and underwent the labeling procedure (off DOX) during the presentation of the backward CS in a novel context. We reactivated the captured ensemble twenty-four hours later with the synthetic DREADD ligand CNO. Consistent with our hypothesis, reactivation of covertly captured hippocampal ensembles drove conditioned freezing in a novel context. In fact, control rats expressing similar levels of activity-dependent hM3Dq labeling, which was driven by mere placement in the retrieval context, failed to show CNO-induced freezing. For the first time, these data revealed that engram ensembles captured at the time of retrieval are sufficient to drive conditioned behavior.

These data suggest that the CS in a backward conditioning paradigm indirectly retrieves a hippocampal-dependent context memory that can be accessed and manipulated without returning the animals to the conditioning context. We next examined whether covertly reactivated hippocampal ensembles are sensitive to disruption. We approached this issue by destabilizing the retrieved context fear memory with a protein synthesis inhibitor shortly after retrieval. Memories require time after acquisition to be consolidated into long-term storage, and during this time period, memories are labile and susceptible to disruption (Frankland and Bontempi 2005; McGaugh 2000). Memories are similarly labile following retrieval (Alberini 2005; Nader et al. 2000). During these windows, memories can be disrupted via a variety of manipulations (Davis and Squire 1984; Haubrich and Nader 2018; Nader et al. 2000), including the use of protein synthesis inhibitors due to the necessity of protein synthesis on memory storage (Davis and Squire 1984; Haubrich and Nader 2018). Because of the specificity of the engram being recruited during backward fear retrieval, pharmacological methods of reconsolidation disruption specifically target the context memory, resulting in decreased fear of the conditioning context if reconsolidation after indirect retrieval is disrupted. Indeed, we found that post-CS retrieval infusion of the protein-synthesis inhibitor rapamycin into the hippocampus resulted in impaired context fear retrieval in backward-, but not forward-, conditioned rats. This indicates that the hippocampal-dependent memory retrieved by exposure to a backward CS is vulnerable to disruption and, therefore, may suggest a future direction for clinical interventions for trauma- and stressor-related disorders.

Covert capture is a novel method for targeting aversive memories. Conditioning memories are generally stronger and more resistant to retrieval failure than extinction memories, making them more attractive targets for therapeutic intervention. The disruption of conditioning memories after covert capture is potentially a more permanent approach to alleviating fear than traditional fear reduction techniques that utilize extinction-based methods. Although disruption of conditioning memories tagged during indirect retrieval is demonstrably effective in rodents, this technology has not been tested in humans, nor have its ethics been discussed. Moreover, although we found that intra-HPC rapamycin impaired reconsolidation of the retrieved context memory, it is unclear if the pharmacological disruption of memory in covert capture results from memory erasure or retrieval failure due to disruption of one part of an integrated memory network. Further work in humans will be necessary to determine if and how reconsolidation disruptions can be used to efficaciously ameliorate maladaptive fear.

## Conclusion

Because of the central role of the hippocampus in contextual memories associated with trauma, understanding the precise roles of the hippocampus in maladaptive fear and anxiety is a major goal for the field. Technologies in cell tagging and manipulation have become exponentially more sophisticated in the past decade, and such technology has allowed researchers to more precisely elucidate the mechanisms by which the hippocampus modulates memories. While the role of the hippocampus and its ensembles is complex, researchers have more ability than ever to paint a more unified picture of what the hippocampus does during learning and memory retrieval.

## References

Akerboom J, Chen T-W, Wardill TJ, Tian L, Marvin JS, Mutlu S, Calderón NC, Esposti F, Borghuis BG, Sun XR, Gordus A, Orger MB, Portugues R, Engert F, Macklin JJ, Filosa A, Aggarwal A, Kerr RA, Takagi R, Kracun S, Shigetomi E, Khakh BS, Baier H, Lagnado L, Wang SS-H, Bargmann CI, Kimmel BE, Jayaraman V, Svoboda K, Kim DS, Schreiter ER, Looger LL (2012) Optimization of a GCaMP calcium indicator for neural activity imaging. J Neurosci 32(40):13819–13840. https://doi.org/10.1523/JNEUROSCI.2601-12.2012

Alberini CM (2005) Mechanisms of memory stabilization: are consolidation and reconsolidation similar or distinct processes? Trends Neurosci 28(1):51–56. https://doi.org/10.1016/j.tins.2004.11.001

Alvarez P, Squire LR (1994) Memory consolidation and the medial temporal lobe: a simple network model. Proc Natl Acad Sci USA 91(15):7041–7045. https://doi.org/10.1073/pnas.91.15.7041

Anagnostaras SG, Maren S, Fanselow MS (1999) Temporally graded retrograde amnesia of contextual fear after hippocampal damage in rats: within-subjects examination. J Neurosci 19(3):1106–1114. https://doi.org/10.1523/JNEUROSCI.19-03-01106.1999

Bakker A, Kirwan CB, Miller M, Stark CEL (2008) Pattern separation in the human hippocampal CA3 and dentate gyrus. Science 319(5870):1640–1642. https://doi.org/10.1126/science.1152882

Bannerman DM, Rawlins JNP, McHugh SB, Deacon RMJ, Yee BK, Bast T, Zhang WN et al (2004) Regional dissociations within the hippocampus—memory and anxiety. Neurosci Biobehav Rev 28(3):273–283. https://doi.org/10.1016/j.neubiorev.2004.03.004

Beeman CL, Bauer PS, Pierson JL, Quinn JJ (2013) Hippocampus and medial prefrontal cortex contributions to trace and contextual fear memory expression over time. Learn Mem 20(6):336–343. https://doi.org/10.1101/lm.031161.113

Bernier BE, Lacagnina AF, Ayoub A, Shue F, Zemelman BV, Krasne FB, Drew MR (2017) Dentate Gyrus contributes to retrieval as well as encoding: evidence from context fear conditioning, recall, and extinction. J Neurosci 37(26):6359–6371. https://doi.org/10.1523/JNEUROSCI.3029-16.2017

Blanchard RJ, Fukunaga KK, Blanchard DC (1976) Environmental control of defensive reactions to footshock. Bull Psychon Soc 8(2):129–130. https://doi.org/10.3758/BF03335103

Bouton ME (2002) Context, ambiguity, and unlearning: sources of relapse after behavioral extinction. Biol Psychiatry 52(10):976–986. https://doi.org/10.1016/s0006-3223(02)01546-9

Bouton ME, King DA (1983) Contextual control of the extinction of conditioned fear: tests for the associative value of the context. J Exp Psychol Anim Behav Process 9(3):248–265. https://doi.org/10.1037/0097-7403.9.3.248

Bouton ME, Peck CA (1989) Context effects on conditioning, extinction, and reinstatement in an appetitive conditioning preparation. Anim Learn Behav 17(2):188–198. https://doi.org/10.3758/BF03207634

Bouton ME, Maren S, McNally GP (2021) Behavioral and neurobiological mechanisms of pavlovian and instrumental extinction learning. Physiol Rev 101(2):611–681. https://doi.org/10.1152/physrev.00016.2020

Bozon B, Davis S, Laroche S (2003) A requirement for the immediate early gene zif268 in reconsolidation of recognition memory after retrieval. Neuron 40(4):695–701. https://doi.org/10.1016/s0896-6273(03)00674-3

Cai DJ, Aharoni D, Shuman T, Shobe J, Biane J, Song W, Wei B, Veshkini M, La-Vu M, Lou J, Flores SE, Kim I, Sano Y, Zhou M, Baumgaertel K, Lavi A, Kamata M, Tuszynski M, Mayford M et al (2016) A shared neural ensemble links distinct contextual memories encoded close in time. Nature 534(7605):115–118. https://doi.org/10.1038/nature17955

Caruana DA, Alexander GM, Dudek SM (2012) New insights into the regulation of synaptic plasticity from an unexpected place: hippocampal area CA2. Learn Mem 19(9):391–400. https://doi.org/10.1101/lm.025304.111

Chang RC, Blaisdell AP, Miller RR (2003) Backward conditioning: mediation by the context. J Exp Psychol Anim Behav Process 29(3):171–183. https://doi.org/10.1037/0097-7403.29.3.171

Chang T, Meyer U, Feldon J, Yee BK (2007) Disruption of the US preexposure effect and latent inhibition in two-way active avoidance by systemic amphetamine in C57BL/6 mice. Psychopharmacology 191(2):211–221. https://doi.org/10.1007/s00213-006-0649-z

Chen BK, Murawski NJ, Cincotta C, McKissick O, Finkelstein A, Hamidi AB, Merfeld E, Doucette E, Grella SL, Shpokayte M, Zaki Y, Fortin A, Ramirez S (2019) Artificially enhancing and suppressing hippocampus-mediated memories. Curr Biol 29(11):1885–1894.e4. https://doi.org/10.1016/j.cub.2019.04.065

Chun MM, Phelps EA (1999) Memory deficits for implicit contextual information in amnesic subjects with hippocampal damage. Nat Neurosci 2(9):844–847. https://doi.org/10.1038/12222

Churchwell JC, Kesner RP (2011) Hippocampal-prefrontal dynamics in spatial working memory: interactions and independent parallel processing. Behav Brain Res 225(2):389–395. https://doi.org/10.1016/j.bbr.2011.07.045

Cohen JD, Bolstad M, Lee AK (2017) Experience-dependent shaping of hippocampal CA1 intracellular activity in novel and familiar environments. elife 6. https://doi.org/10.7554/eLife.23040

Corcoran KA, Desmond TJ, Frey KA, Maren S (2005) Hippocampal inactivation disrupts the acquisition and contextual encoding of fear extinction. J Neurosci 25(39):8978–8987. https://doi.org/10.1523/JNEUROSCI.2246-05.2005

Dani JW, Chernjavsky A, Smith SJ (1992) Neuronal activity triggers calcium waves in hippocampal astrocyte networks. Neuron 8(3):429–440. https://doi.org/10.1016/0896-6273(92)90271-e

Daumas S, Halley H, Francés B, Lassalle J-M (2005) Encoding, consolidation, and retrieval of contextual memory: differential involvement of dorsal CA3 and CA1 hippocampal subregions. Learn Mem 12(4):375–382. https://doi.org/10.1101/lm.81905

Davis HP, Squire LR (1984) Protein synthesis and memory: a review. Psychol Bull 96(3):518–559. https://doi.org/10.1037/0033-2909.96.3.518

Davis S, Bozon B, Laroche S (2003) How necessary is the activation of the immediate early gene zif268 in synaptic plasticity and learning? Behav Brain Res 142(1–2):17–30. https://doi.org/10.1016/s0166-4328(02)00421-7

Denny CA, Kheirbek MA, Alba EL, Tanaka KF, Brachman RA, Laughman KB, Tomm NK, Turi GF, Losonczy A, Hen R (2014) Hippocampal memory traces are differentially modulated by experience, time, and adult neurogenesis. Neuron 83(1):189–201. https://doi.org/10.1016/j.neuron.2014.05.018

Dixsaut L, Gräff J (2021) The medial prefrontal cortex and fear memory: dynamics, connectivity, and engrams. Int J Mol Sci 22(22). https://doi.org/10.3390/ijms222212113

Duncan K, Doll BB, Daw ND, Shohamy D (2018) More than the sum of its parts: a role for the hippocampus in configural reinforcement learning. Neuron 98(3):645–657.e6. https://doi.org/10.1016/j.neuron.2018.03.042

Eichenbaum H (1996) Is the rodent hippocampus just for "place"? Curr Opin Neurobiol 6(2):187–195. https://doi.org/10.1016/S0959-4388(96)80072-9

Eichenbaum H (1999) The hippocampus and mechanisms of declarative memory. Behav Brain Res 103(2):123–133. https://doi.org/10.1016/s0166-4328(99)00044-3

Eichenbaum H (2004) Hippocampus: cognitive processes and neural representations that underlie declarative memory. Neuron 44(1):109–120. https://doi.org/10.1016/j.neuron.2004.08.028
Epsztein J, Brecht M, Lee AK (2011) Intracellular determinants of hippocampal CA1 place and silent cell activity in a novel environment. Neuron 70(1):109–120. https://doi.org/10.1016/j.neuron.2011.03.006
Fanselow MS (1986) Associative vs topographical accounts of the immediate shock-freezing deficit in rats: implications for the response selection rules governing species-specific defensive reactions. Learn Motiv 17(1):16–39. https://doi.org/10.1016/0023-9690(86)90018-4
Fanselow MS (1990) Factors governing one-trial contextual conditioning. Anim Learn Behav 18(3):264–270. https://doi.org/10.3758/BF03205285
Fanselow MS (2000) Contextual fear, gestalt memories, and the hippocampus. Behav Brain Res 110(1–2):73–81. https://doi.org/10.1016/s0166-4328(99)00186-2
Fanselow MS, Dong H-W (2010) Are the dorsal and ventral hippocampus functionally distinct structures? Neuron 65(1):7–19. https://doi.org/10.1016/j.neuron.2009.11.031
Feldman LA, Shapiro ML, Nalbantoglu J (2010) A novel, rapidly acquired and persistent spatial memory task that induces immediate early gene expression. Behav Brain Funct 6:35. https://doi.org/10.1186/1744-9081-6-35
Frankland PW, Bontempi B (2005) The organization of recent and remote memories. Nat Rev Neurosci 6(2):119–130. https://doi.org/10.1038/nrn1607
Frankland PW, Cestari V, Filipkowski RK, McDonald RJ, Silva AJ (1998) The dorsal hippocampus is essential for context discrimination but not for contextual conditioning. Behav Neurosci 112(4):863–874. https://doi.org/10.1037/0735-7044.112.4.863
Fyhn M, Molden S, Witter MP, Moser EI, Moser M-B (2004) Spatial representation in the entorhinal cortex. Science 305(5688):1258–1264. https://doi.org/10.1126/science.1099901
Garner AR, Rowland DC, Hwang SY, Baumgaertel K, Roth BL, Kentros C, Mayford M (2012) Generation of a synthetic memory trace. Science 335(6075):1513–1516. https://doi.org/10.1126/science.1214985
Girven KS, Sparta DR (2017) Probing deep brain circuitry: new advances in in vivo calcium measurement strategies. ACS Chem Neurosci 8(2):243–251. https://doi.org/10.1021/acschemneuro.6b00307
Goode TD, Tanaka KZ, Sahay A, McHugh TJ (2020) An integrated index: engrams, place cells, and hippocampal memory. Neuron 107(5):805–820. https://doi.org/10.1016/j.neuron.2020.07.011
Govindarajan A, Kelleher RJ, Tonegawa S (2006) A clustered plasticity model of long-term memory engrams. Nat Rev Neurosci 7(7):575–583. https://doi.org/10.1038/nrn1937
Grienberger C, Konnerth A (2012) Imaging calcium in neurons. Neuron 73(5):862–885. https://doi.org/10.1016/j.neuron.2012.02.011
Guenthner CJ, Miyamichi K, Yang HH, Heller HC, Luo L (2013) Permanent genetic access to transiently active neurons via TRAP: targeted recombination in active populations. Neuron 78(5):773–784. https://doi.org/10.1016/j.neuron.2013.03.025
Guzowski JF, McNaughton BL, Barnes CA, Worley PF (1999) Environment-specific expression of the immediate-early gene Arc in hippocampal neuronal ensembles. Nat Neurosci 2(12):1120–1124. https://doi.org/10.1038/16046
Han J-H, Kushner SA, Yiu AP, Hsiang H-LL, Buch T, Waisman A, Bontempi B, Neve RL, Frankland PW, Josselyn SA (2009) Selective erasure of a fear memory. Science 323(5920):1492–1496. https://doi.org/10.1126/science.1164139
Haubrich J, Nader K (2018) Memory reconsolidation. Curr Top Behav Neurosci 37:151–176. https://doi.org/10.1007/7854_2016_463
Herman JP, Ostrander MM, Mueller NK, Figueiredo H (2005) Limbic system mechanisms of stress regulation: hypothalamo-pituitary-adrenocortical axis. Prog Neuro-Psychopharmacol Biol Psychiatry 29(8):1201–1213. https://doi.org/10.1016/j.pnpbp.2005.08.006
Hermann A, Stark R, Milad MR, Merz CJ (2016) Renewal of conditioned fear in a novel context is associated with hippocampal activation and connectivity. Soc Cogn Affect Neurosci 11(9):1411–1421. https://doi.org/10.1093/scan/nsw047

Heth CD, Rescorla RA (1973) Simultaneous and backward fear conditioning in the rat. J Comp Physiol Psychol 82(3):434–443. https://doi.org/10.1037/h0034124
Hirsh R (1974) The hippocampus and contextual retrieval of information from memory: a theory. Behav Biol 12(4):421–444. https://doi.org/10.1016/S0091-6773(74)92231-7
Hobin JA, Goosens KA, Maren S (2003) Context-dependent neuronal activity in the lateral amygdala represents fear memories after extinction. J Neurosci 23(23):8410–8416. https://doi.org/10.1523/JNEUROSCI.23-23-08410.2003
Holland PC, Bouton ME (1999) Hippocampus and context in classical conditioning. Curr Opin Neurobiol 9(2):195–202. https://doi.org/10.1016/S0959-4388(99)80027-0
Huckleberry KA, Ferguson LB, Drew MR (2016) Behavioral mechanisms of context fear generalization in mice. Learn Mem 23(12):703–709. https://doi.org/10.1101/lm.042374.116
Igarashi KM, Ito HT, Moser EI, Moser M-B (2014) Functional diversity along the transverse axis of hippocampal area CA1. FEBS Lett 588(15):2470–2476. https://doi.org/10.1016/j.febslet.2014.06.004
Jarrard LE (1993) On the role of the hippocampus in learning and memory in the rat. Behav Neural Biol 60(1):9–26. https://doi.org/10.1016/0163-1047(93)90664-4
Jarrard LE (1995) What does the hippocampus really do? Behav Brain Res 71(1–2):1–10. https://doi.org/10.1016/0166-4328(95)00034-8
Ji J, Maren S (2007) Hippocampal involvement in contextual modulation of fear extinction. Hippocampus 17(9):749–758. https://doi.org/10.1002/hipo.20331
Jimenez JC, Su K, Goldberg AR, Luna VM, Biane JS, Ordek G, Zhou P, Ong SK, Wright MA, Zweifel L, Paninski L, Hen R, Kheirbek MA (2018) Anxiety cells in a hippocampal-hypothalamic circuit. Neuron 97(3):670–683.e6. https://doi.org/10.1016/j.neuron.2018.01.016
Jin J, Maren S (2015) Fear renewal preferentially activates ventral hippocampal neurons projecting to both amygdala and prefrontal cortex in rats. Sci Rep 5:8388. https://doi.org/10.1038/srep08388
Josselyn SA, Tonegawa S (2020) Memory engrams: recalling the past and imagining the future. Science 367(6473). https://doi.org/10.1126/science.aaw4325
Jung MW, McNaughton BL (1993) Spatial selectivity of unit activity in the hippocampal granular layer. Hippocampus 3(2):165–182. https://doi.org/10.1002/hipo.450030209
Kennedy PJ, Shapiro ML (2004) Retrieving memories via internal context requires the hippocampus. J Neurosci 24(31):6979–6985. https://doi.org/10.1523/JNEUROSCI.1388-04.2004
Kesner RP (2007) Behavioral functions of the CA3 subregion of the hippocampus. Learn Mem 14(11):771–781. https://doi.org/10.1101/lm.688207
Khalaf O, Resch S, Dixsaut L, Gorden V, Glauser L, Gräff J (2018) Reactivation of recall-induced neurons contributes to remote fear memory attenuation. Science 360(6394):1239–1242. https://doi.org/10.1126/science.aas9875
Kheirbek MA, Drew LJ, Burghardt NS, Costantini DO, Tannenholz L, Ahmari SE, Zeng H, Fenton AA, Hen R (2013) Differential control of learning and anxiety along the dorsoventral axis of the dentate gyrus. Neuron 77(5):955–968. https://doi.org/10.1016/j.neuron.2012.12.038
Kim JJ, Fanselow MS (1992) Modality-specific retrograde amnesia of fear. Science 256(5057):675–677. https://doi.org/10.1126/science.1585183
Krasne FB, Cushman JD, Fanselow MS (2015) A Bayesian context fear learning algorithm/automaton. Front Behav Neurosci 9:112. https://doi.org/10.3389/fnbeh.2015.00112
Lacagnina AF, Brockway ET, Crovetti CR, Shue F, McCarty MJ, Sattler KP, Lim SC, Santos SL, Denny CA, Drew MR (2019) Distinct hippocampal engrams control extinction and relapse of fear memory. Nat Neurosci 22(5):753–761. https://doi.org/10.1038/s41593-019-0361-z
Lanahan A, Worley P (1998) Immediate-early genes and synaptic function. Neurobiol Learn Mem 70(1–2):37–43. https://doi.org/10.1006/nlme.1998.3836
Landeira-Fernandez J, Fanselow MS, Decola JP, Kim JJ (1995) Effects of handling and context preexposure on the immediate shock deficit. Anim Learn Behav 23(3):335–339. https://doi.org/10.3758/BF03198930

Landeira-Fernandez J, DeCola JP, Kim JJ, Fanselow MS (2006) Immediate shock deficit in fear conditioning: effects of shock manipulations. Behav Neurosci 120(4):873–879. https://doi.org/10.1037/0735-7044.120.4.873

LeDoux JE (2000) Emotion circuits in the brain. Annu Rev Neurosci 23:155–184. https://doi.org/10.1146/annurev.neuro.23.1.155

Lee I, Kesner RP (2002) Differential contribution of NMDA receptors in hippocampal subregions to spatial working memory. Nat Neurosci 5(2):162–168. https://doi.org/10.1038/nn790

Leutgeb S, Leutgeb JK, Treves A, Moser M-B, Moser EI (2004) Distinct ensemble codes in hippocampal areas CA3 and CA1. Science 305(5688):1295–1298. https://doi.org/10.1126/science.1100265

Leutgeb JK, Leutgeb S, Moser M-B, Moser EI (2007) Pattern separation in the dentate gyrus and CA3 of the hippocampus. Science 315(5814):961–966. https://doi.org/10.1126/science.1135801

Li J, Jiang RY, Arendt KL, Hsu Y-T, Zhai SR, Chen L (2020) Defective memory engram reactivation underlies impaired fear memory recall in Fragile X syndrome. elife 9. https://doi.org/10.7554/eLife.61882

Liu X, Ramirez S, Pang PT, Puryear CB, Govindarajan A, Deisseroth K, Tonegawa S (2012) Optogenetic stimulation of a hippocampal engram activates fear memory recall. Nature 484(7394):381–385. https://doi.org/10.1038/nature11028

Liu X, Ramirez S, Tonegawa S (2014) Inception of a false memory by optogenetic manipulation of a hippocampal memory engram. Philos Trans R Soc Lond Ser B Biol Sci 369(1633):20130142. https://doi.org/10.1098/rstb.2013.0142

Mandler G (1980) Recognizing: the judgment of previous occurrence. Psychol Rev 87(3):252–271. https://doi.org/10.1037/0033-295X.87.3.252

Marek R, Strobel C, Bredy TW, Sah P (2013) The amygdala and medial prefrontal cortex: partners in the fear circuit. J Physiol 591(10):2381–2391. https://doi.org/10.1113/jphysiol.2012.248575

Maren S (2000) Auditory fear conditioning increases CS-elicited spike firing in lateral amygdala neurons even after extensive overtraining. Eur J Neurosci 12(11):4047–4054. https://doi.org/10.1046/j.1460-9568.2000.00281.x

Maren S (2001) Neurobiology of Pavlovian fear conditioning. Annu Rev Neurosci 24:897–931. https://doi.org/10.1146/annurev.neuro.24.1.897

Maren S (2003) The amygdala, synaptic plasticity, and fear memory. Ann N Y Acad Sci 985:106–113. https://doi.org/10.1111/j.1749-6632.2003.tb07075.x

Maren S (2008) Pavlovian fear conditioning as a behavioral assay for hippocampus and amygdala function: cautions and caveats. Eur J Neurosci 28(8):1661–1666. https://doi.org/10.1111/j.1460-9568.2008.06485.x

Maren S, Holt W (2000) The hippocampus and contextual memory retrieval in Pavlovian conditioning. Behav Brain Res 110(1–2):97–108. https://doi.org/10.1016/s0166-4328(99)00188-6

Maren S, Quirk GJ (2004) Neuronal signalling of fear memory. Nat Rev Neurosci 5(11):844–852. https://doi.org/10.1038/nrn1535

Maren S, Aharonov G, Fanselow MS (1997) Neurotoxic lesions of the dorsal hippocampus and Pavlovian fear conditioning in rats. Behav Brain Res 88(2):261–274. https://doi.org/10.1016/s0166-4328(97)00088-0

Maren S, Phan KL, Liberzon I (2013) The contextual brain: implications for fear conditioning, extinction and psychopathology. Nat Rev Neurosci 14(6):417–428. https://doi.org/10.1038/nrn3492

Maren S (2016) Neural circuits for context processing in aversive learning and memory. In I Liberzon & KJ Ressler (Eds.), Neurobiology of PTSD: From brain to mind (pp. 85–114). Oxford University Press

Mayford M (2014) The search for a hippocampal engram. Philosophical Transactions of the Royal Society B: Biological Sciences 369(1633):20130161. https://doi.org/10.1098/rstb.2013.0161

Matus-Amat P, Higgins EA, Barrientos RM, Rudy JW (2004) The role of the dorsal hippocampus in the acquisition and retrieval of context memory representations. J Neurosci 24(10):2431–2439. https://doi.org/10.1523/JNEUROSCI.1598-03.2004

McGaugh JL (2000) Memory--a century of consolidation. Science 287(5451):248–251. https://doi.org/10.1126/science.287.5451.248

McHugh TJ, Blum KI, Tsien JZ, Tonegawa S, Wilson MA (1996) Impaired hippocampal representation of space in CA1-specific NMDAR1 knockout mice. Cell 87(7):1339–1349. https://doi.org/10.1016/s0092-8674(00)81828-0

McHugh TJ, Jones MW, Quinn JJ, Balthasar N, Coppari R, Elmquist JK et al (2007) Dentate gyrus NMDA receptors mediate rapid pattern separation in the hippocampal network. Science 317(5834):94–99. https://doi.org/10.1126/science.1140263

Minatohara K, Akiyoshi M, Okuno H (2015) Role of immediate-early genes in synaptic plasticity and neuronal ensembles underlying the memory trace. Front Mol Neurosci 8:78. https://doi.org/10.3389/fnmol.2015.00078

Mizumori SJY, Ragozzino KE, Cooper BG, Leutgeb S (1999) Hippocampal representational organization and spatial context. Hippocampus. https://doi.org/10.1002/(SICI)1098-1063(1999)9:4%3C444::AID-HIPO10%3E3.0.CO;2-Z

Moita MAP, Rosis S, Zhou Y, LeDoux JE, Blair HT (2004) Putting fear in its place: remapping of hippocampal place cells during fear conditioning. J Neurosci 24(31):7015–7023. https://doi.org/10.1523/JNEUROSCI.5492-03.2004

Morgan CA, Grillon C, Southwick SM, Davis M, Charney DS (1995) Fear-potentiated startle in posttraumatic stress disorder. Biol Psychiatry 38(6):378–385. https://doi.org/10.1016/0006-3223(94)00321-S

Moser M-B, Moser EI (1998) Functional differentiation in the hippocampus. Hippocampus 8:608

Moser MB, Moser EI, Forrest E, Andersen P, Morris RG (1995) Spatial learning with a minislab in the dorsal hippocampus. Proc Natl Acad Sci USA 92(21):9697–9701. https://doi.org/10.1073/pnas.92.21.9697

Moser M-B, Rowland DC, Moser EI (2015) Place cells, grid cells, and memory. Cold Spring Harb Perspect Biol 7(2):a021808. https://doi.org/10.1101/cshperspect.a021808

Moyer JR, Deyo RA, Disterhoft JF (1990) Hippocampectomy disrupts trace eye-blink conditioning in rabbits. Behav Neurosci 104(2):243–252. https://doi.org/10.1037/0735-7044.104.2.243

Nader K, Schafe GE, Le Doux JE (2000) Fear memories require protein synthesis in the amygdala for reconsolidation after retrieval. Nature 406(6797):722–726. https://doi.org/10.1038/35021052

Nalloor R, Bunting KM, Vazdarjanova A (2012) Encoding of emotion-paired spatial stimuli in the rodent hippocampus. Front Behav Neurosci 6:27. https://doi.org/10.3389/fnbeh.2012.00027

O'Keefe J, Nadel L (1978) The hippocampus as a cognitive map. Oxford University Press

O'Keefe J, Burgess N, Donnett JG, Jeffery KJ, Maguire EA (1998) Place cells, navigational accuracy, and the human hippocampus. Philos Trans R Soc Lond Ser B Biol Sci 353(1373):1333–1340. https://doi.org/10.1098/rstb.1998.0287

Parfitt GM, Nguyen R, Bang JY, Aqrabawi AJ, Tran MM, Seo DK, Richards BA, Kim JC (2017) Bidirectional control of anxiety-related behaviors in mice: role of inputs arising from the ventral hippocampus to the lateral septum and medial prefrontal cortex. Neuropsychopharmacology 42(8):1715–1728. https://doi.org/10.1038/npp.2017.56

Pevzner A, Miyashita T, Schiffman AJ, Guzowski JF (2012) Temporal dynamics of Arc gene induction in hippocampus: relationship to context memory formation. Neurobiol Learn Mem 97(3):313–320. https://doi.org/10.1016/j.nlm.2012.02.004

Phillips RG, LeDoux JE (1992) Differential contribution of amygdala and hippocampus to cued and contextual fear conditioning. Behav Neurosci 106(2):274–285. https://doi.org/10.1037/0735-7044.106.2.274

Press GA, Amaral DG, Squire LR (1989) Hippocampal abnormalities in amnesic patients revealed by high-resolution magnetic resonance imaging. Nature 341(6237):54–57. https://doi.org/10.1038/341054a0

Quinn JJ, Wied HM, Ma QD, Tinsley MR, Fanselow MS (2008) Dorsal hippocampus involvement in delay fear conditioning depends upon the strength of the tone-footshock association. Hippocampus 18(7):640–654. https://doi.org/10.1002/hipo.20424
Ramirez S, Liu X, Lin P-A, Suh J, Pignatelli M, Redondo RL, Ryan TJ, Tonegawa S (2013) Creating a false memory in the hippocampus. Science 341(6144):387–391. https://doi.org/10.1126/science.1239073
Redondo RL, Kim J, Arons AL, Ramirez S, Liu X, Tonegawa S (2014) Bidirectional switch of the valence associated with a hippocampal contextual memory engram. Nature 513(7518):426–430. https://doi.org/10.1038/nature13725
Reijmers LG, Perkins BL, Matsuo N, Mayford M (2007) Localization of a stable neural correlate of associative memory. Science 317(5842):1230–1233. https://doi.org/10.1126/science.1143839
Ressler RL, Goode TD, Kim S, Ramanathan KR, Maren S (2021) Covert capture and attenuation of a hippocampus-dependent fear memory. Nat Neurosci 24(5):677–684. https://doi.org/10.1038/s41593-021-00825-5
Richmond MA, Yee BK, Pouzet B, Veenman L, Rawlins JN, Feldon J, Bannerman DM (1999) Dissociating context and space within the hippocampus: effects of complete, dorsal, and ventral excitotoxic hippocampal lesions on conditioned freezing and spatial learning. Behav Neurosci 113(6):1189–1203. https://doi.org/10.1037/0735-7044.113.6.1189
Rolls ET (2016) Pattern separation, completion, and categorisation in the hippocampus and neocortex. Neurobiology of Learning and Memory 129:4–28. https://doi.org/10.1016/j.nlm.2015.07.008
Roy DS, Tonegawa S (2017) Manipulating memory in space and time. Curr Opin Behav Sci 17:1–6. https://doi.org/10.1016/j.cobeha.2017.05.020
Ross WN (1989) Changes in intracellular calcium during neuron activity. Annu Rev Physiol 51:491–506. https://doi.org/10.1146/annurev.ph.51.030189.002423
Ross RS, Slotnick SD (2008) The hippocampus is preferentially associated with memory for spatial context. J Cogn Neurosci 20(3):432–446. https://doi.org/10.1162/jocn.2008.20035
Rudy JW, O'Reilly RC (2001) Conjunctive representations, the hippocampus, and contextual fear conditioning. Cogn Affect Behav Neurosci 1(1):66–82. https://doi.org/10.3758/cabn.1.1.66
Runyan JD, Dash PK (2005) Inhibition of hippocampal protein synthesis following recall disrupts expression of episodic-like memory in trace conditioning. Hippocampus 15(3):333–339. https://doi.org/10.1002/hipo.20055
Ryan TJ, Frankland PW (2022) Forgetting as a form of adaptive engram cell plasticity. Nat Rev Neurosci 23(3):173–186. https://doi.org/10.1038/s41583-021-00548-3
Ryan TJ, Roy DS, Pignatelli M, Arons A, Tonegawa S (2015) Memory. Engram cells retain memory under retrograde amnesia. Science 348(6238):1007–1013. https://doi.org/10.1126/science.aaa5542
Sakurai Y (1994) Involvement of auditory cortical and hippocampal neurons in auditory working memory and reference memory in the rat. J Neurosci 14(5 Pt 1):2606–2623. https://doi.org/10.1523/JNEUROSCI.14-05-02606.1994
Sakurai K, Zhao S, Takatoh J, Rodriguez E, Lu J, Leavitt AD, Fu M, Han B-X, Wang F (2016) Capturing and manipulating activated neuronal ensembles with CANE delineates a hypothalamic social-fear circuit. Neuron 92(4):739–753. https://doi.org/10.1016/j.neuron.2016.10.015
Scoville WB, Milner B (1957) Loss of recent memory after bilateral hippocampal lesions. J Neurol Neurosurg Psychiatry 20(1):11–21. https://doi.org/10.1136/jnnp.20.1.11
Semon RW (1921) The Mneme (L. Simon, Trans.). Legare Street Press, p 312
Shapiro ML, Eichenbaum H (1999) Hippocampus as a memory map: synaptic plasticity and memory encoding by hippocampal neurons. Hippocampus 9:365
Sheng M, Greenberg ME (1990) The regulation and function of c-fos and other immediate early genes in the nervous system. Neuron 4(4):477–485. https://doi.org/10.1016/0896-6273(90)90106-p
Smith DM, Bulkin DA (2014) The form and function of hippocampal context representations. Neurosci Biobehav Rev 40:52–61. https://doi.org/10.1016/j.neubiorev.2014.01.005

Smith DM, Mizumori SJY (2006) Hippocampal place cells, context, and episodic memory. Hippocampus 16(9):716–729. https://doi.org/10.1002/hipo.20208

Soeter M, Kindt M (2015) Retrieval cues that trigger reconsolidation of associative fear memory are not necessarily an exact replica of the original learning experience. Front Behav Neurosci 9:122. https://doi.org/10.3389/fnbeh.2015.00122

Spear NE (1973) Retrieval of memory in animals. Psychol Rev 80(3):163–194. https://doi.org/10.1037/h0034326

Svoboda K, Denk W, Kleinfeld D, Tank DW (1997) In vivo dendritic calcium dynamics in neocortical pyramidal neurons. Nature 385(6612):161–165. https://doi.org/10.1038/385161a0

Tanaka KZ, McHugh TJ (2018) The hippocampal engram as a memory index. Journal of Experimental Neuroscience 12:1179069518815942. https://doi.org/10.1177/1179069518815942

Tanaka KZ, He H, Tomar A, Niisato K, Huang AJY, McHugh TJ (2018) The hippocampal engram maps experience but not place. Science 361(6400):392–397. https://doi.org/10.1126/science.aat5397

Tayler KK, Tanaka KZ, Reijmers LG, Wiltgen BJ (2013) Reactivation of neural ensembles during the retrieval of recent and remote memory. Curr Biol 23(2):99–106. https://doi.org/10.1016/j.cub.2012.11.019

Teyler TJ, DiScenna P (1986) The hippocampal memory indexing theory. Behav Neurosci 100(2):147–154. https://doi.org/10.1037/0735-7044.100.2.147

Tischmeyer W, Grimm R (1999) Activation of immediate early genes and memory formation. Cell Mol Life Sci 55(4):564–574. https://doi.org/10.1007/s000180050315

Tonegawa S, Pignatelli M, Roy DS, Ryan TJ (2015) Memory engram storage and retrieval. Current Opinion in Neurobiology 35:101–109. https://doi.org/10.1016/j.conb.2015.07.009

Treves A, Rolls ET (1994) Computational analysis of the role of the hippocampus in memory. Hippocampus 4(3):374–391. https://doi.org/10.1002/hipo.450040319

Tsien JZ, Huerta PT, Tonegawa S (1996) The essential role of hippocampal CA1 NMDA receptor-dependent synaptic plasticity in spatial memory. Cell 87(7):1327–1338. https://doi.org/10.1016/s0092-8674(00)81827-9

Tulving E, Markowitsch HJ (1998) Episodic and declarative memory: role of the hippocampus. Hippocampus 8:198

Tulving E, Thomson DM (1973) Encoding specificity and retrieval processes in episodic memory. Psychol Rev 80(5):352–373. https://doi.org/10.1037/h0020071

Tzakis N, Holahan MR (2019) Social memory and the role of the hippocampal CA2 region. Front Behav Neurosci 13:233. https://doi.org/10.3389/fnbeh.2019.00233

Wang W, Wildes CP, Pattarabanjird T, Sanchez MI, Glober GF, Matthews GA, Tye KM, Ting AY (2017) A light- and calcium-gated transcription factor for imaging and manipulating activated neurons. Nat Biotechnol 35(9):864–871. https://doi.org/10.1038/nbt.3909

Wang Y, DeMarco EM, Witzel LS, Keighron JD (2021) A selected review of recent advances in the study of neuronal circuits using fiber photometry. Pharmacol Biochem Behav 201:173113. https://doi.org/10.1016/j.pbb.2021.173113

Watarai A, Tao K, Wang M-Y, Okuyama T (2021) Distinct functions of ventral CA1 and dorsal CA2 in social memory. Curr Opin Neurobiol 68:29–35. https://doi.org/10.1016/j.conb.2020.12.008

Wiltgen BJ, Sanders MJ, Anagnostaras SG, Sage JR, Fanselow MS (2006) Context fear learning in the absence of the hippocampus. J Neurosci 26(20):5484–5491. https://doi.org/10.1523/JNEUROSCI.2685-05.2006

Wood ER, Dudchenko PA, Eichenbaum H (1999) The global record of memory in hippocampal neuronal activity. Nature 397(6720):613–616. https://doi.org/10.1038/17605

Xu C, Krabbe S, Gründemann J, Botta P, Fadok JP, Osakada F, Saur D, Grewe BF, Schnitzer MJ, Callaway EM, Lüthi A (2016) Distinct hippocampal pathways mediate dissociable roles of context in memory retrieval. Cell 167(4):961–972.e16. https://doi.org/10.1016/j.cell.2016.09.051

Yoshii T, Hosokawa H, Matsuo N (2017) Pharmacogenetic reactivation of the original engram evokes an extinguished fear memory. Neuropharmacology 113(Pt A):1–9. https://doi.org/10.1016/j.neuropharm.2016.09.012

Yuste R, Peinado A, Katz LC (1992) Neuronal domains in developing neocortex. Science 257(5070):665–669. https://doi.org/10.1126/science.1496379

Zaki Y, Mau W, Cincotta C, Monasterio A, Odom E, Doucette E, Grella SL, Merfeld E, Shpokayte M, Ramirez S (2022) Hippocampus and amygdala fear memory engrams re-emerge after contextual fear relapse. Neuropsychopharmacology 47(11):1992–2001. https://doi.org/10.1038/s41386-022-01407-0

Zelikowsky M, Bissiere S, Hast TA, Bennett RZ, Abdipranoto A, Vissel B, Fanselow MS (2013) Prefrontal microcircuit underlies contextual learning after hippocampal loss. Proc Natl Acad Sci USA 110(24):9938–9943. https://doi.org/10.1073/pnas.1301691110

Zhang WN, Bast T, Feldon J (2001) The ventral hippocampus and fear conditioning in rats: different anterograde amnesias of fear after infusion of N-methyl-D-aspartate or its noncompetitive antagonist MK-801 into the ventral hippocampus. Behav Brain Res 126(1–2):159–174. https://doi.org/10.1016/s0166-4328(01)00256-x

Zhang TR, Larosa A, Di Raddo M-E, Wong V, Wong AS, Wong TP (2019) Negative memory engrams in the hippocampus enhance the susceptibility to chronic social defeat stress. J Neurosci 39(38):7576–7590. https://doi.org/10.1523/JNEUROSCI.1958-18.2019

# The Role of Prefrontal Ensembles in Memory Across Time: Time-Dependent Transformations of Prefrontal Memory Ensembles

**Zachary Zeidler and Laura DeNardo**

**Abstract** The medial prefrontal cortex (mPFC) plays a critical role in recalling recent and remote fearful memories. Modern neuroscience techniques, such as projection-specific circuit manipulation and activity-dependent labeling, have illuminated how mPFC memory ensembles are reorganized over time. This chapter discusses the implications of new findings for traditional theories of memory, such as the systems consolidation theory and theories of memory engrams. It also examines the specific contributions of mPFC subregions, like the prelimbic and infralimbic cortices, in fear memory, highlighting how their distinct connections influence memory recall. Further, it elaborates on the cellular and molecular changes within the mPFC that support memory persistence and how these are influenced by interactions with the hippocampus. Ultimately, this chapter provides insights into how lasting memories are dynamically encoded in prefrontal circuits, arguing for a key role of memory ensembles that extend beyond strict definitions of the engram.

**Keywords** Prefrontal cortex · Memory recall · Time-dependent organization · Neuronal ensembles · Engram

## Introduction

For nearly 70 years, the medial prefrontal cortex (mPFC) has been investigated for its role in memory (Lashley 1950). As the field's methods have advanced, our understanding of the role of the mPFC in memory has grown increasingly nuanced. In particular, two recent advances in neuroscience—projection-specific circuit manipulation and activity-dependent labeling—have revolutionized our understanding of how the mPFC functions in acquiring, consolidating, and recalling

Z. Zeidler (✉) · L. DeNardo (✉)
Department of Physiology, University of California Los Angeles, Los Angeles, CA, USA
e-mail: zzeidler@ucla.edu; ldenardo@ucla.edu

 
J. Gräff, S. Ramirez (eds.), *Engrams*, Advances in Neurobiology 38,
https://doi.org/10.1007/978-3-031-62983-9_5

memories. Owing to the anatomical resolution and indelible ensemble tagging, these techniques have helped us understand that mPFC involvement in memory is highly dynamic. As time passes and memories are recalled, the prefrontal ensembles that control memory reorganize along anatomical (e.g. projection-class) and functional dimensions.

In this chapter, we will discuss new work that has advanced our understanding of how mPFC memory ensembles reorganize from recent to remote timepoints. We will then use this understanding to reexamine basic concepts in memory research, including core concepts in the standard theory of systems consolidation of memory and the definition of engram itself. Recent memories are those that are recalled hours or days after the learning event, while remote memories generally refer to recall after 7–30 days or older. While the binarization of memory into recent and remote categories likely oversimplifies a gradient of physiological processes, it nevertheless remains a useful simplification. Cellular and molecular changes occur rapidly (minutes to hours) after the learning event. These changes reinforce the event-activated ensembles, often evident at the synaptic level. Yet over days to weeks, cellular changes accrue that reorganize the mPFC circuits involved in memory recall.

There is a notable tension between this reorganization of mPFC memory ensembles and the core idea of how the brain stores memory across time. At its face, the classical conception of memory as the storing and retaining of information is at odds with a dynamic neural population mediating recall. Yet multiple studies have now reported that only a minority of learning or recent recall-activated mPFC neurons are reactivated during remote recall (DeNardo et al. 2019; Lee et al. 2022). Furthermore, certain mPFC populations critical for recall one day following learning are unnecessary for recall at a remote timepoint and vice versa (Do-Monte et al. 2015). Investigating how the stability of memory is reconciled with the dynamic nature of memory ensembles pushes the forefront of engram research.

Engrams are typically defined with the dual criteria of being active during learning and recall. The reorganization of mPFC ensembles provides a counterpoint to the classical definition of an engram. While the learning-activated population is a crucial area for memory-related insight, limiting our study of mPFC memory to engram cells results in an incomplete understanding of the mPFC's role in memory processes. By examining the evolution of mPFC engram and non-engram, recall-activated neurons across time, a richer understanding of the dynamic nature of memory appears.

## Prefrontal Subdivisions and Fear Memory

The dynamic nature of the mPFC in memory recall may relate to its unique cognitive role. The general function of the mPFC has been described as "the representation and execution of new forms of goal-directed action"(Fuster 2015). This includes control over many complex cognitive and emotional functions like working memory, attention, decision-making, and emotional regulation. Each of these functions intersects with memory in some capacity, but their exact contributions, and at different memory stages, are an area of active investigation.

The diversity of mPFC functions is in part mediated by subregions with distinct connectivity (Lui et al. 2020). The literature reviewed in this chapter largely draws from work using a rodent model species; therefore, we use anatomy and nomenclature from the rodent literature. The mPFC includes the prelimbic (PL), infralimbic (IL), and anterior cingulate cortices (ACC). These have extensive inputs from limbic regions, including the amygdala and hippocampus, which likely influence the representation of contextual cues in the mPFC, biasing action selection toward salient cues. Despite limbic inputs to all regions, PL, IL, and ACC can play distinct yet partially overlapping roles in fear conditioning and fear recall. IL is critical for the extinction of fear memory while PL and ACC are involved in recall and initiation or execution of memory and memory-based action. While the functional dichotomy between IL and PL/ACC is well-characterized in the realm of fear learning and memory, distinctions between the roles of PL and ACC are more nuanced (Almada et al. 2015; Rea et al. 2019). mPFC subregion-specific functions have been thoroughly discussed in recent reviews (Sung and Kaang 2022; Merre et al. 2021; Euston et al. 2012; Giustino and Maren 2015).

PL and ACC are often associated with distinct aspects of fear learning and memory. The ACC has been well-established as central for the recall of remote contextual fear memories (Frankland et al. 2004; Restivo et al. 2009; Vetere et al. 2011; Tulogdi et al. 2012; Wheeler et al. 2013). PL is involved in the recall of recent cued fear memory (Do-Monte et al. 2015; Corcoran and Quirk 2007; Burgos-Robles et al. 2009; Choi et al. 2010; Sierra-Mercado et al. 2011) as well as remote cued fear memory (DeNardo et al. 2019; Do-Monte et al. 2015). PL has also been shown to be involved in the expression and recall of recent (Corcoran and Quirk 2007) and remote contextual fear (Lee et al. 2022; Dixsaut and Gräff 2022; Kitamura et al. 2017).

The difference in connectivity between PL and ACC may direct their roles in different forms of fear memory. While both structures have overlapping afferent anatomical connections, including strong recurrent connectivity within the mPFC itself, the ACC generally receives more sensorimotor input while PL receives more limbic input (Hoover and Vertes 2007). The efferents from these structures likely also affect how they participate in fear memory expression. According to a recent single-neuron projectome of the mouse prefrontal cortex, approximately 3% of ACC neurons have axons in the basolateral amygdala (BLA) or central amygdala (CeA), regions critical for fear expression, while 15% of PL neurons project to the amygdala (Gao et al. 2022). In addition to PL having more efferents to the amygdala than ACC, it also has more afferent amygdalar connections (Hoover and Vertes 2007). While both ACC and PL are crucial for fear memory, PL has generally been reported to be involved in fear expression across a broader range of fear paradigms than ACC, and the high degree of bidirectional connections between PL and amygdala may inform this bias in the literature.

## mPFC Engrams—Cellular and Molecular Plasticity Across Time

Recent studies intersecting activity-dependent labeling and modulation of mPFC activity have suggested that activity during learning is important for later recruitment into the memory ensemble—what is sometimes referred to as an eligibility trace (DeNardo et al. 2019; Jung et al. 2023). In these studies, researchers showed that active mPFC neurons during cued fear conditioning could later drive freezing when optogenetically reactivated. However, if mPFC neurons were silenced during fear learning, those neurons were unable to promote freezing with the same optogenetic manipulation (DeNardo et al. 2019). These data support the idea that excitability—or "tagging"—during memory acquisition is key to future memory trace membership (Jung et al. 2023; Han et al. 2007; Lesburguères et al. 2011; Kim et al. 2013; Yiu et al. 2014). However, changes at the memory ensemble's cellular, molecular, and circuit levels complicate the idea of a static engram from learning to recall.

Immediately after the learning event, plasticity occurs to promote consolidation of the experience. While much of the literature in this area has been done in the amygdala and hippocampus (for reviews, see (Johansen et al. 2011; Orsini and Maren 2012; Tovote et al. 2015; Chaaya et al. 2018)), similar mechanisms may occur for the consolidation of mPFC fear memory ensembles. Lee and colleagues investigated the timecourse of synaptic strengthening between fear conditioning-activated and non-fear-conditioning-activated mPFC neurons, finding increased synaptic strength at 28 days but not at 7 days following learning (Lee et al. 2022). These data align with studies of dendritic spines in the mPFC showing a time-dependent increase in spine density examining across recent and remote memory (Vetere et al. 2011; Kitamura et al. 2017; Aceti et al. 2015). Together, these data show that classic molecular and cellular mechanisms of synaptic potentiation, typically studied in the subcortical limbic regions, are also present in the mPFC. The timecourse of these changes demonstrates that molecular mechanisms of plasticity may not occur or may not be evident immediately following learning in mPFC ensembles.

Despite the protracted timecourse of consolidation, activity immediately after learning may be required to initiate ensemble strengthening. Evidence comes from synapse-specific optogenetic inhibition of cortical fear circuits immediately following auditory fear conditioning, observing impaired remote recall without affecting recent recall (Lee et al. 2021). This work suggests that cortical fear circuits are required to be engaged immediately after learning for intact remote memory. Similar mechanisms may be in play for mPFC ensembles. Prefrontal neurons show changes in firing rate within minutes (Takehara-Nishiuchi et al. 2020) and changes to their transcriptomic and ultrastructural synaptic profile within an hour (Bero et al. 2014) following fear conditioning. This work suggests that within minutes of a fearful experience occurring, changes in gene expression and synaptic signaling begin to occur that influence the long-term ability of mPFC circuits to mediate memory recall independently of the consolidation occurring in recent recall circuits.

While the precise mechanism that guides the strengthening of prefrontal memory ensembles across time requires more investigation, mounting evidence supports the role of the hippocampus in this process. Studies tagged cFos+ cells during fear conditioning in the hippocampus (Lee et al. 2022; Kitamura et al. 2017) or in regions that facilitate hippocampo-prefrontal communication, including the medial entorhinal and restrosplenial cortices. In those mice, ablating or inhibiting synaptic communication of those learning-activated cells throughout consolidation resulted in diminished freezing and fewer cFos+ mPFC neurons at remote recall. Electrophysiological recordings in the hippocampus and mPFC found that mPFC neurons modulated their firing in response to hippocampal theta and sharp wave ripples during both sleep and wake, which may support memory retrieval and updating (Wierzynski et al. 2009; Jadhav et al. 2016). In the 12 hours following a fearful event, CA1 and PL memory encoding ensembles are reactivated and strengthened during non-REM sleep (Miyawaki and Mizuseki 2022). Similarly, CA1-PL ensembles co-activated during a learning experience are maximally reactivated during hippocampal sharp wave ripples (Miyawaki and Mizuseki 2022). The interplay between prefrontal memory and hippocampal activity occurs for both spatial (Kitamura et al. 2017; Goto et al. 2021) and non-spatial (Lee et al. 2022; Sawangjit et al. 2018) memory, implying that prefrontal-hippocampal coordination is a general mechanism of memory consolidation. Even more broadly, hippocampal sharp wave ripples coordinate activity across the neocortex (Siapas and Wilson 1998), pointing toward the reorganization of memory ensembles beyond the mPFC.

## mPFC Engrams—Circuit Mechanisms Across Time

During the acquisition of a fear memory, the mPFC has been shown to be broadly engaged (Lee et al. 2022; Burgos-Robles et al. 2009; Dixsaut and Gräff 2022; Kitamura et al. 2017; Roy et al. 2022; Cummings and Clem 2020). Interestingly, pre-training inactivation of the mPFC during acquisition of fear memory has historically had mixed results on fear acquisition and retrieval across time, often resulting in no effect (for an exhaustive review, see (Courtin et al. 2013)). However, recent work using activity-dependent tagging and circuit-specific inhibition of mPFC input during encoding of fear memory revealed different impairments on recent or remote memory depending on the circuit (Dixsaut and Gräff 2022). More research is needed to fully understand the relationship between mPFC activity during memory encoding, the afferent input driving that activity, and its effects on memory recall.

Just as varied mPFC inputs control consolidation, heterogenous mPFC outputs control recall. The mPFC projects to many cortical and subcortical brain regions, with wide-ranging transcriptomic and task-related activity profiles across projection classes (Lui et al. 2020). This diversity of structure and function makes the mPFC a prime candidate for orchestrating the recruitment of different circuits to support memory recall. In addition to output circuits, the mPFC exhibits recurrent connectivity (Gao et al. 2022, 2023; Gongwer et al. 2022) with important roles for local inhibitory circuits in memory (Cummings and Clem 2020; Cummings et al. 2022), all of which may facilitate shifts in prefrontal memory ensembles across time (Lee et al. 2022).

To dissect the contributions of prefrontal circuits in fear recall has required both anatomical and temporal precision. Recent work demonstrated that prefrontal neurons active during recent memory recall have enhanced functional connectivity with subcortical regions compared to a remote recall ensemble (DeNardo et al. 2019). In support of this, optogenetic inhibition of mPFC projections to the BLA strongly impaired recall of recent cued fear memory (Do-Monte et al. 2015). These results demonstrate that fronto-amygdalar circuits can support recall at early timepoints. However, other work utilizing activity-dependent tagging to exclusively target prefrontal neurons active during fear learning found no effect of fronto-amygdalar circuit inhibition on recall of recent context fear memory (Kitamura et al. 2017). These findings either reflect differences in the prefrontal circuit mechanisms governing recent cued versus contextual fear memory and/or suggest that while the mPFC supports recent memory recall through amygdalar activation, it does so through a different ensemble than the one engaged during memory acquisition.

Indeed, the mPFC exhibits striking reorganization of the ensembles and circuits supporting recall as a memory age. Prefrontal populations active during remote fear memory generally have stronger functional connectivity with high-order sensory cortices compared to recent recall ensembles (DeNardo et al. 2019). Recurrent prefrontal connectivity is strengthened following contextual fear conditioning at a remote compared to recent timepoint (Lee et al. 2022). Also, both learning-activated and remote recall-activated prefrontal neurons strengthen their connections to learning-activated amygdala neurons at a remote timepoint in contextual memory recall (Lee et al. 2022).

In agreement with these results, inhibition of learning-activated prefronto-amygdalar projections during remote recall impaired context memory (Kitamura et al. 2017). However, inhibition of bulk fronto-amygdalar projections during remote-cued fear recall failed to impact memory (Do-Monte et al. 2015). Like recent recall, these findings may indicate differences in prefrontal fear memory circuitry depending on the content of the memory and/or suggest a privileged role for learning-activated fronto-amygdalar neurons in remote memory. Additional work is needed to distinguish the influence of memory content and learning-activated ensembles on prefrontal recall circuits across time.

Certain cortical areas show pronounced recruitment by remote prefrontal memory ensembles (DeNardo et al. 2019) and are involved in remote-cued fear memory, including the auditory association area (Sacco and Sacchetti 2010; Cambiaghi et al. 2016, 2017), retrosplenial cortex (Todd et al. 2016, 2018), and postrhinal cortex (DeAngeli et al. 2022). Other cortical areas similarly activated by remote memory ensembles in the mPFC (DeNardo et al. 2019) participate in remote contextual fear memory, such as the perirhinal and postrhinal cortex (Burwell et al. 2004). While evidence indicates that mPFC circuits are poised to coordinate these cortical areas in remote memory, direct evidence to support this idea is currently lacking (Fig. 1).

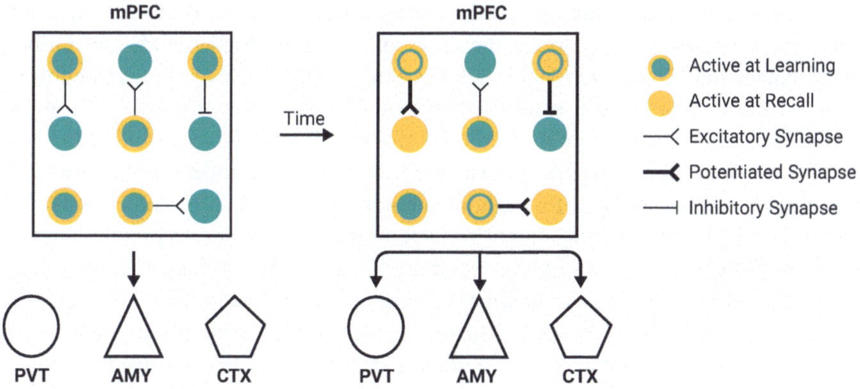


**Fig. 1** Graphical summary of prefrontal circuits supporting recall across time. AMY amygdala; PVT paraventricular thalamus; CTX cortex

Taken together, these studies suggest that the reorganization of mPFC circuits from recent to remote recall may be directed toward the recruitment of specific populations with enhanced functional connectivity to cortical areas (DeNardo et al. 2019). While the mechanism of this dynamic change is an area of active investigation, two pieces of information hint at what may be happening. First, activity-dependent labeling of learning-activated and recall-activated ensembles has consistently found about 30% of the prefrontal memory ensembles to be reactivated upon recall (DeNardo et al. 2019; Lee et al. 2022). This suggests that there is a small population of neurons consistently engaged in memory across time, potentially acting as a memory scaffold. Second, learning-activated prefrontal neurons show progressive strengthening onto other, non-learning-activated prefrontal neurons across time (Lee et al. 2022). These data suggest that a small, stable population of prefrontal neurons could be active throughout memory, progressively recruiting new members to the prefrontal memory ensemble. In parallel to modifications of synaptic weights in other circuits supporting memory, this results in a remotely active ensemble with distinct functional connectivity compared to recent ones (DeNardo et al. 2019).

## mPFC Engrams and Ensembles—Broadening Theories and Definitions

After reviewing the current literature on mPFC engrams, we can revisit two core concepts in memory research: systems consolidation theory of memory (SCM) and the definition of an engram itself. The standard theory of SCM proposes that unstable memories are initially stored in the hippocampus and gradually consolidated into cortical networks, including the mPFC (Müller and Pilzecker 1900; Marr 1970,

1971; Frankland and Bontempi 2005; Tonegawa et al. 2018) (but see important counter-arguments among "trace theories" of memory (Moscovitch and Gilboa 2022; Nadel and Moscovitch 1997; Sekeres et al. 2018). However, human neuroimaging studies report mixed results when examining whether the mPFC is more active during the recall of remote memories (for reviews, see (Cabeza and St Jacques 2007; Dede and Smith 2018)). As discussed, findings from rodent work show that the mPFC can be engaged throughout memory recall via different circuits and ensembles depending on the content and age of the memory. Key synaptic connections in mPFC circuits—including recurrent circuits—are modulated across time, giving the mPFC a persistent yet changing role in recall. The mPFC indeed plays a different role in remote compared to recent memory, in principle agreeing with SCM. Yet a preponderance of the evidence demonstrates that the mPFC can be engaged across time, which is an important update to traditional models of SCM (Frankland and Bontempi 2005).

Just as mPFC engrams force us to reconsider traditional notions of SCM, it also raises questions about traditional denotations of the engram itself. Engram cells are commonly defined as "a population of neurons that are activated by learning, have enduring cellular changes as a consequence of learning, and whose reactivation by a part of the original stimuli delivered during learning results in memory recall" (Tonegawa et al. 2015). A key criterion here is "activated by learning," and while that is an important and interesting aspect of the engram, is not the only dimension to consider when discussing membership in memory ensembles.

The concept of silent versus active engram neurons has emerged to describe engram neurons that are not reactivated during a specific recall event, such as some prefrontal neurons during recent recall (Kitamura et al. 2017). Critically though, silent engram neurons are activated by learning. Yet memory recall in the mPFC (DeNardo et al. 2019; Lee et al. 2022) and other areas such as the amygdala (Cho et al. 2021) is predominantly guided by neurons that were inactive during learning. Conflating "engram" with "memory ensemble" excludes the important class of recall-activated, learning-silent neurons that make up a majority of the memory recall ensemble in the mPFC. Despite being silent during learning (insofar as our activity-dependent technologies can detect), these cells exhibit enduring cellular changes as a consequence of learning and activation by the original stimulus (Fig. 2).

Broadening the notion of memory ensembles beyond the strict definition of an engram is more than arguing semantics. It reflects an updated idea that memories and the systems that carry them are not static. Given the dynamic nature of memory updating (Dudai 2012) and the chaotic nature of neural states (Skarda and Freeman 1987), it is important to not constrain the scope of research questions by constraining our technical discourse. Understanding the role of recall-activated ensembles in memory, and how they may evolve from learning-activated engrams, will provide important insight into how memory functions across time, in the mPFC and beyond.

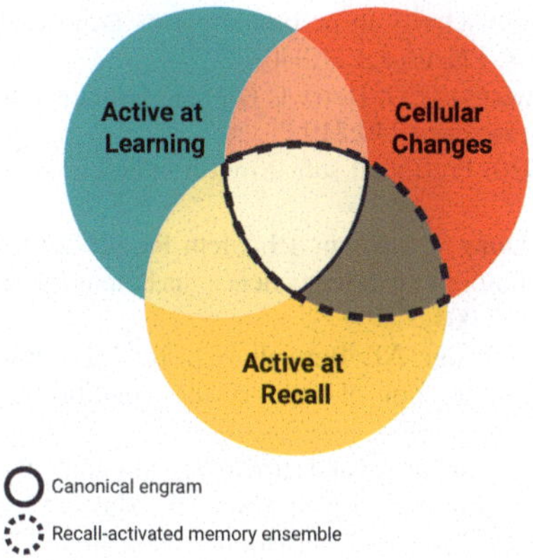


**Fig. 2** Diagram of criteria for canonical engram neurons (solid outline) and recall-activated memory ensembles (dashed outline)

## References

Aceti M, Vetere G, Novembre G, Restivo L, Ammassari-Teule M (2015) Progression of activity and structural changes in the anterior cingulate cortex during remote memory formation. Neurobiol Learn Mem 123:67

Almada RC, Coimbra NC, Brandão ML (2015) Medial prefrontal cortex serotonergic and GABAergic mechanisms modulate the expression of contextual fear: Intratelencephalic pathways and differential involvement of cortical subregions. Neuroscience 284:988

Bero AW, Meng J, Cho S, Shen AH, Canter RG, Ericsson M, Tsai L-H (2014) Early remodeling of the neocortex upon episodic memory encoding. Proc Natl Acad Sci 111:11852

Burgos-Robles A, Vidal-Gonzalez I, Quirk GJ (2009) Sustained conditioned responses in prelimbic prefrontal neurons are correlated with fear expression and extinction failure. J Neurosci 29:8474

Burwell RD, Bucci DJ, Sanborn MR, Jutras MJ (2004) Perirhinal and postrhinal contributions to remote memory for context. J Neurosci 24:11023

Cabeza R, St Jacques P (2007) Functional neuroimaging of autobiographical memory. Trends Cogn Sci 11:219

Cambiaghi M, Grosso A, Likhtik E, Mazziotti R, Concina G, Renna A, Sacco T, Gordon JA, Sacchetti B (2016) Higher-order sensory cortex drives basolateral amygdala activity during the recall of remote, but not recently learned fearful memories. J Neurosci 36:1647

Cambiaghi M, Renna A, Milano L, Sacchetti B (2017) Reversible inactivation of the higher order auditory cortex during fear memory consolidation prevents memory-related activity in the basolateral amygdala during remote memory retrieval. Front Behav Neurosci 11:138

Chaaya N, Battle AR, Johnson LR (2018) An update on contextual fear memory mechanisms: transition between amygdala and hippocampus. Neurosci Biobehav Rev 92:43

Cho H-Y, Shin W, Lee H-S, Lee Y, Kim M, Oh J-P, Han J, Jeong Y, Suh B, Kim E, Han J-H (2021) Turnover of fear engram cells by repeated experience. Curr Biol. https://doi.org/10.1016/j.cub.2021.10.004

Choi DC, Maguschak KA, Ye K, Jang S-W, Myers KM, Ressler KJ (2010) Prelimbic cortical BDNF is required for memory of learned fear but not extinction or innate fear. Proc Natl Acad Sci 107:2675

Corcoran KA, Quirk GJ (2007) Activity in prelimbic cortex is necessary for the expression of learned, but not innate, fears. J Neurosci 27:840
Courtin J, Bienvenu TCM, Einarsson EÖ, Herry C (2013) Medial prefrontal cortex neuronal circuits in fear behavior. Neuroscience 240:219
Cummings KA, Clem RL (2020) Prefrontal somatostatin interneurons encode fear memory. Nat Neurosci 23:61
Cummings KA, Bayshtok S, Dong TN, Kenny PJ, Clem RL (2022) Control of fear by discrete prefrontal GABAergic populations encoding valence-specific information. Neuron. https://doi.org/10.1016/j.neuron.2022.07.004
DeAngeli NE, Fournier DI, Gulledge AT, Burwell RD, Todd TP, Bucci DJ (2022) Postrhinal cortex contributions to the expression of auditory fear conditioning. Neurobiol Learn Mem 191:107609
Dede AJO, Smith CN (2018) The functional and structural neuroanatomy of systems consolidation for autobiographical and semantic memory. In: Clark RE, Martin SJ (eds) Behavioral neuroscience of learning and memory. Springer International Publishing, Cham, pp 119–150. https://doi.org/10.1007/7854_2016_452
DeNardo LA, Liu CD, Allen WE, Adams EL, Friedmann D, Fu L, Guenthner CJ, Tessier-Lavigne M, Luo L (2019) Temporal evolution of cortical ensembles promoting remote memory retrieval. Nat Neurosci 22:460
Dixsaut L, Gräff J (2022) Brain-wide screen of prelimbic cortex inputs reveals a functional shift during early fear memory consolidation. elife 11:e78542
Do-Monte FH, Quiñones-Laracuente K, Quirk GJ (2015) A temporal shift in the circuits mediating retrieval of fear memory. Nature 519:460
Dudai Y (2012) The restless engram: consolidations never end. Annu Rev Neurosci 35:227
Euston DR, Gruber AJ, McNaughton BL (2012) The role of medial prefrontal cortex in memory and decision making. Neuron 76:1057
Frankland PW, Bontempi B (2005) The organization of recent and remote memories. Nat Rev Neurosci 6:119
Frankland PW, Bontempi B, Talton LE, Kaczmarek L, Silva AJ (2004) The involvement of the anterior cingulate cortex in remote contextual fear memory. Science 304:881
Fuster JM (2015) Chapter 1 - Introduction. In: Fuster JM (ed) The prefrontal cortex, 5th edn. Academic Press, San Diego, pp 1–8. https://doi.org/10.1016/B978-0-12-407815-4.00001-5
Gao L, Liu S, Gou L, Hu Y, Liu Y, Deng L, Ma D, Wang H, Yang Q, Chen Z, Liu D, Qiu S, Wang X, Wang D, Wang X, Ren B, Liu Q, Chen T, Shi X, Yao H, Xu C, Li CT, Sun Y, Li A, Luo Q, Gong H, Xu N, Yan J (2022) Single-neuron projectome of mouse prefrontal cortex. Nat Neurosci 25:515
Gao L, Liu S, Wang Y, Wu Q, Gou L, Yan J (2023) Single-neuron analysis of dendrites and axons reveals the network organization in mouse prefrontal cortex. Nat Neurosci 26:1111
Giustino TF, Maren S (2015) The role of the medial prefrontal cortex in the conditioning and extinction of fear. Front Behav Neurosci 9:298
Gongwer MW, Klune CB, Enos AS, Chen R, Friedmann D, DeNardo LA (2022) Mapping whole-brain projections of anatomically-defined prefrontal neurons using combined 3D convolutional networks. bioRxiv 2022.03.31.486619. https://doi.org/10.1101/2022.03.31.486619
Goto A, Bota A, Miya K, Wang J, Tsukamoto S, Jiang X, Hirai D, Murayama M, Matsuda T, McHugh TJ, Nagai T, Hayashi Y (2021) Stepwise synaptic plasticity events drive the early phase of memory consolidation. Science 374:857
Han J-H, Kushner SA, Yiu AP, Cole CJ, Matynia A, Brown RA, Neve RL, Guzowski JF, Silva AJ, Josselyn SA (2007) Neuronal competition and selection during memory formation. Science 316:457
Hoover WB, Vertes RP (2007) Anatomical analysis of afferent projections to the medial prefrontal cortex in the rat. Brain Struct Funct 212:149
Jadhav SP, Rothschild G, Roumis DK, Frank LM (2016) Coordinated excitation and inhibition of prefrontal ensembles during awake hippocampal sharp-wave ripple events. Neuron 90:113

Johansen JP, Cain CK, Ostroff LE, LeDoux JE (2011) Molecular mechanisms of fear learning and memory. Cell 147:509

Jung JH, Wang Y, Mocle AJ, Zhang T, Köhler S, Frankland PW, Josselyn SA (2023) Examining the engram encoding specificity hypothesis in mice. Neuron 111:1830

Kim D, Paré D, Nair SS (2013) Assignment of model amygdala neurons to the fear memory trace depends on competitive synaptic interactions. J Neurosci 33:14354

Kitamura T, Ogawa SK, Roy DS, Okuyama T, Morrissey MD, Smith LM, Redondo RL, Tonegawa S (2017) Engrams and circuits crucial for systems consolidation of a memory. Science 356:73

Lashley KS (1950) In search of the engram, [i]Society of Experimental Biology Symposium, No. 4: psychological mechanisms in animal behavior[/i]. Cambridge University Press, Cambridge, p 454–455, 468–473, 477–480, (1988). https://doi.org/10.7551/mitpress/4943.003.0007

Lee Y, Oh J-P, Han J-H (2021) Dissociated role of thalamic and cortical input to the lateral amygdala for consolidation of long-term fear memory. J Neurosci. https://doi.org/10.1523/jneurosci.1167-21.2021

Lee J-H, Kim WB, Park EH, Cho J-H (2022) Neocortical synaptic engrams for remote contextual memories. Nat Neurosci 1. https://doi.org/10.1038/s41593-022-01223-1

Lesburguères E, Gobbo OL, Alaux-Cantin S, Hambucken A, Trifilieff P, Bontempi B (2011) Early tagging of cortical networks is required for the formation of enduring associative memory. Science 331:924

Lui JH, Nguyen ND, Grutzner SM, Darmanis S, Peixoto D, Wagner MJ, Allen WE, Kebschull JM, Richman EB, Ren J, Newsome WT, Quake SR, Luo L (2020) Differential encoding in prefrontal cortex projection neuron classes across cognitive tasks. Cell. https://doi.org/10.1016/j.cell.2020.11.046

Marr D (1970) A theory for cerebral neocortex. Proc R Soc Lond B Biol Sci 176:161

Marr D (1971) Simple memory: a theory for archicortex. Philos Trans R Soc Lond Ser B Biol Sci 262:23

Merre PL, Ährlund-Richter S, Carlén M (2021) The mouse prefrontal cortex: Unity in diversity. Neuron. https://doi.org/10.1016/j.neuron.2021.03.035

Miyawaki H, Mizuseki K (2022) De novo inter-regional coactivations of preconfigured local ensembles support memory. Nat Commun 13:1272

Moscovitch M, Gilboa A (2022) Has the concept of systems consolidation outlived its usefulness? Identification and evaluation of premises underlying systems consolidation. F1000Prime Rep 11:33

Müller GE, Pilzecker A (1900) Experimentelle Beiträge zur Lehre vom Gedächtnis. Zeitschrift für Psychologie. Ergänzungsband 1:1–300

Nadel L, Moscovitch M (1997) Memory consolidation, retrograde amnesia and the hippocampal complex. Curr Opin Neurobiol 7:217

Orsini CA, Maren S (2012) Neural and cellular mechanisms of fear and extinction memory formation. Neurosci Biobehav Rev 36:1773

Rea K, McGowan F, Corcoran L, Roche M, Finn DP (2019) The prefrontal cortical endocannabinoid system modulates fear–pain interactions in a subregion-specific manner. Br J Pharmacol 176:1492

Restivo L, Vetere G, Bontempi B, Ammassari-Teule M (2009) The formation of recent and remote memory is associated with time-dependent formation of dendritic spines in the hippocampus and anterior cingulate cortex. J Neurosci 29:8206

Roy DS, Park Y-G, Kim ME, Zhang Y, Ogawa SK, DiNapoli N, Gu X, Cho JH, Choi H, Kamentsky L, Martin J, Mosto O, Aida T, Chung K, Tonegawa S (2022) Brain-wide mapping reveals that engrams for a single memory are distributed across multiple brain regions. Nat Commun 13:1799

Sacco T, Sacchetti B (2010) Role of secondary sensory cortices in emotional memory storage and retrieval in rats. Science 329:649

Sawangjit A, Oyanedel CN, Niethard N, Salazar C, Born J, Inostroza M (2018) The hippocampus is crucial for forming non-hippocampal long-term memory during sleep. Nature 564:109

Sekeres MJ, Winocur G, Moscovitch M (2018) The hippocampus and related neocortical structures in memory transformation. Neurosci Lett 680:39

Siapas AG, Wilson MA (1998) Coordinated interactions between hippocampal ripples and cortical spindles during slow-wave sleep. Neuron 21:1123

Sierra-Mercado D, Padilla-Coreano N, Quirk GJ (2011) Dissociable roles of prelimbic and infralimbic cortices, ventral hippocampus, and basolateral amygdala in the expression and extinction of conditioned fear. Neuropsychopharmacology 36:529

Skarda CA, Freeman WJ (1987) How brains make chaos in order to make sense of the world. Behav Brain Sci 10:161

Sung Y, Kaang B-K (2022) The three musketeers in the medial prefrontal cortex: subregion-specific structural and functional plasticity underlying fear memory stages. Exp Neurobiol 31:221

Takehara-Nishiuchi K, Morrissey MD, Pilkiw M (2020) Prefrontal neural ensembles develop selective code for stimulus associations within minutes of novel experiences. J Neurosci 40:8355

Todd TP, Mehlman ML, Keene CS, DeAngeli NE, Bucci DJ (2016) Retrosplenial cortex is required for the retrieval of remote memory for auditory cues. Learn Mem 23:278

Todd TP, Jiang MY, DeAngeli NE, Bucci DJ (2018) A functional circuit for the retrieval of remote cued fear memory. Behav Neurosci 132:403

Tonegawa S, Liu X, Ramirez S, Redondo R (2015) Memory engram cells have come of age. Neuron 87:918

Tonegawa S, Morrissey MD, Kitamura T (2018) The role of engram cells in the systems consolidation of memory. Nat Rev Neurosci 19:485

Tovote P, Fadok JP, Lüthi A (2015) Neuronal circuits for fear and anxiety. Nat Rev Neurosci 16:317

Tulogdi Á, Sörös P, Tóth M, Nagy R, Biró L, Aliczki M, Klausz B, Mikics É, Haller J (2012) Temporal changes in c-Fos activation patterns induced by conditioned fear. Brain Res Bull 88:359

Vetere G, Restivo L, Cole CJ, Ross PJ, Ammassari-Teule M, Josselyn SA, Frankland PW (2011) Spine growth in the anterior cingulate cortex is necessary for the consolidation of contextual fear memory. Proc Natl Acad Sci 108:8456

Wheeler AL, Teixeira CM, Wang AH, Xiong X, Kovacevic N, Lerch JP, McIntosh AR, Parkinson J, Frankland PW (2013) Identification of a functional connectome for long-term fear memory in mice. PLoS Comput Biol 9:e1002853

Wierzynski CM, Lubenov EV, Gu M, Siapas AG (2009) State-dependent spike-timing relationships between hippocampal and prefrontal circuits during sleep. Neuron 61:587

Yiu AP, Mercaldo V, Yan C, Richards B, Rashid AJ, Hsiang H-LL, Pressey J, Mahadevan V, Tran MM, Kushner SA, Woodin MA, Frankland PW, Josselyn SA (2014) Neurons are recruited to a memory trace based on relative neuronal excitability immediately before training. Neuron 83:722

# Part III
# Insights into Intra-Engram Functioning

# All IEGs Are Not Created Equal—Molecular Sorting Within the Memory Engram

**Tushar D. Yelhekar, Meizhen Meng, Joslyn Doupe, and Yingxi Lin**

**Abstract** When neurons are recruited to form the memory engram, they are driven to activate the expression of a series of immediate-early genes (IEGs). While these IEGs have been used relatively indiscriminately to identify the so-called engram neurons, recent research has demonstrated that different IEG ensembles can be physically and functionally distinct within the memory engram. This inherent heterogeneity of the memory engram is driven by the diversity in the functions and distributions of different IEGs. This process, which we call molecular sorting, is analogous to sorting the entire population of engram neurons into different sub-engrams molecularly defined by different IEGs. In this chapter, we will describe the molecular sorting process by systematically reviewing published work on engram ensemble cells defined by the following four major IEGs: Fos, Npas4, Arc, and Egr1. By comparing and contrasting these likely different components of the memory engram, we hope to gain a better understanding of the logic and significance behind the molecular sorting process for memory functions.

---

Tushar D. Yelhekar, Meizhen Meng and Joslyn Doupe contributed equally with all other contributors.

---

T. D. Yelhekar · Y. Lin (✉)
Department of Psychiatry, O'Donnell Brain Institute, University of Texas Southwestern Medical Center, Dallas, TX, USA
e-mail: Yingxi.Lin@UTSouthwestern.edu

M. Meng
Department of Psychiatry, O'Donnell Brain Institute, University of Texas Southwestern Medical Center, Dallas, TX, USA

Neuroscience Graduate Program, University of Texas Southwestern Medical Center, Dallas, TX, USA

J. Doupe
Neuroscience Graduate Program, SUNY Upstate Medical University, Syracuse, NY, USA

 
J. Gräff, S. Ramirez (eds.), *Engrams*, Advances in Neurobiology 38,
https://doi.org/10.1007/978-3-031-62983-9_6

**Keywords** cFos · Arc · Npas4 · Egr1 · Molecular sorting

## Introduction

It is generally accepted that active ensemble neurons, the groups of neurons that are activated by memory experience, are the prime candidates for sites of the memory engram (Josselyn and Frankland 2018; Lashley 1950; Semon 1921; Tonegawa et al. 2015). However, as a field we are still at an early stage in terms of our understanding of how and what information is encoded in the active ensemble neurons.

One way to mark activated neurons is to use the activation of immediate-early genes (IEGs), such as Fos, Npas4, Arc and Egr1, in those neurons (Denny et al. 2014; Guenthner et al. 2013; Ramirez et al. 2013; Sorensen et al. 2016; Kawashima et al. 2009; Reijmers et al. 2007). The IEGs, in this case, are used as neuronal activity markers, as they have been for many decades, because their expression is turned on robustly and rapidly after neurons are activated by sensory and behavioral experiences, often within minutes (Greenberg and Ziff 1984; Lanahan and Worley 1998; Nedivi et al. 1993; Qian et al. 1993; Allen et al. 2017). Many decades of painstaking research has established that the expression of many IEGs in regions involved in learning and memory, such as the hippocampus, is correlated with experiences involving learning (Denny et al. 2014; Tischmeyer et al. 1990; Sun and Lin 2016; Sun et al. 2020; Ramamoorthi et al. 2011; Nikolaev et al. 1992; Guzowski et al. 2000). With the development of tools that are based on IEG activation to capture activated neuronal ensembles, whose activity can now be readily manipulated optogenetically or chemogenetically, the IEG ensembles have been shown to possess certain features of the memory engram (Tonegawa et al. 2015; Denny et al. 2014; Sun et al. 2020; Liu et al. 2012; Brito et al. 2022).

Until now, all the IEGs have often been and still are used interchangeably to mark engram cells. However, although their functions have not been compared systematically, one can easily see from published studies that these IEGs have different circuit functions, ranging from regulating excitatory to inhibitory synapses and from controlling synaptic plasticity to homeostasis (Sun et al. 2020; Choi and Kaang 2022; Nambu et al. 2022; Ryan and Frankland 2022; Fleischmann et al. 2003; Lin et al. 2008; Ryan et al. 2015; Shepherd and Bear 2011; Shepherd et al. 2006; Weng et al. 2018; Yap et al. 2021).

If the active neuronal ensembles defined by the different IEGs are all part of the memory engram, we raised the question a few years ago of whether there is functional heterogeneity within the memory engram that can be defined by these different IEGs. We went on to show that this is indeed the case for the contextual fear memory engram in the dentate gyrus (DG), where Fos and Npas4 define functionally distinct neuronal ensembles (Sun et al. 2020). While the Fos ensemble promotes the generalization of contextual fear memory, the Npas4 ensemble is required for its discrimination. We also showed that the Fos and Npas4 neuronal ensembles likely acquired their distinct functionality through respective Fos- and

Npas4-dependent synaptic mechanisms triggered by contextual learning. Our findings support the existence of functional heterogeneity within the memory engram.

While functional heterogeneity within the memory engram is a relatively new concept at the molecular level, at the systems level, as demonstrated by in vivo recording or imaging, the active ensemble neurons have been shown to be quite diverse and to encode different information. Heterogeneous population activities have been observed during learning (Grewe et al. 2017; Grosmark and Buzsaki 2016; Herry et al. 2008; Tanaka et al. 2018). From a theoretical point of view, such heterogeneity is beneficial for efficient information encoding (Chelaru and Dragoi 2008; Osborne et al. 2008). Furthermore, heterogeneous encoding has been hypothesized to allow different aspects of the memory to be individually represented and regulated, which would allow for flexible memory expression in an ever-changing environment (Ghandour et al. 2019; Mallory and Giocomo 2018; Richter et al. 2016; Tanaka and McHugh 2018). To experimentally test and validate the theoretical framework of how memory is encoded in the brain, we need to understand how functionally distinct neuronal ensembles within the memory engram can be defined at the genetic and molecular level. While our previous work provided the first such example, more work is clearly needed to dissect individual functional units within the memory engram.

Here we want to put forward the idea that there is a molecular sorting process (Fig. 1) that distributes the encoded memory in functionally distinct ensembles that constitute the memory engram. Accordingly, information is sorted into different ensembles to drive different behavioral outputs. The functionality of these neuronal ensembles is determined by distinct molecular pathways triggered within each ensemble by the learning experiences.

In this chapter, we will systematically review published work on engram ensemble cells defined by the following four major IEGs: Fos, Npas4, Arc, and Egr1. We will survey the approaches developed to capture the Fos, Npas4, Arc, and Egr1 ensembles, their unique functions in memory expression, and the unique cellular and synaptic properties of the engram cells defined by these IEGs. We will also survey known synaptic and circuit functions of each IEG that, in our opinion, determine the functions of the engram cells they define. Our goal is to compare and contrast these components within the engram to understand the logic and significance behind the molecular sorting events for memory functions.

## Fos: The Prototypical, Best-Known, and Most Widely Used IEG

Fos, also known as c-Fos, was first identified as a proto-oncogene. It belongs to the AP-1 (Activator protein 1) transcription factor family. FOS forms a heterodimer, the AP-1 complex, with one of the JUN family of transcription factors, such as c-Jun, to regulate gene expression. In the 1980s, Fos transcription was found in in vitro studies to be activated rapidly and transiently following depolarization-induced

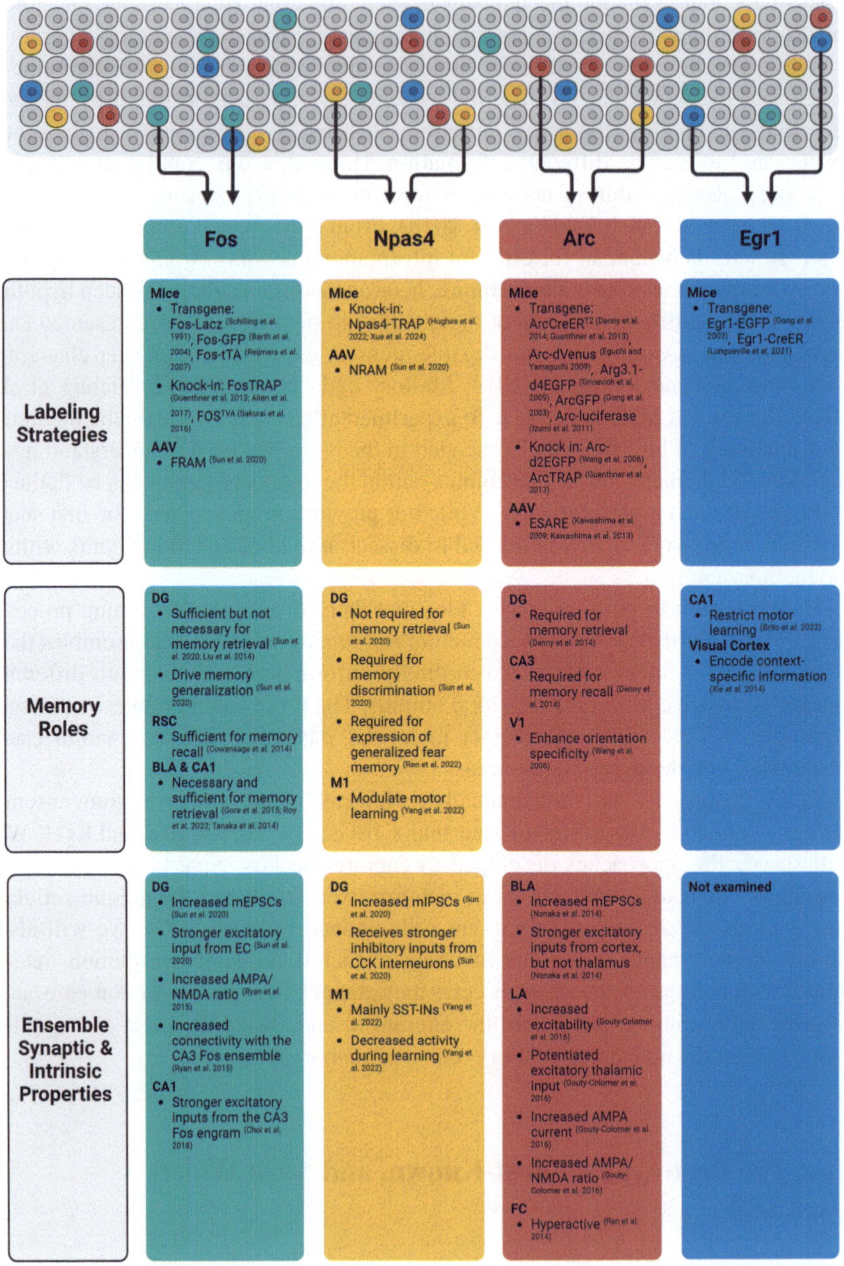


**Fig. 1** Schematic presentation of molecular sorting within the memory engram. The functions and properties of each IEG ensemble that is discussed in the chapter are outlined here. Please note that, while the functions of each IEG are distinct, depending on brain areas and/or memory tasks, engrams labeled with different IEGs can sometimes overlap

calcium influx in neuronally differentiated PC12 cells (Greenberg et al. 1986; Morgan and Curran 1986). In vivo, Fos was first found to be strongly induced in the central nervous system by seizure (Morgan et al. 1987). Elevated Fos expression was subsequently found in specific regions after physiological stimulation (Dragunow and Robertson 1987; Hunt et al. 1987) and behavior training (Tischmeyer et al. 1990; Nikolaev et al. 1992). Since then, Fos has been used as a neuronal activity marker to label behaviorally activated neuronal ensembles. Consistent with its speculative roles in learning and memory, selective deletion of Fos in the central nerve system impairs hippocampus-dependent spatial and associative learning, which is correlated with a reduction of long-term potentiation (LTP) at the hippocampal CA3–CA1 synapses (Fleischmann et al. 2003). These early findings suggested that Fos may be involved in the LTP mechanisms critical for hippocampus-dependent learning. However, whether learning specifically induces synaptic plasticity on the Fos ensemble neurons was not directly examined until recently, with our ability to capture the Fos ensemble neurons.

## *Labeling of the Fos Engram Ensemble*

With the development of engram labeling and manipulation techniques in recent years, the function of Fos-expressing ensemble neurons in the memory engram and the underlying synaptic plasticity have been studied in several brain areas. Here, we summarize the strategies used to label and manipulate Fos-expressed neuronal ensembles. As similar strategies are also used to capture other IEG engram ensembles, the pros and cons of each strategy will be thoroughly discussed here.

Early on, a transgene that harbored a piece of the murine Fos genomic DNA was used to generate several very valuable transgenic animal models to study the Fos ensemble. The transgene Fos-lacZ was first developed by the Curran group (Schilling et al. 1991). It spans from about 1 kb upstream from the Fos transcription start site to its fourth (and last) exon. It encodes a fusion protein containing the first 315 (out of 380) amino acids of the FOS protein and LACZ. The resulting Fos-LacZ transgenic mice and rats (Kasof et al. 1995; Smeyne et al. 1992) were used to examine recently activated neurons in fixed tissue. In addition, these transgenic animals were used to ablate the Fos ensemble neurons, taking advantage of the ability of the LacZ product beta-galactosidase to convert the prodrug Daun02 into Daunorubicin, which inactivates neurons through apoptosis or blockade of voltage-gated calcium channels (Engeln et al. 2016; Koya et al. 2009). This approach of manipulating ensemble neurons was rather innovative before the era of optogenetics and chemogenetics.

The same Fos genomic DNA from the Curran lab (Schilling et al. 1991) was used to generate the Fos-EGFP transgenic mouse (Barth et al. 2004), in which the EGFP coding sequence was fused in frame with the last exon of the Fos gene, resulting in the FOS-EGFP fusion protein. Probably due to the short half-life of the FOS protein, the half-life of the FOS-EGFP protein is shortened to 4–6 hours, allowing the visualization of the Fos ensemble neurons activated by experience a few hours previously. The Fos-EGFP transgenic mice have been used for ex vivo electrophysiology studies (Barth et al. 2004; Clem and Barth 2006) and in vivo imaging (Jouhanneau et al. 2014). A potential caveat for the Fos-LacZ and Fos-EGFP transgenes is the potential artifacts associated with the overexpression of a functional portion of the FOS protein, which have not been carefully examined.

The Fos-tTA transgenic mouse (JAX 018306) (Reijmers et al. 2007) developed by the Mayford group is probably the most widely used Fos transgenic mouse for engram studies. The transgene has two components, presumably in tandem, a Fos promoter-driven tetracycline transactivator (tTA) and a Fos promoter-driven short half-life (~2 hours) GFP (shGFP), but the exact sequence of the transgene was not described in detail in the original publication. It is half of the TetTag mouse (JAX 008344) (Reijmers et al. 2007), which has a second TetO-lacZ transgene. lacZ expression in the TetTag mouse is under the control of tetracycline responsive regulatory element (TRE) and is suppressed in the presence of doxycycline (Dox), which prevents tTA from binding to TRE to drive transcription. The Fos ensemble can be labeled and studied during the temporal window in which Dox is removed (Dox-off). In addition to providing temporal control, the Dox-off system provides the flexibility to express a variety of reporter molecules, such as optogenetic and chemogenetic actuators and genetically encoded $Ca^{2+}$ sensors, under the control of TRE, in the Fos ensemble neurons. Nowadays, the Fos-tTA transgenic mouse is widely used to label, manipulate, and monitor the activity of Fos ensemble neurons in vivo.

Instead of the transgene approach using a truncated Fos promoter, the FosTRAP (Targeted Recombination in Active Populations) mouse (JAX 021882) (Guenthner et al. 2013) developed by Dr. Liqun Luo's lab, replaces the Fos open reading frame (ORF) with the tamoxifen (TM)-dependent Cre recombinase iCreER$^{T2}$ (Feil et al. 1997). The FosTRAP mouse is used together with a Cre-dependent reporter, either transgenic or viral, to permanently express the reporter molecule in Fos ensemble neurons. Since all the transcriptional regulatory elements outside the Fos ORF are intact, activation of expression of iCreER$^{T2}$ should closely resemble that of Fos. Temporal control is provided by the TM-regulated translocation of the iCreER$^{T2}$ protein from the cytosol to the nucleus, where it can activate a Cre-dependent reporter. Since the original FosTRAP mouse disrupts the endogenous Fos expression, an improved version, FosTRAP2 (JAX 030323) (Allen et al. 2017) has since been developed by an in-frame insertion of a 2A-iCreER$^{T2}$ cassette at the end of the Fos ORF. The FosTRAP strategy provides great flexibility to express a wide range of reporter proteins in the Fos ensemble using the plethora of Cre-dependent reporters currently available.

The FosTRAP strategy overcomes some of the technical limitations associated with the earlier Fos transgene strategies. First, tagging of the Fos ensemble is triggered by the endogenous Fos promoter. Second, the TM-dependent tagging window, which can be as short as 6 hours, has a better temporal resolution than the 24–48 hour window provided by the Dox-off system. Third, the tagging is permanent, while the tagging from the Dox-off system lasts 2–4 weeks, depending on the half-life of the reporter protein. However, the Cre-dependent reporter protein usually takes time to accumulate, making the FosTRAP mouse only suited for investigating events some time (typically 1–2 weeks) after the initial tagging event; the transcription amplification provided by the Dox-off system allows the Fos ensemble tagged in the Fos-tTA mouse to be investigated and manipulated as soon as 12–24 hours later. In addition, the tagging efficiency of the FosTRAP mouse is typically lower than that of the Fos-tTA mouse; this is an inherent limitation of the CreER$^{T2}$ system, which typically captures only a subset of the ensemble neurons.

Another knock-in mouse, the Fos$^{TVA}$ mouse (JAX 027831), was generated to be used in the CANE (Capturing and manipulating Activated Neural Ensembles) strategy (Sakurai et al. 2016). Similar to the FosTRAP2 mouse, a 2A-dsTVA (destabilized TVA) cassette is inserted at the end of the Fos ORF such that the Fos ensemble neurons express the TVA receptor and can be infected by a lentivirus pseudotyped with the avian envelope protein EnvA (EnvA-LV). Pseudotyped monosynaptic rabies viruses (EnvA-RVs) can also be used to reveal synaptic inputs to the Fos ensemble neurons. Temporal control is provided by the short half-life of dsTVA and a mutant EnvA, EnvAM21, which has reduced TVA-binding potency, such that only the Fos ensemble neurons activated within a short time before the EnvA-LV injection will be tagged. There are some inherent limitations associated with the CANE strategy. First, mice need to undergo stereotaxic surgeries for virus injection about 1 hour after the behavioral experience, which is not only operationally demanding but can also confound some learning and memory tasks. In addition, like the FosTRAP strategy, the Fos ensemble can only be investigated weeks after tagging.

Our lab developed the Robust Activity Marking (RAM) system (Sorensen et al. 2016) to improve the throughput and signal-to-noise ratio of engram cell capturing. At the core of a RAM reporter is a synthetic activity promoter that is designed to be activated by a specific IEG pathway (described later). For temporal control, the RAM system uses an improved Dox-off system with a short-life tTA, destabilized tTA or d2-tTA, which significantly reduces the accumulation of tTA from previous (irrelevant) activations and improves the signal-to-noise ratio of ensemble tagging. The RAM reporters are compact enough to be packaged into AAV particles, bypassing the need to generate transgenic animals and thus significantly increasing throughput. In addition, the viral-based RAM system can be applied to species other than just the mouse (Sorensen et al. 2016; Dolzani et al. 2018).

Based on the RAM system, we developed the F-RAM (FOS-dependent Robust Activity Marking) reporter to target the Fos ensemble (Sun et al. 2020). Instead of using the Fos promoter, the F-RAM promoter contains the AP-1 DNA binding element that is activated by the endogenous Fos protein. This design offers several advantages. First, it fully reflects the endogenous activation of Fos. Second, since Fos is a transcription factor, exerting its functions through the activation of downstream transcriptional targets, F-RAM captures neurons expressing Fos, where its activation not only occurs but also results in meaningful transcription outputs. Neurons tagged by F-RAM are likely to encode behaviorally induced structural and functional changes mediated by FOS. Third, the synthetic F-RAM promoter is designed to have a high signal-to-noise ratio and high tagging efficiency. It is also compact enough to be packaged into AAVs.

## *Roles of Fos Ensemble Neurons in Memory Expression*

To assess whether Fos (and also other IEG) ensemble neurons are part of a memory engram, loss-of-function and gain-of-function experiments have been typically used to examine if they are necessary and sufficient for memory expression, respectively. In loss-of-function studies, the activity of the Fos ensemble is inhibited during memory recall, while in gain-of-function studies, the Fos ensemble is artificially activated to elicit memory expression. Neither necessity nor sufficiency can prove, or the lack of both rule out, that the Fos ensemble neurons are the physical location of the encoded memory. Nevertheless, published data from the last decade consistently showed that memory expression can indeed be modulated by the manipulation of the Fos ensemble.

The Fos ensemble, like other IEG ensembles, has been best studied in the hippocampus, especially in the DG, in the context of contextual fear conditioning (CFC). In one of the first demonstrations (Liu et al. 2012; Garner et al. 2012) of engram-like properties of the Fos ensemble, Liu et al. (2012, 2014) expressed ChR2 in the Fos ensemble neurons in the DG of Fos-tTA transgenic mice during the acquisition of contextual fear memory. Subsequently, when the Fos ensemble was activated by blue light, mice displayed freezing behavior in a neutral context without being exposed to the conditioned context, suggesting that activation of the Fos ensemble triggered the recall of the contextual fear memory. A similar finding was reported in a later study we conducted (Sun et al. 2020). Interestingly, while necessity was not examined in the 2012 study, chemogenetic inhibition of the Fos ensemble, captured by the F-RAM reporter, showed that the Fos ensemble in the DG is not required for contextual fear memory recall (Sun et al. 2020). Nevertheless, while it is not required in the DG for memory recall, the Fos ensemble has been shown to promote memory generalization (Sun et al. 2020; Pignatelli et al. 2019).

In another example showing the DG Fos ensemble's ability to modulate the expression of contextual fear memory, Ramirez et al. (2013) created false memories by optogenetically manipulating the Fos ensemble neurons (Ramirez et al. 2013). The Fos ensemble neurons were first captured when exposed to a neutral context (without foot shock). These Fos ensemble neurons were then optically activated during fear conditioning in a different context. These treatments caused mice to show a fear response (i.e., freezing) in the original, neutral context. A plausible explanation for the "memory inception" is that the contextual fear memory is preferentially allocated to the Fos ensemble representing the neutral context because optogenetic activation of those ensemble neurons increases their chance of being recruited to form the memory engram (Miller et al. 2023). However, this finding differs significantly from an earlier study (Garner et al. 2012), where a similar behavioral task resulted in the formation of a synthetic contextual fear memory combining both contexts rather than a simple "implanted" fear memory associated only with the neutral context. Interestingly, the study by Garner et al. (2012), also contrasts with that of Liu et al. (2012), in that the activation of the Fos ensemble linked to CFC alone failed to retrieve fear memory afterward. This discrepancy might stem from the Fos ensemble captured in the Garner study, which encompassed the entire brain rather than being limited to the DG.

The Fos ensemble in other brain regions has also been shown to be sufficient to drive learned behaviors. For example, in the retrosplenial cortex (RSC), fear memory can be recalled in a neutral context when the Fos ensemble that encoded fear memory is optogenetically activated (Cowansage et al. 2014). The Fos ensemble can be associated not only with negative (fear) memories but also with positive memories, as shown in the DG (Ramirez et al. 2015) and in the BLA (Gore et al. 2015; Redondo et al. 2014). In addition, the Fos ensembles in different brain regions, when activated simultaneously, confer a greater level of memory recall than the activation of a single Fos ensemble (Roy et al. 2022). Furthermore, synchronous activation of Fos ensembles in different brain regions that encode unrelated memories can create a new artificial memory, as shown by Ohkawa et al. (2015).

In both CA1 and BLA, the Fos ensemble has been shown to be both necessary and sufficient for memory retrieval (Gore et al. 2015; Roy et al. 2022; Tanaka et al. 2014). Recently, Xing et al. (2021) used FosTRAP transgenic mice to capture Fos ensembles associated with social investigation in mPFC. There too, optogenetic inhibition of the Fos ensemble impaired social memory without affecting social motivation (Xing et al. 2021).

## *Synaptic Plasticity on the Fos Ensemble Neurons*

Long-term physiological changes, especially at synapses, have long been believed to underlie the formation and storage of memories, but direct evidence for the causal relationship between experience-dependent synaptic changes and memory only started to emerge recently (Nabavi et al. 2014). With the ability to capture engram neurons, the field is ready to directly examine the hypothesis that learning-induced synaptic changes on engram neurons underlie memory and learned behavioral adaptation. Here, we summarize synaptic changes on the Fos ensemble neurons that are triggered by learning and are responsible for memory recall.

Synaptic changes on the Fos ensemble neurons have been best examined in the hippocampus, again in the context of contextual fear conditioning. In general, classical synaptic plasticity mechanisms, such as NMDA receptor-dependent LTP, are found to occur on Fos ensemble neurons. Fos ensemble neurons in the DG captured during contextual fear conditioning using the Fos-tTA mouse were found to receive stronger excitatory input from the upstream entorhinal cortex and also displayed a higher AMPA/NMDA ratio than neighboring, "non-engram," control neurons (Ryan et al. 2015). Structural analysis also showed that these Fos ensemble neurons had more dendritic spines, consistent with strengthened excitatory drive. Application of the protein synthesis inhibitor anisomycin immediately after learning not only abolished these synaptic enhancements on the Fos ensemble neurons, but it also blocked memory retrieval. However, even without the synaptic enhancement, memory retrieval could be restored by optogenetic activation of the DG Fos ensemble (Ryan et al. 2015). These results suggest that learning-induced synaptic changes on the Fos ensemble neurons are important for memory retrieval but not required for memory storage, at least not in the DG.

Similar, but not identical, synaptic changes were found on DG Fos ensemble neurons captured by the F-RAM reporter during contextual fear conditioning (Sun et al. 2020). These Fos ensemble neurons were found to receive higher excitatory postsynaptic currents from the medial entorhinal cortex (MEC) but not the lateral entorhinal cortex (LEC), but with a similar AMPA/NMDA current ratio to the "non-engram" control neurons (Sun et al. 2020). The strengthened MEC input on the Fos ensemble neurons may drive the Fos ensemble to promote memory generalization because we found that the input of the MEC, but not LEC, in the DG was required for memory generalization (Sun et al. 2020).

In CA1, the Fos ensemble also undergoes similar learning-dependent synaptic changes. Using the dual-eGRASP (Green Fluorescent Protein Reconstitution Across Synaptic Partners) technique, Choi et al. (2018) found that the number and size of spines are significantly increased on the CA1 Fos ensemble neurons receiving input from the CA3 Fos ensemble neurons (Choi et al. 2018), suggesting that contextual fear learning potentiated Schaffer collateral synapses between the Fos ensemble neurons in CA3 and CA1. This finding was confirmed by electrophysiological recordings showing that the CA1 ensemble neurons received more excitatory postsynaptic current from the CA3 ensemble neurons, and these potentiated excitatory synapses between ensemble neurons were occluded from further potentiation (Choi et al. 2018).

The studies from both Choi et al. (2018) and Sun et al. (2020) demonstrate that the magnitude of learning-induced synaptic changes on the Fos ensemble neurons is correlated with memory strength. The excitatory inputs from MEC to DG Fos ensemble neurons significantly increased as the shock intensity increased, which increased freezing levels (Sun et al. 2020). The same was found on the synapses between CA3 and CA1 ensemble neurons using structural analysis (Choi et al. 2018). Similar learning-induced synaptic changes on the Fos ensemble neurons were observed outside the hippocampus (Zhou et al. 2019; Kim and Cho 2020). These findings suggest that synaptic changes on the Fos ensemble neurons may encode the strength of memory.

These learning-induced synaptic changes on the Fos ensemble neurons typically involve only excitatory synapses. However, a recent study reported bidirectional inhibitory plasticity on the Fos ensemble neurons in the CA1 that are activated by a prolonged (2–9 day) exploration of a novel enriched environment (Yap et al. 2021). There is an increase in perisomatic inhibition by parvalbumin-expressing interneurons on the Fos ensemble neurons, accompanied by weakened inhibition from cholecystokinin-expressing interneurons. This bidirectional synaptic plasticity is mediated by the neuropeptide SCG2, a downstream target gene of Fos (Yap et al. 2021). Nevertheless, it remains to be determined whether these bi-directional inhibitory synaptic changes result from memory encoding or are a homeostatic response to sustained neural network activation from prolonged environmental stimulation.

Besides synaptic changes, studies have demonstrated a transient elevation in the intrinsic excitability of Fos ensemble neurons during the hours following memory encoding (in CA1) (Cai et al. 2016) and retrieval (in DG) (Pignatelli et al. 2019). This experience-triggered metaplasticity, although temporary, appears crucial for associating memories formed in close temporal proximity (Cai et al. 2016) and enhancing the precision of memory retrieval (Pignatelli et al. 2019). The alteration in excitability can manifest as quickly as 5 minutes after the (re)activation of Fos ensemble neurons, as illustrated by Pignatelli et al. (2019), likely through post-transcriptional and post-translational mechanisms that impact ion channel localization and properties.

## Npas4: A Neuronal-Specific and Activity Selective IEG

Neuronal PAS Domain Protein 4 (Npas4) is a newcomer, compared to Fos and other IEGs, that was initially found to play a role in developing inhibitory synapses on excitatory neurons (Lin et al. 2008). Compared to other IEGs, Npas4 is unique in that it is only expressed in neurons and is selectively induced by neuronal activity. It can be expressed in both inhibitory and excitatory neurons, but its roles appear to differ in different neuronal types. During development, it recruits inhibitory synapses on excitatory neurons, but it also strengthens excitatory synapses on inhibitory GABAergic neurons (Lin ct al. 2008; Spiegel et al. 2014; Yoshihara et al. 2014). In CA3 pyramidal neurons, Npas4 selectively restricts the excitatory mossy fiber input from the DG granule cells (Weng et al. 2018). Npas4 plays important roles in learning and memory. It is required for contextual memory formation in CA3 (Ramamoorthi et al. 2011; Weng et al. 2018), fear memory formation and reconsolidation in the lateral amygdala (LA) (Ploski et al. 2011), consolidation of safety memory in the LA (Ko et al. 2023), and motor learning in the primary motor cortex (Yang et al. 2022). Unlike Fos, the functions of the Npas4 ensemble neurons in learning and memory have not been explored extensively.

### *Labeling of the Npas4 Engram Ensemble*

Based on our RAM system (Sorensen et al. 2016), we created an NPAS4-dependent RAM reporter, N-RAM, to capture activated ensemble neurons defined by Npas4 expression (Sun et al. 2020). The activity-dependent promoter in the N-RAM reporter consists of DNA-binding elements of NPAS4. Instead of the transcription activation of the Npas4 gene, the N-RAM reporter reflects the transcriptional output of the Npas4 protein, like the F-RAM reporter. Like all RAM reporters, the N-RAM reporter uses the built-in, improved Dox-off system for temporal control (Sorensen et al. 2016).

Several groups, including ours, have very recently generated Npas4-CreER$^{T2}$ or Npas4TRAP mice, in which a P2A-CreER$^{T2}$ cassette is inserted into the end of the Npas4 ORF (Hughes et al. 2022; Xue et al. 2024). The Npas4TRAP mice have yet to be used for engram studies.

## *Role of Npas4 Ensemble Neurons in Learning and Memory*

Being a relatively new IEG, the Npas4-defined active neuronal ensembles have not been studied as widely as those defined by Fos. However, based on the existing data, it is clear that the Npas4 and Fos ensembles play rather different roles in the memory engram.

In the DG, the Npas4 ensemble captured using the N-RAM reporter during contextual fear conditioning is physically and functionally distinct from the Fos ensemble (Sun et al. 2020). When the Npas4 ensemble is chemogenetically inhibited during memory recall, the animal's ability to discriminate between two similar but distinct contexts is abolished, suggesting that the Npas4 ensemble is required for contextual memory discrimination. Furthermore, activation of the Npas4 ensemble during memory recall does not promote contextual memory generalization in a neutral context, a phenomenon seen when the Fos ensemble is activated. These data provide the first evidence suggesting distinct memory functions for the Npas4 and Fos ensembles: while the DG Npas4 ensemble is required for contextual memory discrimination, the DG Fos ensemble promotes contextual memory generalization.

Npas4 and Fos ensembles were also shown to have distinct functions in the DG in another study by Ren et al. (2022). In a stress-induced memory generalization paradigm, mice are trained in CFC for 3 days before undergoing a 4-day social defeat (SD) stress treatment. This results in a severe over-generalization of contextual fear memory demonstrated by high freezing levels in both conditioned and neutral contexts. Inhibition of the Npas4 ensemble, but not the Fos ensemble, captured at fear memory recall (not at CFC training) before SD, reverses the stress-induced memory generalization. This finding suggests that the Npas4 ensemble associated with the original contextual fear memory retrieval is somehow required for triggering the memory in an unrelated, neutral context, generalizing the fear response. This finding is intriguing given that the Npas4 ensemble captured at CFC encoding is required for memory discrimination (Sun et al. 2020). Thus, it is possible that the normal function of the Npas4 ensemble is altered under pathological conditions caused by SD. Future studies will be needed to fully understand this phenomenon.

The Npas4 ensemble has been found to be critical in the motor cortex for motor learning (Yang et al. 2022). Unlike the one-trial CFC paradigm, the motor learning paradigm used in the study of Yang et al. (2022) is progressive, requiring multi-day training sessions before the motor output is stabilized. Surprisingly, the Npas4 ensemble activated during the learning process consists mostly of somatostatin (SST)-expressing GABAergic interneurons (SST-IN). Furthermore, the activity of the SST-IN Npas4 ensemble neurons continues to decrease as learning progresses, and this reduction in the ensemble activity is required for motor learning; elevating the activity of the SST-IN Npas4 ensemble during learning completely blocked learning. The study did not examine the function of the Fos ensemble. But, if the Fos ensemble is mostly made of pyramidal neurons like in other cortical areas and given SST-IN's role in modulating the function of pyramidal neurons, it is possible that the SST-IN Npas4 ensemble modulates the motor output by modulating the output of the Fos ensemble neurons in the motor cortex. Such a scenario would provide additional strong evidence for functionally distinct neuronal ensembles within the learning circuits.

## *Synaptic Plasticity on Npas4 Ensemble Neurons*

In the DG, learning-induced synaptic changes on Npas4 ensemble neurons are very different from those on the Fos ensemble neurons. The Npas4 ensemble neurons activated during CFC selectively recruit GABAergic inhibitory inputs from CCK GABAergic interneurons, while their excitatory inputs remain unchanged (Sun et al. 2020). This is distinct from the Fos ensemble neurons, whose GABAergic inputs are not altered while their excitatory inputs from the MEC are strengthened (Sun et al. 2020). The level of CCK inhibition on the Npas4 ensemble neurons increases with the strength of CFC training (foot shock intensity). Furthermore, artificial induction of Npas4 expression in a random, sparse population of DG granule neurons is sufficient to selectively recruit CCK inhibition onto those neurons, suggesting that learning-induced Npas4 expression is responsible for recruiting CCK inhibition onto these neurons. Consistent with the idea that this learning-induced CCK-inhibition controls the memory discrimination function of the Npas4 ensemble neurons, inhibiting CCK interneurons during memory recall leads to abolition of memory discrimination, therefore phenocopying inhibition of the Npas4 ensemble.

Learning-induced synaptic changes on the Npas4 ensemble neurons were not examined in the studies by Ren et al. (2022) and Yang et al. (2022). However, the continuously decreased activity observed in the Npas4 SST-IN ensemble neurons during motor learning (Yang et al. 2022) could imply a decrease in the excitatory drive or an increase in the inhibitory drive on those neurons. Furthermore, when the activity of the Npas4 SST-IN ensemble was kept elevated chemogenetically during motor training sessions, excitatory synapse reorganization (measured by spine density and turnover changes), which correlated closely with the acquisition of the motor task, was completely blocked (Yang et al. 2022). This finding indicates that the SST-INs forming the Npas4 ensemble directly or indirectly modulate learning-associated excitatory synaptic plasticity on the pyramidal neurons that are likely part of the motor memory engram.

## Arc: A Non-transcription Factor IEG with Complex Functions

Activity-regulated cytoskeleton-associated protein (Arc) (Lyford et al. 1995), also known as activity-regulated gene (Arg3.1) (Link et al. 1995), is another IEG that has been widely used to capture or visualize memory engrams. Its expression is induced by many sensory and behavioral experiences, such as fear-based learning (Denny et al. 2014; Lacagnina et al. 2019; Perusini et al. 2017). While it is expressed primarily in excitatory neurons under normal physiological conditions, it can be induced in inhibitory neurons by strong stimuli such as seizures (Vazdarjanova et al. 2006). Unlike most other IEGs, Arc is not a transcription factor; rather, it acts as a cytosolic signaling protein.

Given its association with the cytoskeleton and localization at dendritic spines, Arc has been extensively studied for its roles in synaptic plasticity processes such as LTP and long-term depression (LTD), as well as in homeostasis (Guzowski et al. 2000; Shepherd and Bear 2011; Shepherd et al. 2006; Kyrke-Smith et al. 2021; Messaoudi et al. 2007; Plath et al. 2006). Arc gene deletion in mice leads to deficits in circuit plasticity (Shepherd and Bear 2011; McCurry et al. 2010) and impairments in learning and memory (Guzowski et al. 2000; Messaoudi et al. 2007; Plath et al. 2006).

In addition to being transcriptionally regulated as an IEG, Arc mRNA can be transported to active synapses to be translated locally (Moga et al. 2004; Steward et al. 1998; Steward and Worley 2001a, b), where it is involved in excitatory synaptic plasticity (Shepherd et al. 2006; Plath et al. 2006; Waung et al. 2008; Chowdhury et al. 2006; Park et al. 2008). Furthermore, the ARC protein, which structurally resembles retroviral Gag proteins, can self-assemble into viral-like capsids to transfer genetic materials between neurons (Pastuzyn et al. 2018). While these additional functions of ARC are important for brain function, we will focus here on the active neuronal ensembles defined by Arc transcription.

### *Labeling the Arc Engram Ensemble*

The first mouse model to allow capturing and monitoring the Arc active neuronal ensemble was a knock-in mouse in which the coding sequence of the destabilized fluorescent protein d2EGFP replaces the entire Arc ORF (Wang et al. 2006). d2EGFP was chosen to match the short half-life of the ARC protein. The Arc-d2EGFP mouse has been successfully used to monitor Arc activation in vivo in various cortical regions during learning and sensory processing (Wang et al. 2006; Cao et al. 2015; Mastwal et al. 2014; Ren et al. 2014). However, the d2EGFP signal is relatively weak for imaging because it is only expressed from one endogenous Arc allele. The more recent BAC transgenic mouse models Arc-dVenus (Eguchi and Yamaguchi 2009) and TgArc/Arg3.1-d4EGFP (Grinevich et al. 2009) significantly improved reporter signal intensity thanks to multiple copies of the transgene. These mice are widely used for in vivo 2-photon imaging (Wang et al. 2006; Grinevich et al. 2009; Cao et al. 2013), while the ArcGFP BAC transgenic mouse developed by the GENSAT BAC transgenic resource (Gong et al. 2003) has only been used in in vitro settings (Yassin et al. 2010). Additionally, there is also an Arc-luciferase BAC transgenic mouse, which can be used to monitor the magnitude of Arc activation in vivo by detecting bioluminescence signals (Izumi et al. 2011).

Two mouse models have been developed to capture and manipulate Arc ensembles. Both used the Arc promoter to drive the expression of tamoxifen-inducible Cre recombinase CreER$^{T2}$. One is the ArcCreER$^{T2}$ BAC transgenic mouse generated by Denny et al. (2014). The other is the ArcTRAP mouse generated by the Luo lab, which replaces the Arc ORF with CreER$^{T2}$ (Guenthner et al. 2013). Furthermore, the development of a synthetic Arc promoter, Enhanced Activity-Responsive Element (ESARE) (Kawashima et al. 2009, 2013), has allowed the research community to develop powerful and versatile viral tools to study the Arc engram ensembles in vivo.

One note of caution: since all the Arc engram reporters capture cells that have undergone activity-dependent Arc transcription but not those engaged in local translation of Arc, the properties of the Arc ensemble neurons captured in this way may not encompass all the functions that ARC mediates.

## *Roles of the Arc Ensemble in the Memory Engram*

The Arc ensemble has been used in the same manner as the Fos ensemble as a proxy for the memory engram. Traditionally, it was believed that the Arc ensembles behave very similarly to the Fos ensembles, and there is evidence that ARC and FOS proteins overlap substantially in vivo (Denny et al. 2014; Jaeger et al. 2018; Jiang and VanDongen 2021), where both Arc and Fos are reported to play roles in synaptic plasticity processes such as LTP (Fleischmann et al. 2003; Shepherd and Bear 2011). Although manipulation of the Arc and Fos engrams results in some similar behavioral changes (Denny et al. 2014; Liu et al. 2014; Kitamura et al. 2017), closer examination nevertheless shows some differences between them.

Here we highlight the study by Denny et al. (2014), which used the ArcCreER$^{T2}$ BAC transgenic mouse to capture the Arc ensemble during CFC. Optogenetic inhibition of the Arc ensemble in either CA3 or DG during memory recall significantly reduced freezing levels, suggesting that the Arc ensemble is required for contextual memory retrieval in those two subregions of the hippocampus (Denny et al. 2014). No other engram ensembles have been functionally examined in CA3, so it is not known whether the Arc ensemble is unique in being required in CA3 for contextual fear memory expression. It is intriguing, however, to find the Arc ensemble in the DG is required for contextual memory recall since recall is not affected even when the whole DG is inhibited during recall (Bernier et al. 2017). Another study (Lacagnina et al. 2019), using the same ArcCreER$^{T2}$ BAC transgenic mouse, demonstrated the same requirement for the Arc ensemble in the DG. A computational model proposed by the Drew lab suggests that, while contextual memory can be recalled through the perforant input to the CA3 and bypass the whole DG, silencing only the Arc ensemble DG neurons results in abnormal DG activity, which interferes with the recall (Bernier et al. 2017). However, the computational model cannot explain why inhibition of either the Fos or Npas4 ensemble doesn't affect contextual memory retrieval like the Arc ensemble does (Sun et al. 2020). Nevertheless, the necessity of the Arc ensemble in the DG suggests that it may be functionally different from the Fos or Npas4 ensembles.

## *Synaptic Plasticity on the Arc Ensemble*

While the roles of the Arc gene in hippocampal synaptic plasticity have been extensively investigated, no studies examined the changes in neuronal and synaptic properties occurring on the hippocampal Arc ensemble neurons after learning. Outside the hippocampus, synaptic properties of the Arc ensemble neurons, captured in the Arc-dVenus mouse during CFC, were examined in the basolateral amygdala (BLA). These Arc ensemble neurons exhibit an increased miniature excitatory postsynaptic current (mEPSC) frequency compared to non-ensemble neurons (Nonaka et al. 2014a). Furthermore, these strengthened excitatory inputs on the Arc ensemble neurons are a result of presynaptic potentiation, shown by a decrease in the paired-pulse ratio (PPR) from the cortical but not the thalamic pathway.

The Arc ensemble was also examined in the lateral amygdala (LA) after auditory fear conditioning (AFC) (Gouty-Colomer et al. 2016), using the Arc-dVenus mouse. Interestingly, the Arc-expressing neurons in the LA display enhanced intrinsic excitability at baseline and are preferentially recruited into newly encoded memory ensembles. Glutamatergic thalamic inputs on the TFC-induced Arc ensemble neurons were found to be selectively potentiated, primarily through postsynaptic mechanisms, with an increased AMPA/NMDA ratio.

Additionally, Ren et al. (2014) discovered that, after motor learning, the Arc ensemble neurons in the frontal cortex are hyperactive compared to the neurons that do not express Arc. This hyperactivity is not observed in untrained mice (Ren et al. 2014). ARC-positive neurons also displayed higher rates of spontaneous firing in somatosensory neurons, even in untrained animals (Yassin et al. 2010).

These studies collectively indicate the potentiation of excitatory inputs onto Arc ensemble neurons, which is reminiscent of the Fos ensemble neurons in the DG (Sun et al. 2020; Ryan et al. 2015). The Arc ensemble neurons, in some cases, display higher excitability, which is not seen in the Fos or Npas4 ensemble; in fact the Npas4 ensemble neurons are hypoexcitable (Sun et al. 2020). Although it has been suggested that the Fos and Arc ensembles largely overlap in the DG, as approximately 80% of the Arc engram populations are reported to also express FOS (Denny et al. 2014), the behavioral data suggest otherwise (see above). Further investigations are warranted to delineate the Fos and Arc ensembles.

## Egr1: An Old-Timer IEG That Has Not Received as Much Attention

Early growth response 1 (Egr1) (Sukhatme et al. 1988), also called NGFI-A (Milbrandt 1987), Krox-24 (Lemaire et al. 1988), TIS8 (Lim et al. 1987, 1989), or Zif268 (Christy et al. 1988), was identified as an IEG only a few years after Fos. While initially found to be induced by growth factors such as nerve growth factor (NGF) (Milbrandt 1987), it was soon shown to be robustly activated by neuronal activity in vivo, in a manner that's very similar to Fos (Sukhatme et al. 1988). It is a zinc finger protein and belongs to the EGR family of transcription factors (Christy et al. 1988). Interestingly, Arc was identified as one of the potential direct transcriptional targets of EGR1 (Li et al. 2005).

In the brain, EGR1 expression is not present embryonically (McMahon et al. 1990; Crosby et al. 1992), but it peaks around the second postnatal week, which coincides with synaptogenesis and the onset of synaptic plasticity (Watson and Milbrandt 1990; Beckmann and Wilce 1997). Egr1 knockout mice exhibit normal early LTP but are defective in late LTP. Consistently, mutant Egr1 mice, despite having intact short-term memory across various tasks, have impaired long-term memory (Jones et al. 2001; Bozon et al. 2003). Egr1 knockdown in the whole amygdala impairs contextual fear memory, whereas specific knockdown in the basolateral amygdala impairs auditory fear memory (Malkani et al. 2004; Maddox et al. 2011). Interestingly, mice with Egr1 knockdown in the dorsal hippocampus or Egr1 heterozygous knockout mice (Besnard et al. 2013) are impaired in memory reconsolidation but not consolidation, while Egr1 total knockout mice are defective in both consolidation and reconsolidation of the contextual fear memory (Besnard et al. 2013). On the other hand, during an object-place recognition memory task, EGR1 overexpression improved long-term memory of the spatial location of objects but not memory of objects (Penke et al. 2014). Taken together, these studies suggest that the strength of memory might be modulated by the Egr1 gene dosage.

### *Labeling the Egr1 Engram Ensemble*

The Egr1-EGFP BAC transgenic mouse was generated from the GENSAT project (Gong et al. 2003; Heintz 2004). A recent study cautioned that the expression of EGFP is not consistent with the activity-dependent expression of Egr1 in the Egr1-EGFP mouse (Demchuk et al. 2020), highlighting the limitations of transgenic approaches. However, studies from Ji-Song Guan's group found a linear correlation between the Egr1 and EGFP mRNA levels, suggesting that Egr1-EGFP transcription mimics that of the native Egr1 gene expression (Wang et al. 2021; Xie et al. 2014). In addition, following careful monitoring of neural activities and EGR1-EGFP signals, it was confirmed that the EGR1-EGFP signal correlates with neural activity (Mahringer et al. 2022).

Another BAC transgenic mouse, the Egr1-CreER$^{T2}$ mouse, was recently generated to permanently tag neurons activated to express Egr1 (Longueville et al. 2021). It has been shown to tag neurons activated by seizure and cocaine administration, which both elicit very strong levels of neuronal activity in the brain. Researchers have just started to use this method to tag Egr1 ensembles to study memory engrams (Brito et al. 2022).

## *Roles of Egr1 Ensemble Neurons in Memory Expression*

Brito et al. (2022) used the Egr1-CreER$^{T2}$ BAC transgenic mouse (Longueville et al. 2021) to tag the Egr1 ensemble in the CA1 during motor learning (Brito et al. 2022). Surprisingly, depletion of the Egr1 engram neurons or shRNA knockdown of Egr1 in the CA1 improves motor learning, whereas chemogenetic activation of the Egr1 engram leads to impaired motor performance. These results suggest that the CA1 Egr1 ensemble activated during motor learning negatively regulates motor learning. Although it is hard to compare the active ensembles tagged during motor learning to those during contextual fear conditioning, the Egr1 ensemble's ability to restrict learning appears to be the opposite of what we would expect from other IEG ensembles. This suggests that the Egr1 ensemble is distinct from the Fos, Npas4, and Arc ensembles.

The Egr1 ensemble in the Egr1-EGFP BAC transgenic mice was monitored using 2-photon [$Ca^{2+}$] in vivo imaging (Xie et al. 2014). The authors observed Egr1 ensembles in layer II of multiple cortical areas, including the motor cortex, the visual cortex, the somatosensory cortex, and the retrosplenial cortex, that are specific to stimulus context (Xie et al. 2014). This observation is reminiscent of the Arc ensemble encoding different oriented visual stimuli in the primary visual cortex (Wang et al. 2006).

Although Egr1 is one the first identified IEGs in the nervous system, and its expression has been widely used to map memory engrams, we are still at a very early stage in exploring the functions and features of the Egr1 ensembles in learning and memory. There is a complete lack of knowledge of the neuronal and synaptic properties associated with the Egr1 ensembles.

# Discussion

In this chapter, we have described the process of molecular sorting during the formation of memory engrams (Fig. 1). Sensory and behavioral experiences activate ensembles of neurons that are defined by the activation of different IEGs. These IEG ensembles are broadly considered part of the memory engram as they possess the capacity to regulate the expression of the memory. Yet these IEG ensembles are often functionally distinct, resulting in a functional heterogeneity within the memory engram. The unique gene function of each IEG, which drives specific neuronal and synaptic changes triggered by the learning experience, is likely responsible for its ensemble's unique function within the memory engram. However, the driving force(s) behind the molecular sorting process remains elusive. Here, we discuss some of the major factors that likely determine the distribution and makeup of the memory engram in vivo.

## *Signaling Pathways for IEG Transcription Activation*

At the transcription level, the activation of different IEGs involves many overlapping pathways. It is typically triggered by $Ca^{2+}$ influx through $Ca^{2+}$-permeable channels such as voltage-gated calcium channels, NMDA receptors, and AMPA receptors, or by the release of intracellular $Ca^{2+}$ storage (Berridge 1998; Jonas and Burnashev 1995; Lyons and West 2011). $Ca^{2+}$ serves as a second messenger to directly or indirectly regulate a cascade of signaling pathways, mediated by the Ras-mitogen-associated protein kinases (MAPKs), the calcium/calmodulin-dependent protein kinases (CaMKs), PI3 Kinase/Akt, calcineurin, and others. These signaling pathways subsequently activate transcription factors that are pre-bound to the promoters of IEGs, such as cyclic adenosine monophosphate (cAMP)-responsive element binding protein (CREB), serum response factor (SRF), and myocyte enhancer factor 2 (MEF2), unleashing IEG transcription (Greer and Greenberg 2008). Although they are all essential upstream components for IEG induction, disruption of the transcriptional functions of CREB, SRF, and MEF2 each causes unique defects in circuit function in vivo (Pulimood et al. 2017), highlighting that there is non-uniform activation of signaling pathways, and therefore non-uniform IEG expression in vivo.

Non-uniform activation of different IEGs is likely due to their variable sensitivities to their shared upstream signaling pathways (Ramamoorthi et al. 2011; Lin et al. 2008). As we showed in cultured hippocampal neurons (Fig. 2), Npas4 is selectively activated by neuronal activity but insensitive to the activation of signaling pathways that robustly activate Fos and Arc. Fos activation, in contrast, is more sensitive to Forskolin (PKA activation) than Arc, whereas Arc transcription is more robustly driven by neurotrophic factors than Fos. Activation of the NMDA receptor, for example, by chemically induced LTP and LTD in hippocampal slices, engages distinct downstream signaling pathways to activate Arc and Npas4 (Coba et al. 2008). In vivo, in striatal medium spiny neurons, Npas4 activation is very distinct from Fos, Arc, and other IEGs, partly because it is insensitive to dopaminergic input, which activates the PKA signaling pathway (Lissek et al. 2021). Overall, the different signaling pathways downstream of synaptic activity are likely responsible for sorting experience-generated circuit activity into different active neuronal ensembles.

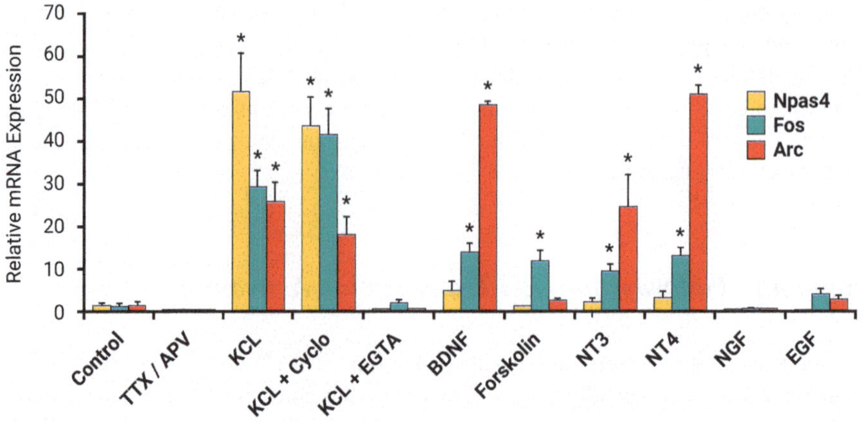


**Fig. 2** The induction of Npas4, Fos, and Arc by various extracellular stimuli in cultured hippocampal neurons. Note the different selectivity displayed by each IEG. (Modified from Ramamoorthi et al. Science 2011)

## *Post-transcriptional Regulation*

It is important to emphasize that transcription activation is only the first of many steps before the functional output of an IEG. Post-transcriptional regulation determines whether mRNAs are translated into functional proteins, whether the appropriate signaling pathways are activated to modify proteins post-translationally, and whether the necessary protein complexes are formed for specific functional outputs. For these reasons, it is important to recognize that the ensembles defined by an IEG mRNA, protein, and functional output can be very different. For example, while Npas4 mRNA is quite abundant in CA1 after CFC (our unpublished observations), NPAS4 protein is largely absent there (Ramamoorthi et al. 2011), suggesting potential post-transcriptional regulation in CA1 that limits Npas4 expression. Furthermore, while FOS and NPAS4 proteins show substantial overlap in the DG after CFC (our unpublished observation), their transcriptional outputs are detected in separate populations (Sun et al. 2020).

In our opinion, for IEGs that are also transcription factors, such as FOS, NPAS4, and EGR1, it is particularly important to define their engram ensembles based on their transcriptional outputs. This was a major reason for us to design our RAM reporters to report the transcriptional outputs of IEG transcription factors. Most engram capturing tools utilize an IEG's promoter and therefore report the equivalent of the IEG mRNA population. It is important to draw biological distinctions among the engram populations defined by the IEG mRNA, protein and transcriptional output if we are to reconstitute the entirety of the memory engram.

### *Intrinsic and Extrinsic Factors*

An important unanswered question is whether the recruitment of the engram neurons is determined intrinsically or extrinsically to the neurons. Work from the Silva and Josselyn labs suggests that the probability of recruitment, i.e., memory allocation, is at least partly determined by the intrinsic excitability of the neurons (Josselyn and Frankland 2018; Silva et al. 2009). Artificially increasing neuron excitability by CREB overexpression or chemogenetic/optogenetic activation significantly increases their probability of being recruited into the memory trace (Han et al. 2007, 2009; Ramsaran et al. 2023; Zhou et al. 2009). There are other intrinsic factors that could also affect engram recruitment but have not been explored, for example, synaptic connections (Castello-Waldow et al. 2020), epigenetic states (Marco et al. 2020), and neuronal age (Kee et al. 2007). It also remains to be seen whether any of the above-mentioned intrinsic factors differentially affect the recruitment of different IEG ensembles.

The Fos, Npas4, and Arc engram ensembles all show remarkable input-specific, learning-induced synaptic plasticity (Sun et al. 2020; Ryan et al. 2015; Choi et al. 2018; Nonaka et al. 2014b). Because these input-specific synaptic modifications (on the Fos and Npas4 ensembles at least) have been shown to be not pre-existing but induced by experience (Sun et al. 2020; Choi et al. 2018), it suggests the possibility that the recruitment is driven pre-synaptically, not cell-autonomously. Therefore, input-specific extrinsic factors could be responsible for the heterogenous distribution of different IEG engram ensembles.

## Closing Remarks

Here we propose the concept of molecular sorting to explain heterogeneity within the memory engram. This is supported by an array of existing data (Fig. 1) showing distinct functionalities in engram ensembles defined by different IEGs (Denny et al. 2014; Sun et al. 2020). We would like to close this chapter with the following open questions to stimulate future research:

1. The Fos and Npas4 ensembles appear to be physically and functionally distinct, at least in the contextual fear memory engram in the DG. Are they parts of a larger orthogonal set of ensembles within the memory engram? What is their relationship to the Arc and Egr1 ensembles? How many orthogonal ensembles are there?
2. The Fos ensembles in different subregions of the hippocampus have been shown to be preferentially connected (Ryan et al. 2015). The same was found with the Arc ensembles in the orbitofrontal cortex and dorsal striatum (Wall et al. 2019). This raises the question of whether such hyperconnectivity exists only between ensembles defined by the same IEG or among all different ensembles of a memory engram.

3. What is the biological meaning of the different IEG ensembles in an engram? Do they represent different aspects of the memory? Or do they represent various regulations of the learning-induced behavioral adaptation (e.g., discrimination vs. generalization)?
4. Are there other, non-IEG-based approaches to decomposing a memory engram?

## References

Allen WE et al (2017) Thirst-associated preoptic neurons encode an aversive motivational drive. Science 357:1149–1155. https://doi.org/10.1126/science.aan6747

Barth AL, Gerkin RC, Dean KL (2004) Alteration of neuronal firing properties after in vivo experience in a FosGFP transgenic mouse. J Neurosci 24:6466–6475. https://doi.org/10.1523/JNEUROSCI.4737-03.2004

Beckmann AM, Wilce PA (1997) Egr transcription factors in the nervous system. Neurochem Int 31:477–510; discussion 517–476. https://doi.org/10.1016/s0197-0186(96)00136-2

Bernier BE et al (2017) Dentate gyrus contributes to retrieval as well as encoding: evidence from context fear conditioning, recall, and extinction. J Neurosci 37:6359–6371. https://doi.org/10.1523/JNEUROSCI.3029-16.2017

Berridge MJ (1998) Neuronal calcium signaling. Neuron 21:13–26. https://doi.org/10.1016/s0896-6273(00)80510-3

Besnard A, Caboche J, Laroche S (2013) Recall and reconsolidation of contextual fear memory: differential control by ERK and Zif268 expression dosage. PLoS One 8:e72006. https://doi.org/10.1371/journal.pone.0072006

Bozon B et al (2003) MAPK, CREB and zif268 are all required for the consolidation of recognition memory. Philos Trans R Soc Lond Ser B Biol Sci 358:805–814. https://doi.org/10.1098/rstb.2002.1224

Brito V et al (2022) Hippocampal Egr1-dependent neuronal ensembles negatively regulate motor learning. J Neurosci 42:5346–5360. https://doi.org/10.1523/JNEUROSCI.2258-21.2022

Cai DJ et al (2016) A shared neural ensemble links distinct contextual memories encoded close in time. Nature 534:115–118. https://doi.org/10.1038/nature17955

Cao VY et al (2013) In vivo two-photon imaging of experience-dependent molecular changes in cortical neurons. J Vis Exp. https://doi.org/10.3791/50148

Cao VY et al (2015) Motor learning consolidates arc-expressing neuronal ensembles in secondary motor cortex. Neuron 86:1385–1392. https://doi.org/10.1016/j.neuron.2015.05.022

Castello-Waldow TP et al (2020) Hippocampal neurons with stable excitatory connectivity become part of neuronal representations. PLoS Biol 18:e3000928. https://doi.org/10.1371/journal.pbio.3000928

Chelaru MI, Dragoi V (2008) Efficient coding in heterogeneous neuronal populations. Proc Natl Acad Sci USA 105:16344–16349. https://doi.org/10.1073/pnas.0807744105

Choi DI, Kaang BK (2022) Interrogating structural plasticity among synaptic engrams. Curr Opin Neurobiol 75:102552. https://doi.org/10.1016/j.conb.2022.102552

Choi JH et al (2018) Interregional synaptic maps among engram cells underlie memory formation. Science 360:430–435. https://doi.org/10.1126/science.aas9204

Chowdhury S et al (2006) Arc/Arg3.1 interacts with the endocytic machinery to regulate AMPA receptor trafficking. Neuron 52:445–459. https://doi.org/10.1016/j.neuron.2006.08.033

Christy BA, Lau LF, Nathans D (1988) A gene activated in mouse 3T3 cells by serum growth factors encodes a protein with "zinc finger" sequences. Proc Natl Acad Sci USA 85:7857–7861. https://doi.org/10.1073/pnas.85.21.7857

Clem RL, Barth A (2006) Pathway-specific trafficking of native AMPARs by in vivo experience. Neuron 49:663–670. https://doi.org/10.1016/j.neuron.2006.01.019

Coba MP, Valor LM, Kopanitsa MV, Afinowi NO, Grant SG (2008) Kinase networks integrate profiles of N-methyl-D-aspartate receptor-mediated gene expression in hippocampus. J Biol Chem 283:34101–34107. https://doi.org/10.1074/jbc.M804951200

Cowansage KK et al (2014) Direct reactivation of a coherent neocortical memory of context. Neuron 84:432–441. https://doi.org/10.1016/j.neuron.2014.09.022

Crosby SD et al (1992) Neural-specific expression, genomic structure, and chromosomal localization of the gene encoding the zinc-finger transcription factor NGFI-C. Proc Natl Acad Sci USA 89:4739–4743. https://doi.org/10.1073/pnas.89.10.4739

Demchuk AM, Dube ST, Mesina L, McNaughton BL (2020) Limitations of the GENSAT Egr1-EGFP transgenic mouse strain for neural circuit activity mapping. Neurosci Lett 732:135072. https://doi.org/10.1016/j.neulet.2020.135072

Denny CA et al (2014) Hippocampal memory traces are differentially modulated by experience, time, and adult neurogenesis. Neuron 83:189–201. https://doi.org/10.1016/j.neuron.2014.05.018

Dolzani SD et al (2018) Inhibition of a descending prefrontal circuit prevents ketamine-induced stress resilience in females. eNeuro 5. https://doi.org/10.1523/ENEURO.0025-18.2018

Dragunow M, Robertson HA (1987) Kindling stimulation induces c-fos protein(s) in granule cells of the rat dentate gyrus. Nature 329:441–442. https://doi.org/10.1038/329441a0

Eguchi M, Yamaguchi S (2009) In vivo and in vitro visualization of gene expression dynamics over extensive areas of the brain. NeuroImage 44:1274–1283. https://doi.org/10.1016/j.neuroimage.2008.10.046

Engeln M et al (2016) Selective inactivation of striatal FosB/DeltaFosB-expressing neurons alleviates L-DOPA-induced Dyskinesia. Biol Psychiatry 79:354–361. https://doi.org/10.1016/j.biopsych.2014.07.007

Feil R, Wagner J, Metzger D, Chambon P (1997) Regulation of Cre recombinase activity by mutated estrogen receptor ligand-binding domains. Biochem Biophys Res Commun 237:752–757. https://doi.org/10.1006/bbrc.1997.7124

Fleischmann A et al (2003) Impaired long-term memory and NR2A-type NMDA receptor-dependent synaptic plasticity in mice lacking c-Fos in the CNS. J Neurosci 23:9116–9122. https://doi.org/10.1523/JNEUROSCI.23-27-09116.2003

Garner AR et al (2012) Generation of a synthetic memory trace. Science 335:1513–1516. https://doi.org/10.1126/science.1214985

Ghandour K et al (2019) Orchestrated ensemble activities constitute a hippocampal memory engram. Nat Commun 10:2637. https://doi.org/10.1038/s41467-019-10683-2

Gong S et al (2003) A gene expression atlas of the central nervous system based on bacterial artificial chromosomes. Nature 425:917–925

Gore F et al (2015) Neural representations of unconditioned stimuli in basolateral amygdala mediate innate and learned responses. Cell 162:134–145. https://doi.org/10.1016/j.cell.2015.06.027

Gouty-Colomer LA et al (2016) Arc expression identifies the lateral amygdala fear memory trace. Mol Psychiatry 21:1153. https://doi.org/10.1038/mp.2016.91

Greenberg ME, Ziff EB (1984) Stimulation of 3T3 cells induces transcription of the c-fos proto-oncogene. Nature 311:433–438. https://doi.org/10.1038/311433a0

Greenberg ME, Ziff EB, Greene LA (1986) Stimulation of neuronal acetylcholine receptors induces rapid gene transcription. Science 234:80–83. https://doi.org/10.1126/science.3749894

Greer PL, Greenberg ME (2008) From synapse to nucleus: calcium-dependent gene transcription in the control of synapse development and function. Neuron 59:846–860. https://doi.org/10.1016/j.neuron.2008.09.002

Grewe BF et al (2017) Neural ensemble dynamics underlying a long-term associative memory. Nature 543:670–675. https://doi.org/10.1038/nature21682

Grinevich V et al (2009) Fluorescent Arc/Arg3.1 indicator mice: a versatile tool to study brain activity changes in vitro and in vivo. J Neurosci Methods 184:25–36. https://doi.org/10.1016/j.jneumeth.2009.07.015

Grosmark AD, Buzsaki G (2016) Diversity in neural firing dynamics supports both rigid and learned hippocampal sequences. Science 351:1440–1443. https://doi.org/10.1126/science.aad1935

Guenthner CJ, Miyamichi K, Yang HH, Heller HC, Luo L (2013) Permanent genetic access to transiently active neurons via TRAP: targeted recombination in active populations. Neuron 78:773–784. https://doi.org/10.1016/j.neuron.2013.03.025

Guzowski JF et al (2000) Inhibition of activity-dependent arc protein expression in the rat hippocampus impairs the maintenance of long-term potentiation and the consolidation of long-term memory. J Neurosci 20:3993–4001. https://doi.org/10.1523/JNEUROSCI.20-11-03993.2000

Han JH et al (2007) Neuronal competition and selection during memory formation. Science 316:457–460. https://doi.org/10.1126/science.1139438

Han JH et al (2009) Selective erasure of a fear memory. Science 323:1492–1496. https://doi.org/10.1126/science.1164139

Heintz N (2004) Gene expression nervous system atlas (GENSAT). Nat Neurosci 7:483. https://doi.org/10.1038/nn0504-483

Herry C et al (2008) Switching on and off fear by distinct neuronal circuits. Nature 454:600–606. https://doi.org/10.1038/nature07166

Hughes BW et al (2022) NPAS4 supports drug-cue associations and relapse-like behavior through regulation of the cell type-specific activation balance in the nucleus accumbens. bioRxiv 2022.2009.2004.506434. https://doi.org/10.1101/2022.09.04.506434

Hunt SP, Pini A, Evan G (1987) Induction of c-fos-like protein in spinal cord neurons following sensory stimulation. Nature 328:632–634. https://doi.org/10.1038/328632a0

Izumi H, Ishimoto T, Yamamoto H, Nishijo H, Mori H (2011) Bioluminescence imaging of Arc expression enables detection of activity-dependent and plastic changes in the visual cortex of adult mice. Brain Struct Funct 216:91–104. https://doi.org/10.1007/s00429-010-0297-2

Jaeger BN et al (2018) A novel environment-evoked transcriptional signature predicts reactivity in single dentate granule neurons. Nat Commun 9:3084. https://doi.org/10.1038/s41467-018-05418-8

Jiang Y, VanDongen AMJ (2021) Selective increase of correlated activity in Arc-positive neurons after chemically induced long-term potentiation in cultured hippocampal neurons. eNeuro 8. https://doi.org/10.1523/ENEURO.0540-20.2021

Jonas P, Burnashev N (1995) Molecular mechanisms controlling calcium entry through AMPA-type glutamate receptor channels. Neuron 15:987–990. https://doi.org/10.1016/0896-6273(95)90087-x

Jones MW et al (2001) A requirement for the immediate early gene Zif268 in the expression of late LTP and long-term memories. Nat Neurosci 4:289–296. https://doi.org/10.1038/85138

Josselyn SA, Frankland PW (2018) Memory allocation: mechanisms and function. Annu Rev Neurosci 41:389–413. https://doi.org/10.1146/annurev-neuro-080317-061956

Jouhanneau JS et al (2014) Cortical fosGFP expression reveals broad receptive field excitatory neurons targeted by POm. Neuron 84:1065–1078. https://doi.org/10.1016/j.neuron.2014.10.014

Kasof GM et al (1995) Kainic acid-induced neuronal death is associated with DNA damage and a unique immediate-early gene response in c-fos-lacZ transgenic rats. J Neurosci 15:4238–4249. https://doi.org/10.1523/JNEUROSCI.15-06-04238.1995

Kawashima T et al (2009) Synaptic activity-responsive element in the Arc/Arg3.1 promoter essential for synapse-to-nucleus signaling in activated neurons. Proc Natl Acad Sci USA 106:316–321. https://doi.org/10.1073/pnas.0806518106

Kawashima T et al (2013) Functional labeling of neurons and their projections using the synthetic activity-dependent promoter E-SARE. Nat Methods 10:889–895. https://doi.org/10.1038/nmeth.2559

Kee N, Teixeira CM, Wang AH, Frankland PW (2007) Preferential incorporation of adult-generated granule cells into spatial memory networks in the dentate gyrus. Nat Neurosci 10:355–362. https://doi.org/10.1038/nn1847

Kim WB, Cho JH (2020) Encoding of contextual fear memory in hippocampal-amygdala circuit. Nat Commun 11:1382. https://doi.org/10.1038/s41467-020-15121-2
Kitamura T et al (2017) Engrams and circuits crucial for systems consolidation of a memory. Science 356:73–78. https://doi.org/10.1126/science.aam6808
Ko B et al (2023) Npas4-mediated dopaminergic regulation of safety memory consolidation. Cell Rep 42:112678. https://doi.org/10.1016/j.celrep.2023.112678
Koya E et al (2009) Targeted disruption of cocaine-activated nucleus accumbens neurons prevents context-specific sensitization. Nat Neurosci 12:1069–1073. https://doi.org/10.1038/nn.2364
Kyrke-Smith M et al (2021) The immediate early gene arc is not required for hippocampal long-term potentiation. J Neurosci 41:4202–4211. https://doi.org/10.1523/JNEUROSCI.0008-20.2021
Lacagnina AF et al (2019) Distinct hippocampal engrams control extinction and relapse of fear memory. Nat Neurosci 22:753–761. https://doi.org/10.1038/s41593-019-0361-z
Lanahan A, Worley P (1998) Immediate-early genes and synaptic function. Neurobiol Learn Mem 70:37–43. https://doi.org/10.1006/nlme.1998.3836
Lashley KS (1950) In search of the engram. In: Society for Experimental Biology, Physiological mechanisms in animal behavior. (Society's symposium IV.), p 454–482
Lemaire P, Revelant O, Bravo R, Charnay P (1988) Two mouse genes encoding potential transcription factors with identical DNA-binding domains are activated by growth factors in cultured cells. Proc Natl Acad Sci USA 85:4691–4695. https://doi.org/10.1073/pnas.85.13.4691
Li L, Carter J, Gao X, Whitehead J, Tourtellotte WG (2005) The neuroplasticity-associated arc gene is a direct transcriptional target of early growth response (Egr) transcription factors. Mol Cell Biol 25:10286–10300. https://doi.org/10.1128/MCB.25.23.10286-10300.2005
Lim RW, Varnum BC, Herschman HR (1987) Cloning of tetradecanoyl phorbol ester-induced 'primary response' sequences and their expression in density-arrested Swiss 3T3 cells and a TPA non-proliferative variant. Oncogene 1:263–270
Lim RW, Varnum BC, O'Brien TG, Herschman HR (1989) Induction of tumor promotor-inducible genes in murine 3T3 cell lines and tetradecanoyl phorbol acetate-nonproliferative 3T3 variants can occur through protein kinase C-dependent and -independent pathways. Mol Cell Biol 9:1790–1793. https://doi.org/10.1128/mcb.9.4.1790-1793.1989
Lin Y et al (2008) Activity-dependent regulation of inhibitory synapse development by Npas4. Nature 455:1198–1204. https://doi.org/10.1038/nature07319
Link W et al (1995) Somatodendritic expression of an immediate early gene is regulated by synaptic activity. Proc Natl Acad Sci USA 92:5734–5738. https://doi.org/10.1073/pnas.92.12.5734
Lissek T et al (2021) Npas4 regulates medium spiny neuron physiology and gates cocaine-induced hyperlocomotion. EMBO Rep 22:e51882. https://doi.org/10.15252/embr.202051882
Liu X et al (2012) Optogenetic stimulation of a hippocampal engram activates fear memory recall. Nature 484:381–385. https://doi.org/10.1038/nature11028
Liu X, Ramirez S, Tonegawa S (2014) Inception of a false memory by optogenetic manipulation of a hippocampal memory engram. Philos Trans R Soc Lond Ser B Biol Sci 369:20130142. https://doi.org/10.1098/rstb.2013.0142
Longueville S et al (2021) Long-lasting tagging of neurons activated by seizures or cocaine administration in Egr1-CreER(T2) transgenic mice. Eur J Neurosci 53:1450–1472. https://doi.org/10.1111/ejn.15060
Lyford GL et al (1995) Arc, a growth factor and activity-regulated gene, encodes a novel cytoskeleton-associated protein that is enriched in neuronal dendrites. Neuron 14:433–445. https://doi.org/10.1016/0896-6273(95)90299-6
Lyons MR, West AE (2011) Mechanisms of specificity in neuronal activity-regulated gene transcription. Prog Neurobiol 94:259–295. https://doi.org/10.1016/j.pneurobio.2011.05.003
Maddox SA, Monsey MS, Schafe GE (2011) Early growth response gene 1 (Egr-1) is required for new and reactivated fear memories in the lateral amygdala. Learn Mem 18:24–38. https://doi.org/10.1101/lm.1980211

Mahringer D, Zmarz P, Okuno H, Bito H, Keller GB (2022) Functional correlates of immediate early gene expression in mouse visual cortex. Peer Community J 2(e45). https://doi.org/10.24072/pcjournal.156

Malkani S, Wallace KJ, Donley MP, Rosen JB (2004) An egr-1 (zif268) antisense oligodeoxynucleotide infused into the amygdala disrupts fear conditioning. Learn Mem 11:617–624. https://doi.org/10.1101/lm.73104

Mallory CS, Giocomo LM (2018) Heterogeneity in hippocampal place coding. Curr Opin Neurobiol 49:158–167. https://doi.org/10.1016/j.conb.2018.02.014

Marco A et al (2020) Mapping the epigenomic and transcriptomic interplay during memory formation and recall in the hippocampal engram ensemble. Nat Neurosci 23:1606–1617. https://doi.org/10.1038/s41593-020-00717-0

Mastwal S et al (2014) Phasic dopamine neuron activity elicits unique mesofrontal plasticity in adolescence. J Neurosci 34:9484–9496. https://doi.org/10.1523/JNEUROSCI.1114-14.2014

McCurry CL et al (2010) Loss of Arc renders the visual cortex impervious to the effects of sensory experience or deprivation. Nat Neurosci 13:450–457. https://doi.org/10.1038/nn.2508

McMahon AP, Champion JE, McMahon JA, Sukhatme VP (1990) Developmental expression of the putative transcription factor Egr-1 suggests that Egr-1 and c-fos are coregulated in some tissues. Development 108:281–287. https://doi.org/10.1242/dev.108.2.281

Messaoudi E et al (2007) Sustained Arc/Arg3.1 synthesis controls long-term potentiation consolidation through regulation of local actin polymerization in the dentate gyrus in vivo. J Neurosci 27:10445–10455. https://doi.org/10.1523/JNEUROSCI.2883-07.2007

Milbrandt J (1987) A nerve growth factor-induced gene encodes a possible transcriptional regulatory factor. Science 238:797–799. https://doi.org/10.1126/science.3672127

Miller AMP et al (2023) Emergence of a predictive model in the hippocampus. Neuron 111:1952–1965 e1955. https://doi.org/10.1016/j.neuron.2023.03.011

Moga DE et al (2004) Activity-regulated cytoskeletal-associated protein is localized to recently activated excitatory synapses. Neuroscience 125:7–11. https://doi.org/10.1016/j.neuroscience.2004.02.004

Morgan JI, Curran T (1986) Role of ion flux in the control of c-fos expression. Nature 322:552–555. https://doi.org/10.1038/322552a0

Morgan JI, Cohen DR, Hempstead JL, Curran T (1987) Mapping patterns of c-fos expression in the central nervous system after seizure. Science 237:192–197. https://doi.org/10.1126/science.3037702

Nabavi S et al (2014) Engineering a memory with LTD and LTP. Nature 511:348–352. https://doi.org/10.1038/nature13294

Nambu MF, Lin YJ, Reuschenbach J, Tanaka KZ (2022) What does engram encode?: heterogeneous memory engrams for different aspects of experience. Curr Opin Neurobiol 75:102568. https://doi.org/10.1016/j.conb.2022.102568

Nedivi E, Hevroni D, Naot D, Israeli D, Citri Y (1993) Numerous candidate plasticity-related genes revealed by differential cDNA cloning. Nature 363:718–722. https://doi.org/10.1038/363718a0

Nikolaev E, Kaminska B, Tischmeyer W, Matthies H, Kaczmarek L (1992) Induction of expression of genes encoding transcription factors in the rat brain elicited by behavioral training. Brain Res Bull 28:479–484. https://doi.org/10.1016/0361-9230(92)90050-8

Nonaka A et al (2014a) Synaptic plasticity associated with a memory engram in the basolateral amygdala. J Neurosci 34:9305–9309. https://doi.org/10.1523/JNEUROSCI.4233-13.2014

Nonaka M et al (2014b) Region-specific activation of CRTC1-CREB signaling mediates long-term fear memory. Neuron 84:92–106. https://doi.org/10.1016/j.neuron.2014.08.049

Ohkawa N et al (2015) Artificial association of pre-stored information to generate a qualitatively new memory. Cell Rep 11:261–269. https://doi.org/10.1016/j.celrep.2015.03.017

Osborne LC, Palmer SE, Lisberger SG, Bialek W (2008) The neural basis for combinatorial coding in a cortical population response. J Neurosci 28:13522–13531. https://doi.org/10.1523/JNEUROSCI.4390-08.2008

Park S et al (2008) Elongation factor 2 and fragile X mental retardation protein control the dynamic translation of Arc/Arg3.1 essential for mGluR-LTD. Neuron 59:70–83. https://doi.org/10.1016/j.neuron.2008.05.023
Pastuzyn ED et al (2018) The neuronal gene Arc encodes a repurposed retrotransposon gag protein that mediates intercellular RNA transfer. Cell 172:275–288 e218. https://doi.org/10.1016/j.cell.2017.12.024
Penke Z et al (2014) Zif268/Egr1 gain of function facilitates hippocampal synaptic plasticity and long-term spatial recognition memory. Philos Trans R Soc Lond Ser B Biol Sci 369:20130159. https://doi.org/10.1098/rstb.2013.0159
Perusini JN et al (2017) Optogenetic stimulation of dentate gyrus engrams restores memory in Alzheimer's disease mice. Hippocampus 27:1110–1122. https://doi.org/10.1002/hipo.22756
Pignatelli M et al (2019) Engram cell excitability state determines the efficacy of memory retrieval. Neuron 101:274–284 e275. https://doi.org/10.1016/j.neuron.2018.11.029
Plath N et al (2006) Arc/Arg3.1 is essential for the consolidation of synaptic plasticity and memories. Neuron 52:437–444. https://doi.org/10.1016/j.neuron.2006.08.024
Ploski JE, Monsey MS, Nguyen T, DiLeone RJ, Schafe GE (2011) The neuronal PAS domain protein 4 (Npas4) is required for new and reactivated fear memories. PLoS One 6:e23760. https://doi.org/10.1371/journal.pone.0023760
Pulimood NS, Rodrigues WDSJ, Atkinson DA, Mooney SM, Medina AE (2017) The role of CREB, SRF, and MEF2 in activity-dependent neuronal plasticity in the visual cortex. J Neurosci 37:6628–6637. https://doi.org/10.1523/JNEUROSCI.0766-17.2017
Qian Z, Gilbert ME, Colicos MA, Kandel ER, Kuhl D (1993) Tissue-plasminogen activator is induced as an immediate-early gene during seizure, kindling and long-term potentiation. Nature 361:453–457. https://doi.org/10.1038/361453a0
Ramamoorthi K et al (2011) Npas4 regulates a transcriptional program in CA3 required for contextual memory formation. Science 334:1669–1675. https://doi.org/10.1126/science.1208049
Ramirez S et al (2013) Creating a false memory in the hippocampus. Science 341:387–391. https://doi.org/10.1126/science.1239073
Ramirez S et al (2015) Activating positive memory engrams suppresses depression-like behaviour. Nature 522:335–339. https://doi.org/10.1038/nature14514
Ramsaran AI et al (2023) A shift in the mechanisms controlling hippocampal engram formation during brain maturation. Science 380:543–551. https://doi.org/10.1126/science.ade6530
Redondo RL et al (2014) Bidirectional switch of the valence associated with a hippocampal contextual memory engram. Nature 513:426–430. https://doi.org/10.1038/nature13725
Reijmers LG, Perkins BL, Matsuo N, Mayford M (2007) Localization of a stable neural correlate of associative memory. Science 317:1230–1233. https://doi.org/10.1126/science.1143839
Ren M, Cao V, Ye Y, Manji HK, Wang KH (2014) Arc regulates experience-dependent persistent firing patterns in frontal cortex. J Neurosci 34:6583–6595. https://doi.org/10.1523/JNEUROSCI.0167-14.2014
Ren LY et al (2022) Stress-induced changes of the cholinergic circuitry promote retrieval-based generalization of aversive memories. Mol Psychiatry 27:3795–3805. https://doi.org/10.1038/s41380-022-01610-x
Richter FR, Cooper RA, Bays PM, Simons JS (2016) Distinct neural mechanisms underlie the success, precision, and vividness of episodic memory. elife 5. https://doi.org/10.7554/eLife.18260
Roy DS et al (2022) Brain-wide mapping reveals that engrams for a single memory are distributed across multiple brain regions. Nat Commun 13:1799. https://doi.org/10.1038/s41467-022-29384-4
Ryan TJ, Frankland PW (2022) Forgetting as a form of adaptive engram cell plasticity. Nat Rev Neurosci 23:173–186. https://doi.org/10.1038/s41583-021-00548-3
Ryan TJ, Roy DS, Pignatelli M, Arons A, Tonegawa S (2015) Memory. Engram cells retain memory under retrograde amnesia. Science 348:1007–1013. https://doi.org/10.1126/science.aaa5542

Sakurai K et al (2016) Capturing and manipulating activated neuronal ensembles with CANE delineates a hypothalamic social-fear circuit. Neuron 92:739–753. https://doi.org/10.1016/j.neuron.2016.10.015
Schilling K, Luk D, Morgan JI, Curran T (1991) Regulation of a fos-lacZ fusion gene: a paradigm for quantitative analysis of stimulus-transcription coupling. Proc Natl Acad Sci USA 88:5665–5669. https://doi.org/10.1073/pnas.88.13.5665
Semon RW (1921) The mneme. G. Allen & Unwin ltd./The Macmillan Company
Shepherd JD, Bear MF (2011) New views of Arc, a master regulator of synaptic plasticity. Nat Neurosci 14:279–284. https://doi.org/10.1038/nn.2708
Shepherd JD et al (2006) Arc/Arg3.1 mediates homeostatic synaptic scaling of AMPA receptors. Neuron 52:475–484. https://doi.org/10.1016/j.neuron.2006.08.034
Silva AJ, Zhou Y, Rogerson T, Shobe J, Balaji J (2009) Molecular and cellular approaches to memory allocation in neural circuits. Science 326:391–395. https://doi.org/10.1126/science.1174519
Smeyne RJ et al (1992) fos-lacZ transgenic mice: mapping sites of gene induction in the central nervous system. Neuron 8:13–23. https://doi.org/10.1016/0896-6273(92)90105-m
Sorensen AT et al (2016) A robust activity marking system for exploring active neuronal ensembles. elife 5. https://doi.org/10.7554/eLife.13918
Spiegel I et al (2014) Npas4 regulates excitatory-inhibitory balance within neural circuits through cell-type-specific gene programs. Cell 157:1216–1229. https://doi.org/10.1016/j.cell.2014.03.058
Steward O, Worley PF (2001a) A cellular mechanism for targeting newly synthesized mRNAs to synaptic sites on dendrites. Proc Natl Acad Sci USA 98:7062–7068. https://doi.org/10.1073/pnas.131146398
Steward O, Worley PF (2001b) Selective targeting of newly synthesized Arc mRNA to active synapses requires NMDA receptor activation. Neuron 30:227–240. https://doi.org/10.1016/s0896-6273(01)00275-6
Steward O, Wallace CS, Lyford GL, Worley PF (1998) Synaptic activation causes the mRNA for the IEG Arc to localize selectively near activated postsynaptic sites on dendrites. Neuron 21:741–751. https://doi.org/10.1016/s0896-6273(00)80591-7
Sukhatme VP et al (1988) A zinc finger-encoding gene coregulated with c-fos during growth and differentiation, and after cellular depolarization. Cell 53:37–43. https://doi.org/10.1016/0092-8674(88)90485-0
Sun X, Lin Y (2016) Npas4: linking neuronal activity to memory. Trends Neurosci 39:264–275. https://doi.org/10.1016/j.tins.2016.02.003
Sun X et al (2020) Functionally distinct neuronal ensembles within the memory engram. Cell 181:410–423 e417. https://doi.org/10.1016/j.cell.2020.02.055
Tanaka KZ, McHugh TJ (2018) The hippocampal engram as a memory index. J Exp Neurosci 12:1179069518815942. https://doi.org/10.1177/1179069518815942
Tanaka KZ et al (2014) Cortical representations are reinstated by the hippocampus during memory retrieval. Neuron 84:347–354. https://doi.org/10.1016/j.neuron.2014.09.037
Tanaka KZ et al (2018) The hippocampal engram maps experience but not place. Science 361:392–397. https://doi.org/10.1126/science.aat5397
Tischmeyer W, Kaczmarek L, Strauss M, Jork R, Matthies H (1990) Accumulation of c-fos mRNA in rat hippocampus during acquisition of a brightness discrimination. Behav Neural Biol 54:165–171. https://doi.org/10.1016/0163-1047(90)91366-j
Tonegawa S, Liu X, Ramirez S, Redondo R (2015) Memory engram cells have come of age. Neuron 87:918–931. https://doi.org/10.1016/j.neuron.2015.08.002
Vazdarjanova A et al (2006) Spatial exploration induces ARC, a plasticity-related immediate-early gene, only in calcium/calmodulin-dependent protein kinase II-positive principal excitatory and inhibitory neurons of the rat forebrain. J Comp Neurol 498:317–329. https://doi.org/10.1002/cne.21003

Wall NR et al (2019) Complementary genetic targeting and monosynaptic input mapping reveal recruitment and refinement of distributed Corticostriatal ensembles by cocaine. Neuron 104:916–930 e915. https://doi.org/10.1016/j.neuron.2019.10.032

Wang KH et al (2006) In vivo two-photon imaging reveals a role of arc in enhancing orientation specificity in visual cortex. Cell 126:389–402. https://doi.org/10.1016/j.cell.2006.06.038

Wang G et al (2021) Egr1-EGFP transgenic mouse allows in vivo recording of Egr1 expression and neural activity. J Neurosci Methods 363:109350. https://doi.org/10.1016/j.jneumeth.2021.109350

Watson MA, Milbrandt J (1990) Expression of the nerve growth factor-regulated NGFI-A and NGFI-B genes in the developing rat. Development 110:173–183. https://doi.org/10.1242/dev.110.1.173

Waung MW, Pfeiffer BE, Nosyreva ED, Ronesi JA, Huber KM (2008) Rapid translation of Arc/Arg3.1 selectively mediates mGluR-dependent LTD through persistent increases in AMPAR endocytosis rate. Neuron 59:84–97. https://doi.org/10.1016/j.neuron.2008.05.014

Weng FJ et al (2018) Npas4 is a critical regulator of learning-induced plasticity at Mossy Fiber-CA3 synapses during contextual memory formation. Neuron 97:1137–1152 e1135. https://doi.org/10.1016/j.neuron.2018.01.026

Xie H et al (2014) In vivo imaging of immediate early gene expression reveals layer-specific memory traces in the mammalian brain. Proc Natl Acad Sci USA 111:2788–2793. https://doi.org/10.1073/pnas.1316808111

Xing B et al (2021) A subpopulation of prefrontal cortical neurons is required for social memory. Biol Psychiatry 89:521–531. https://doi.org/10.1016/j.biopsych.2020.08.023

Xue J et al (2024) Spatiotemporal mapping and molecular basis of whole-brain circuit maturation. bioRxiv 2024.01.03.572456. https://doi.org/10.1101/2024.01.03.572456

Yang J et al (2022) Functionally distinct NPAS4-expressing somatostatin interneuron ensembles critical for motor skill learning. Neuron 110:3339–3355 e3338. https://doi.org/10.1016/j.neuron.2022.08.018

Yap EL et al (2021) Bidirectional perisomatic inhibitory plasticity of a Fos neuronal network. Nature 590:115–121. https://doi.org/10.1038/s41586-020-3031-0

Yassin L et al (2010) An embedded subnetwork of highly active neurons in the neocortex. Neuron 68:1043–1050. https://doi.org/10.1016/j.neuron.2010.11.029

Yoshihara S et al (2014) Npas4 regulates Mdm2 and thus Dcx in experience-dependent dendritic spine development of newborn olfactory bulb interneurons. Cell Rep 8:843–857. https://doi.org/10.1016/j.celrep.2014.06.056

Zhou Y et al (2009) CREB regulates excitability and the allocation of memory to subsets of neurons in the amygdala. Nat Neurosci 12:1438–1443. https://doi.org/10.1038/nn.2405

Zhou Y et al (2019) A ventral CA1 to nucleus accumbens core engram circuit mediates conditioned place preference for cocaine. Nat Neurosci 22:1986–1999. https://doi.org/10.1038/s41593-019-0524-y

# Unveiling Transcriptional and Epigenetic Mechanisms Within Engram Cells: Insights into Memory Formation and Stability

**Miguel Fuentes-Ramos and Ángel Barco**

**Abstract** Memory traces for behavioral experiences, such as fear conditioning or taste aversion, are believed to be stored through biophysical and molecular changes in distributed neuronal ensembles across various brain regions. These ensembles are known as engrams, and the cells that constitute them are referred to as engram cells. Recent advancements in techniques for labeling and manipulating neural activity have facilitated the study of engram cells throughout different memory phases, including acquisition, allocation, long-term storage, retrieval, and erasure. In this chapter, we will explore the application of next-generation sequencing methods to engram research, shedding new light on the contribution of transcriptional and epigenetic mechanisms to engram formation and stability.

**Keywords** Epigenetics · Learning and memory · Engram · Transcriptomics · Consolidation · Chromatin

## Introduction

The representation of a memory in the brain is commonly known as a memory trace. Memory processes can be divided into different phases, from the acquisition and allocation of the memory trace to its long-term storage (Fig. 1). These traces are believed to be stored through biophysical and molecular changes in neuronal ensembles called engrams, which are distributed across different brain regions. The specific molecular and structural changes that occur in engram cells during the acquisition, consolidation, maintenance, retrieval and elimination of memory are still not fully understood. However, recent advancements and refinements in techniques for labeling and manipulating neural activity have allowed for precise

M. Fuentes-Ramos · Á. Barco (✉)
Instituto de Neurociencias, Universidad Miguel Hernández – Consejo Superior de Investigaciones Científicas, Alicante, Spain
e-mail: abarco@umh.es



J. Gräff, S. Ramirez (eds.), *Engrams*, Advances in Neurobiology 38,
https://doi.org/10.1007/978-3-031-62983-9_7

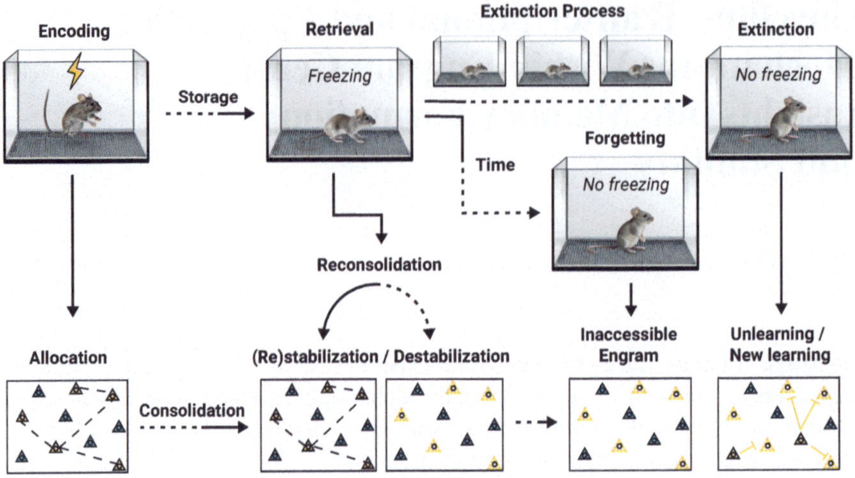


**Fig. 1** Engram neurons and memory phases. Memory phases are represented at two levels: behavioral (upper part) and engram-related mechanisms (bottom part). During memory encoding, new information is acquired and incorporated into a memory trace. At this moment, memory allocation occurs, which refers to the process by which some neurons (triangles) are selected to become part of the engram (bottom left, dashed lines and yellow triangles). Memory storage or consolidation is the phase during which memory stabilization takes place through multi-level brain reorganization, involving a flow of information among regions related to memory traces. This phase can last an hour to days (arrows indicate the time intervals between different processes, and dashed lines denote the lower stability of the original engram). The original engram can be reactivated, and the memory can be retrieved through cues from the previous experience. Retrieval can destabilize engram cells, and new information can be incorporated into the memory trace through a process known as reconsolidation (bottom middle). Over time, memories can decay, leading to forgetting where the engram enters an inaccessible state in which reactivation is no longer possible. Conversely, memories can be modified through an extinction process wherein the original memory is weakened (unlearning), and a new inhibitory engram emerges (new learning)

investigations of engram cells, including examining changes in gene expression and chromatin associated with each memory phase.

Transcriptional and epigenetic mechanisms have been postulated as part of the molecular substrate for memory. The relevance of de novo transcription in memory consolidation was first established almost 60 years ago using transcription blockers (Barondes and Jarvik 1964; Squire and Barondes 1970). Hundreds of studies have later investigated the specific contribution of many individual transcription factors (TF) to learning and memory (Alberini and Kandel 2015; Benito and Barco 2015). Comparatively, the investigation of epigenetic mechanisms in mnemonic processes is more recent but has been intense during the last two decades (Gräff et al. 2011; Lopez-Atalaya and Barco 2014; Sweatt 2013). The intricate mechanisms of epigenetic regulation encompass a variety of modifications, such as covalent changes in the DNA and histone proteins, as well as chromatin interactions. These mechanisms

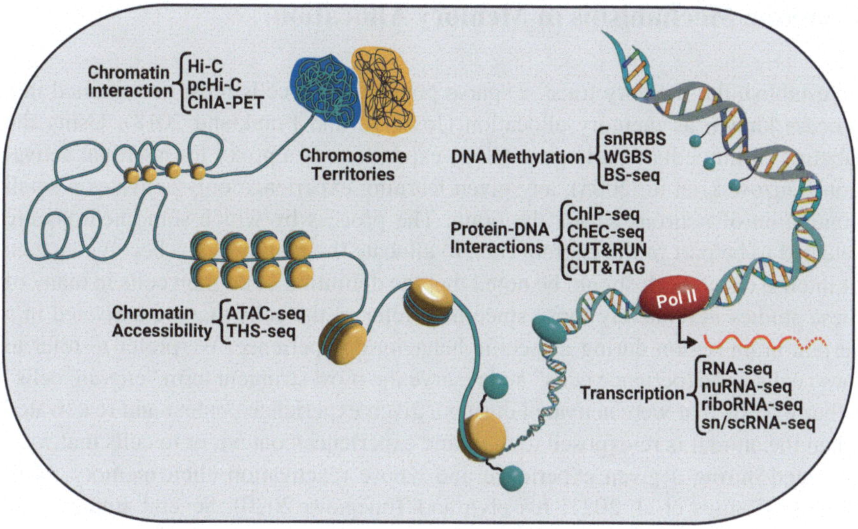


**Fig. 2** New methods to explore transcriptional and epigenetic changes associated with mnemonic processes. Two chromosomes are depicted occupying distinct regions of the nucleus (green and purple). One of them reveals unfolded chromatin, displaying various genomic mechanisms that regulate molecular processes within the nucleus. The catalogue of methodologies for studying each of these mechanisms encompasses (i) chromatin interactions profiling through high-throughput chromosome conformation capture (Hi-C), promoter capture Hi-C (pcHi-C) and chromatin interaction analysis by paired-end tag sequencing (ChIA-PET); (ii) chromatin accessibility through assay for transposase accessible chromatin with high-throughput sequencing (ATAC-seq) and transposome hypersensitive site sequencing (THS-seq); (iii) protein-DNA interaction analysis through chromatin immunoprecipitation (ChIP)-seq, chromatin endogenous cleavage with high-throughput sequencing (ChEC-seq), cleavage under targets and release using nuclease (CUT&RUN) and cleavage under targets and tagmentation (CUT&Tag); (iv) transcriptome analysis by RNA sequencing (RNA-seq), nuclear (nu)RNA-seq, ribosomal RNA immunoprecipitation sequencing (riboRNA-seq), single nucleus (sn)RNA-seq and single cell (sc)RNA-seq; and (v) DNA methylation profiling through bisulfite sequencing (BS-seq), whole-genome bisulfite sequencing (WGBS) and single-nucleus reduced representation bisulfite sequencing (snRRBS). (Modified from Fuentes-Ramos et al. 2021)

interplay with transcriptional and translation regulation to control gene expression. Thanks to technological developments in sequencing methods achieved in recent years, it is now possible to investigate transcriptional and epigenomic changes occurring in a small population of cells with unprecedented sensitivity (Fig. 2). As a result, several recent studies have tackled the characterization of experience-dependent changes in the transcriptome and epigenome in the sparse population of engram cells and even in individual neurons (Fuentes-Ramos et al. 2021). In this chapter, we will discuss the contribution of transcriptional and epigenetic mechanisms in the nucleus of engram cells to the different memory processes.

## Genomic Mechanisms in Memory Allocation

To establish the memory trace, a sparse population of cells must be recruited in a process known as memory allocation (Josselyn and Frankland 2018). Using the labeling of immediate-early gene (IEG) expression as a proxy for neuronal activation (Guzowski et al. 2005), any given learning experience only activates a small population of neurons across the brain. The process by which some neurons are selected to be part of the engram (i.e., to allocate the memory) has become an area of intense research. It should be noted that the definition of engram cells in many of these studies is relatively loose since they refer to the cells that are activated in a certain brain region during a specific behavioral experience. We prefer to refer to these cells as "experience cells" and reserve the most stringent term "engram cells" either to cells that were activated during a given experience/context and reactivated when the animal is re-exposed to the same experience/context, or to cells that were activated during a given experience and whose reactivation elicit memory recall (Fuentes-Ramos et al. 2021; Josselyn and Tonegawa 2020). Several studies have shown that reactivated cells represent only a modest percentage of the experience cells (DeNardo et al. 2019; Roy et al. 2022), although the exact percentage of reactivated cells in relation to the overall population of experience cells may vary depending on the specific memory system and the experimental techniques used to identify and label engram cells. For consistency with the original sources, when discussing some of these studies in the next sections, we will use the term engram cells to refer to what we defined above as experience cells.

### *CREB-Driven Mechanisms in Memory Allocation*

The first findings on the molecular basis of memory allocation came from studies investigating the role of the cAMP response element-binding protein (CREB). Numerous studies had identified this TF as a key factor in long-term memory formation and synaptic plasticity (Bartsch et al. 1995; Benito and Barco 2010; Bourtchuladze et al. 1994; Yin et al. 1994) when Han and colleagues first investigated the role of CREB in memory allocation. They overexpressed CREB in a random subpopulation of neurons (about 20%) in the lateral amygdala (LA) (Han et al. 2007), trained the mice in auditory fear conditioning, and 24 hours later presented the mice to the conditioned tone which, as expected, elicited a fear response. Interestingly, when they mapped the neurons activated during retrieval using antibodies that recognize the activity-regulated cytoskeleton-associated protein Arc, which labels recently activated neurons (Guzowski et al. 1999), they observed a greater-than-chance overlap between the neurons that overexpress CREB and those expressing Arc. This result indicates that the neurons with high levels of CREB were more likely to be recruited as part of the engram. CREB-driven mechanisms for memory allocation have also been observed in other brain regions (Choi et al. 2011; Czajkowski et al. 2014; Sano et al. 2014).

A recent study using a technique that allows a better spatial and temporal control of CREB function revealed that increasing CREB activity in the LA a few minutes before auditory fear conditioning but not 24 hours before or just after training produced an improvement in memory (Park et al. 2020). This memory enhancement was consistent with the idea that the relative levels of CREB in the minutes prior to training play a critical role in memory allocation. In addition to these gain-of-function experiments, the ablation or silencing of CREB-overexpressing neurons impaired fear memory in the amygdala (Han et al. 2009; Zhou et al. 2009) and in other brain regions (Park et al. 2016). Based on all these findings, the process of neuronal allocation during engram formation has been described as a competitive process between neurons, in which CREB plays a key role in influencing the probability that a neuron will be recruited as part of the memory engram.

CREB is known to regulate the intrinsic excitability of neurons in different brain regions. For example, the upregulation of CREB led to increased neuronal excitability in pyramidal CA1 neurons (Lopez de Armentia et al. 2007), whereas a dominant-negative form of CREB produced a decrease in the excitability of neurons in the nucleus accumbens (Dong et al. 2006). Zhou et al. found that cells that overexpressed CREB had lower thresholds for action potentials than cells that did not overexpress it (Zhou et al. 2009). On a different line of evidence, it has been found that neurons with naturally increased excitability during memory encoding tended to be part of the engram (Gouty-Colomer et al. 2015). Further supporting this view, when the exogenous CREB was co-expressed with Kir2.1, an inwardly rectifying K+ channel that decreases evoked action potential firing (Dong et al. 2006), the effect of CREB overexpression on memory allocation was abolished (Yiu et al. 2014). These results indicate that both CREB levels and the intrinsic excitability of the neuron (downstream of CREB action) are determining factors in the selection of neurons to be part of the memory trace (Jacob et al. 2020). But what determines that some neurons present greater excitability or CREB levels than others? Is it a stochastic process or is there an underlying logic? Future studies should address these questions and clarify the mechanisms by which CREB modulates intrinsic excitability in different neuronal types.

## *Epigenetic Mechanisms That Modulate Memory Allocation*

Research on the mechanisms of memory allocation suggests that, at any given time, there is a limited number of neurons in a primed state, characterized by greater intrinsic excitability than other neurons in the same population, which favors their recruitment into an engram (Gouty-Colomer et al. 2015). One intrinsic factor that could impact the cell status, and thus influence excitability, is their epigenetic landscape. For instance, Odell and colleagues have suggested that pre-existing methylation states could influence cell excitability and thus the likelihood that those cells will be recruited to become part of the engram. In that study, the authors analyzed the methylation status of DNA in active (Fos-expressing) and non-active dentate gyrus (DG) granule cells. They found a number of genes that were differentially expressed between these two populations and were also differently methylated, thereby defining a gene network potentially implicated in the regulation of neuronal excitability and structural plasticity (Odell et al. 2020). However, this study could not rule out the possibility that the differences were due to de novo methylation or demethylation resulting from the activation process itself. More so considering that neuronal activation induces a transcriptional program that includes genes encoding for TFs and epigenetic regulators (Jaeger et al. 2018). The hypothesis that memory allocation is not a stochastic process but is based on variations—including of epigenetic nature—of the intrinsic properties of neurons, making some cells more likely to be part of the engram than others, is appealing. Still, the specific mechanisms that govern this priming process remain to be elucidated.

## Genomic Mechanisms Within Engram Cells During Memory Consolidation

After memory allocation, a series of biological processes are thought to occur in engram neurons that contribute to the stabilization and storage of memory (Tonegawa et al. 2018). This phase, known as consolidation, involves synaptic plasticity mechanisms that rely on intracellular molecular signaling cascades activated by learning, including de novo transcription and changes in the chromatin. The consolidation process can span from hours to days (McGaugh 2000). During the first stages of consolidation, which take place at the cellular or synaptic level, there are both post-translational modification and de novo synthesis of synaptic proteins driven by the activation of intracellular signaling cascades that initiate activity-driven transcriptional responses and alter the epigenomic landscape (Day and Sweatt 2011; Gräff and Tsai 2013; Kyrke-Smith and Williams 2018). In addition to these relatively rapid cellular events, a slower form of consolidation that occurs at a larger scale has been also described. In this phase, known as systems consolidation, there is a reorganization of the memory trace across brain regions so that the relevance of certain brain regions changes. For example, initially hippocampus-dependent episodic memories are rendered independent of the hippocampus, whereas cortical circuits become more important (Frankland et al. 2004; Han et al. 2022; Kitamura et al. 2017; Squire 1986).

### *Activity-Driven Transcription During Memory Formation*

The neurons activated by a behavioral experience initiate a complex sequence of transcriptional changes (Benito and Barco 2015; Sullivan et al. 2021). There is a first transcriptional wave elicited within a few minutes of the learning experience that includes the expression of IEGs, many of which encode for TFs (e.g., *Fos* or *Npas4*). This is followed by a second wave of transcription that comprises synaptic plasticity effector genes downstream of the IEGs. Then, there is a third transcriptional wave with a slower dynamic that includes changes in gene expression that can be sustained for days (Sullivan et al. 2021). However, it should be noted that the transcriptional programs induced upon activation can greatly differ depending on cell type and stimuli (Hrvatin et al. 2017). For instance, different patterns of neuronal activity can produce different transcriptional responses (Tyssowski et al. 2018). Moreover, the transcriptional response and long-range wiring can vary depending on the emotional valence of the stimulus that activates the neuronal assemblies (Shpokayte et al. 2022; Ye et al. 2016). Despite the recent progress in this area, the nature of the transcriptional modifications within these neuronal assemblies and the molecular mechanisms underlying the persistence of the changes are still not fully understood (Fuentes-Ramos et al. 2021).

*Fos* is one of the most extensively studied IEG induced by behavioral experiences during the first transcriptional wave. Experience triggers the expression of this TF in a specific group of neurons that play a crucial role in memory acquisition and recall (Garner et al. 2012; Khalaf et al. 2018; Yap and Greenberg 2018). It has been proposed that the neurons that induce Fos expression in response to a learning experience undergo a series of molecular modifications that enable the encoding of different aspects of the experience. In fact, the definition or identification of engram cells in most of the studies in this field is based either on Fos expression or the activity of the Fos promoter. For instance, the reactivation of the neurons tagged using a Fos-based tagging system is sufficient to evoke the memory of the acquired experience (Josselyn and Tonegawa 2020; Liu et al. 2012). The study of the transcriptional response of these neurons has been established as a starting point to elucidate the underlying molecular mechanisms of memory consolidation. Lacar and colleagues conducted one of the first transcriptomic studies of neuronal assemblies activated by an experience (Lacar et al. 2016). These assemblies were composed of neurons that had expressed Fos in the DG during the exploration of a new environment. They compared the transcriptional profiles of neurons expressing Fos with those that do not express it using single nucleus (sn)RNA-seq. As expected, they detected an increase in IEGs in context-activated neurons, as well as genes encoding proteins that are located at the postsynaptic density and genes involved in the MAPK pathway. In addition, they detected transcriptional heterogeneity in the activated neurons that could indicate different patterns of temporal activation.

Similarly, another study reported different transcriptional responses between cell types. Jaeger and colleagues observed a stronger transcriptional response in $Fos^+$ neurons in the DG compared to $Fos^+$ pyramidal neurons in CA1 and to vasoactive intestinal polypeptide (VIP) interneurons (Jaeger et al. 2018). Furthermore, they found that the early activity-induced signature in DG contained genes related to epigenetic regulation, supporting the notion that neuronal activation triggers epigenetic changes (Belgrad and Fields 2018). The late response signature in DG neurons was composed of genes related to neurite outgrowth and the extracellular matrix. These gene programs are likely to underlie the molecular changes responsible for memory encoding and consolidation.

Most of these studies focused on detecting Fos as a hallmark of neuronal assemblies activated by the behavioral experience. However, Fos is not a mere indicator of neuronal activity. A recent study has suggested that Fos expression has an instructive role in engram cells orchestrating persistent circuit modifications (Yap et al. 2021). This study unveiled a bidirectional perisomatic inhibitory plasticity mechanism in which Fos-activated circuits are selectively rearranged through Scg2 expression. This gene encodes multiple neuropeptides to coordinate the changes in inhibition. Consistent with recent experiments underscoring the importance of Fos in spatial learning (Pettit et al. 2022), these findings raise the possibility that Fos-dependent circuit reorganization is a necessary step for encoding and recalling memories in response to experience.

But do all the neurons belonging to an engram express the same IEGs? Can an engram include neurons that do not express Fos? Are there different types of engrams, encoding for different aspects of the same experience and expressing different IEGs? In a recent study, two functionally different engrams were observed within the same memory: a Fos-dependent ensemble that promoted memory generalization and a Npas4-dependent ensemble that promoted memory discrimination (Sun et al. 2020a) (Nps4, like Fos, is an IEG that encodes for a TF). Both ensembles received information from different neuronal types and brain regions. Future studies should explore the interaction between neuronal ensembles associated with the same memory trace but differ in their transcriptional signature during memory encoding.

## *Transcriptional Response in Engram Cells During Memory Consolidation*

The detection and analysis of engram/experience cells through IEG mapping is constrained to a few hours after neuronal activation. To explore later changes it is necessary to develop methods to label the activated cells for longer periods of time. In recent years, several genetic strategies have allowed bypassing this temporal constraint and made it possible to mark the neurons activated during a certain behavioral experience (e.g., learning of a task) for longer periods of time or even permanently (DeNardo and Luo 2017). These strategies allow for the analysis of second-wave effector genes and late processes related to memory consolidation (Fuentes-Ramos et al. 2021).

This is the case of the Arc-dVenus reporter mouse line in which dVenus, a short-life destabilized fluorescent reporter, is expressed under the control of the Arc promoter. This allows neurons activated during encoding or early memory consolidation to be monitored for a period of time longer than the expression of endogenous Arc itself (Eguchi and Yamaguchi 2009). Using this model, Rao-Ruiz and colleagues detected memory consolidation-specific transcriptional changes in DG neurons

24 hours after being activated during CFC (Rao-Ruiz et al. 2019). The analysis of the differentially expressed genes revealed a CREB-dependent transcriptional network, including many genes involved in synaptic function and structural plasticity (i.e., processes required for memory consolidation). To confirm the key role of CREB in early memory consolidation, they found that disruption of CREB exclusively in engram cells through the expression of a dominant negative mutant version of CREB (CREBS133A) using Fos::tTA transgenic mice interfered with memory consolidation. In addition, CREB involvement in consolidation has also been observed in other brain regions, such as the medial prefrontal cortex (mPFC) (Matos et al. 2019). These studies indicate that in addition to its role in memory allocation, CREB-dependent transcription is necessary for memory consolidation (Alberini and Kandel 2015; Benito and Barco 2010). Interestingly, both roles could rely on the same mechanism. If CREB activation is associated with a transient increase of intrinsic excitability, those neurons in which CREB was activated during or by the learning experience would be more likely to be reactivated within a time range. This form of sensitization could contribute to the consolidation of the memory trace during post-training resting or sleep (Benito and Barco 2010).

Besides activity-regulated TFs, such as CREB, Fos and Npas4, other activity-induced genes also play relevant roles in early consolidation. For example, in a recent study, Shen and colleagues found that the C-C chemokine receptor type 5 (Ccr5) gene, which codifies for an immune receptor, was overexpressed 12–24 hours after contextual fear training in dorsal CA1. This upregulation was more pronounced in the neurons that were identified as part of the assembly activated during training (Shen et al. 2022). This mechanism resulted in a transient decrease in excitability that limited the overlap between memory assemblies, thereby reducing the typical time window for memory linking (i.e., the process in which two memories acquired in a narrow period became linked) (Cai et al. 2016; Rashid et al. 2016). These studies evidence the dynamic transcriptional process in the neurons that encode the memory during the first 24 hours after learning. These transcriptional changes are necessary for memory consolidation, probably to perpetuate experience-dependent changes in synaptic plasticity and neural connectivity and to prevent abnormal memory linking.

The next question is whether the transcriptional changes detected shortly after neuronal activation are maintained or modified after the first 24 hours of consolidation. Recent research indicates that the consolidation of memory systems needs the reinforcement of synapses in regions responsible for long-term memory and the weakening of synapses in regions responsible for short-term memory (Kitamura et al. 2017; Lee et al. 2022). These late changes in memory consolidation have been considerably less studied than the initial transcriptional response to activation, but new tools have become available in recent years. In a recent study, the transcriptome of hippocampal engram neurons activated during contextual fear conditioning (CFC) was analyzed at different time points after training (Marco et al. 2020). To do this, they used an activity-dependent labeling strategy known as targeted recombination in active populations (TRAP). This is a binary system in which the expression of a TMX-dependent recombinase creERT2 is driven by the promoter of an

IEG, Fos or Arc (DeNardo and Luo 2017; Guenthner et al. 2013). Two hours after CFC, Marco and colleagues did not observe large changes in the transcriptome of hippocampal neurons activated during memory encoding, a result that contrasts with those obtained in previous transcriptomics analysis in recently activated neurons (Lacar et al. 2016). The differences could be explained by differences in the behavioral paradigm used in each experiment, the region of the brain studied, or the different sensitivities and leakiness of the specific TRAP reporter (Guenthner et al. 2013). Marco et al. identified stronger transcriptional changes in the hippocampal neurons activated by learning 5 days after training. Differentially expressed transcripts at this time were enriched in intronic regions, suggesting that the transcriptomic profile of late memory consolidation reflects immature transcripts (Marco et al. 2020).

In another study exploring late changes associated with memory consolidation, neurons that were activated in the medial prefrontal cortex (mPFC) during learning of contextual fear conditioning were analyzed one month after the training. The upregulated genes included genes encoding presynaptic proteins, indicating a possible growth of presynaptic terminals triggered by learning (Dillingham et al. 2019). This preprint study provides additional evidence for the existence of a particular transcriptional signature in neurons activated during learning that endures long after the experience and may depend on epigenetic mechanisms.

## *Epigenetic Mechanisms Within Memory Engrams*

These data demonstrate that dynamic transcriptional responses are activated both during learning and consolidation to support the functional and structural changes that occur in engram neurons (Ryan et al. 2015). But are these transcriptional cascades permanently altered in engram cells? Epigenetic processes are postulated as a key mechanism for maintaining the functional state of the engram cell (Marco 2022). However, it is still challenging to investigate changes in the chromatin in a sparse cell assembly due to the large input required in most epigenomic techniques.

The assessment of chromatin accessibility through assay for transposase-accessible chromatin with high-throughput sequencing (ATAC-seq) represents an exception since this technique requires only a few thousand cells to obtain reliable results and has even been scaled down to single cells (Buenrostro et al. 2015). Fernández-Albert et al. obtained hippocampal neurons expressing Fos 1 hour after exploration of a new environment using a reporter mouse line in which creERT2 is driven by the promoter of calcium/calmodulin-dependent kinase 2 alpha (Camk2a), allowing for the expression of a fluorescent reporter (Sun1-GFP) specifically in the nuclear envelope of forebrain excitatory neurons. Then, they compared the $Fos^{+}$ and $Fos^{-}$ populations of excitatory neurons by ATAC-seq and found that the activation of these neurons during the experience produced large changes in chromatin accessibility and TF occupancy of genes related to memory and neuronal plasticity (Fernandez-Albert et al. 2019).

Similarly, Marco and colleagues used ATAC-seq and the ArcTRAP mouse strain to characterize the initial phases of memory formation and detected changes in chromatin conformation, predominantly within enhancer regions, that lasted for at least 5 days after learning (Marco et al. 2020). Nuclear 3D architecture is therefore also emerging as a key factor in the dynamic regulation of gene expression in mnemonic processes since both studies reported chromatin rearrangements in the form of novel enhancer-promoter interactions associated with neuronal activation. These results reinforce the idea that there may be a genomic memory for previous activations in the form of chromatin changes, akin to the idea of the genomic action potential (Clayton et al. 2019).

DNA methylation is another epigenetic mechanism that has been related to the acquisition and maintenance of memory (Halder et al. 2016). Still, the studies on that epigenetic mark are based on bulk tissue or sorted neurons rather than in engram cells. However, a recent study compared the methylation of CpG sites between $Fos^{+}$ and $Fos^{-}$ populations of DG granule neurons obtained 1 hour after exposure to a new environment (Odell et al. 2020). Notably, this screen identified genes that displayed both differential methylation and transcription and defined a gene network related to neuronal excitability and structural plasticity. Another recent study investigated the effects of manipulating DNA methylation in DG excitatory neurons through overexpression of DNA methyltransferase 3a2 (Dnmt3a2) during consolidation of fear conditioning. Remarkably, they found that this intervention strengthened engram stability and significantly enhanced memory (Gulmez Karaca et al. 2020).

Lastly, a recent genome editing strategy based on the clustered regularly interspaced short palindromic repeats (CRISPR)–associated endonuclease Cas9 allowed for the specific knock-down of the lysine acetyltransferase CREB binding protein (CBP) in activated neurons of the prelimbic region during remote memory. This decreased fear memory when evaluated one week after the manipulation (Sun et al. 2020b). Overall, these studies underscore the relevance of epigenetic events, such as DNA methylation, 3D chromatin organization and histone acetylation, in regulating gene transcription during memory formation and maintenance.

## Genomic Mechanisms of Engrams During Memory Recall

The next stage of the memory process involves memory retrieval, during which stored information in the brain is accessed. The molecular changes that occur during memory consolidation likely contribute to the process of ecphory, which is the reactivation of an engram when appropriate cues trigger memory recall or retrieval. It is believed that the molecular changes associated with memory consolidation strengthen specific synaptic connections, possibly between engram cells, in a long-lasting or permanent manner, facilitating the reactivation of the neuronal ensemble during memory retrieval (Asok et al. 2019; Ortega-de San Luis and Ryan 2022). Currently, only two studies have explored the transcriptional response and epigenomic changes in cells that were activated during memory encoding and reactivated during recall.

In the first study, mice were exposed to a specific context, and after 4 hours, they were either re-exposed to the same context or to a different one and sacrificed 1 hour later. By examining the expression of two IEGs with different kinetics—Arc and Fos—the authors used flow cytometry to distinguish between "reactivated" and "newly activated" cells. The study found a higher population of reactivated cells in mice re-exposed to the same context, highlighting the presence of an engram. Additionally, the authors observed a distinct transcriptional signature for reactivated cells and noted that the burst of IEG expression was weaker in reactivated cells compared to newly activated cells (Jaeger et al. 2018).

While Jaeger and colleagues' study focused solely on analyzing the transcriptional response of engram cells, the second study simultaneously examined the transcriptional response and chromatin conformational changes in reactivated neurons 90 minutes after fear memory recall, five days after acquisition. The study found that regions that gained chromatin accessibility during memory encoding remained stable during consolidation and were still detected in reactivated cells during memory recall. Furthermore, the study demonstrated that promoter-enhancer interactions were enriched during reactivation, and enhancers primed during memory encoding interacted with their respective promoters, enhancing the transcription of genes related to local translation in the synaptic compartments (Marco et al. 2020). Thus, it appears that epigenetic changes occurring during memory consolidation influence the subsequent transcriptional response of memory-coding cells.

Most studies investigating cellular reactivation during memory recall have used fear and contextual memory paradigms at specific time points, revealing a significant proportion of reactivated cells during the recall process. However, a recent study found that the reactivation pattern of Fos ensembles in the DG region during the daily execution of a Morris Water Maze (MWM) was highly variable (Lamothe-Molina et al. 2022). The authors attributed these changes to the accumulation of ΔFosB, a long-lived splice variant of FosB, in these cells, which interferes with Fos induction and prevents the repeated expression of Fos in DG granule cells. Therefore, there could be a transcriptional response in these cells that differs from their response during memory encoding. These findings suggest that studying engram cells based solely on a single IEG marker likely limits their detection. They also raise the question of whether cells expressing a different marker during memory retrieval still belong to the original memory trace or if the trace has been modified.

## Update, Removal and Fading of Memory Traces

Memories are known to be dynamic, and various processes can shape and modify them. One such process is reconsolidation in which re-exposure to stimuli associated with a memory can destabilize the memory. During reconsolidation, protein synthesis plays a role in the re-stabilization of the memory and allows for the incorporation of new information into the memory trace (Lee et al. 2017; Nader 2003). Re-exposure to stimuli can also lead to memory extinction, a process where a conditioned stimulus is repeatedly presented without an unconditioned stimulus. This results in modifications in the internal representation of the conditioned association, gradually reducing the initial conditioned response. Two mechanisms have been proposed to explain the functionality of engram cells during extinction (Lee and Kaang 2023). The first is the weakening of the original engram, often referred to as "unlearning." The second is the emergence of a new inhibitory engram, known as "new learning" (Lacagnina et al. 2019). The reactivation of engram cells is necessary for both the reconsolidation process and the attenuation of remote memories during extinction (Khalaf et al. 2018). However, the molecular mechanisms within engram cells that underlie these processes are still largely unknown.

Previous studies have demonstrated that gene expression and protein synthesis are required for both reconsolidation and extinction in distinct brain regions (Mamiya et al. 2009). However, the specific transcriptional response within engram cells in these two processes has not been described. Chen et al. (2020) used a FosTRAP2 mouse to permanently label mPFC neurons that were activated during the retrieval of a fear memory encoded 16 days earlier. Then, they investigated the transcriptional signature in these cells by performing a single cell (sc)RNA-seq analysis 9 days after retrieval. Like in the transcriptional profiles obtained after learning, there was heterogeneity in the transcriptional programs elicited during memory reconsolidation in various neuronal types, including non-neuronal cells such as astrocytes and microglia (Chen et al. 2020). A follow-up study by the same laboratory confirmed the persistence of transcriptional changes in cells involved in long-term memory consolidation using spatial and single-cell transcriptomics. Specifically, they focused on the basolateral amygdala (BLA). They detected changes in genes involved in the mitogen-activated protein kinase (MAPK) pathway and CREB, BDNF and neuropeptide signaling (Sun et al. 2024). Along with previous seminal studies (Gräff et al. 2014), these investigations represent the first efforts to tackle the genomic changes triggered by reconsolidation and long-term memory storage.

While the reactivation of engram cells is required for the processes of reconsolidation or extinction, the mechanisms involved in forgetting appear to be different. Forgetting refers to the decreased ability to retrieve information that was previously encoded. It involves circuit remodeling that changes engram cells from an accessible state, where they can be reactivated by recall cues, to an inaccessible state (O'Leary et al. 2023; Ryan and Frankland 2022). Only a few studies have focused on the changes that take place in engram cells during forgetting. For example, a recent study identified Rac1 signaling and α2-chimaerin as key memory molecules regulating long-term memory maintenance and forgetting (Lv et al. 2019). Further research is still needed to uncover whether epigenetic mechanisms within engram cells also contribute to forgetting.

## Conclusions

As shown in this chapter and other articles in this book, the study of cell biology and molecular mechanisms within engram cells has significantly contributed to a better understanding of memory in its various stages. However, there are still numerous fundamental questions that remain unanswered. For example, we have yet to fully comprehend the molecular mechanisms underlying memory allocation beyond the understanding that variations in the intrinsic excitability of neurons play a crucial role in this process. While CREB levels and changes in the epigenomic landscape have been proposed as key factors influencing memory allocation by affecting the excitability of neurons, the specific mechanisms determining why a particular cell, embedded within an apparently homogeneous population, is more or less likely to become part of an engram during learning remains unknown.

Furthermore, we now know that neurons included in the memory trace undergo epigenomic changes related to their activation state during learning, which subsequently will influence both immediate and long-term transcriptional responses. However, there are significant gaps in our understanding of the crucial steps that connect these nuclear changes to the reinforcement of specific synaptic connections. We also ignore if there are specific transcriptional events or epigenetic modifications responsible for creating and maintaining a memory trace. Our knowledge of the contribution of genomic mechanisms to accessing the memory trace for retrieval, modifying the trace during reconsolidation, or erasing it during forgetting is also limited. It has been hypothesized that different transcriptional cascades regulating synaptic plasticity determine the connectivity between engram cells and facilitate memory accessibility by eliminating or creating new synapses. For example, modification of the epigenomic landscape could serve as a priming event to facilitate the reactivation of engram cells during retrieval. Still, there is limited experimental evidence supporting this model.

Researchers in the field of neuroepigenetics are actively exploring the hypothesis that genomic changes play a role in memory consolidation and long-term memory storage, as well as in the gradual fading of epigenetic changes that could potentially underlie forgetting. The ongoing development and optimization of new technologies for studying engram cells hold great promise for addressing these fundamental questions. Furthermore, these advancements in technology have the potential to revolutionize our understanding and treatment of memory-related disorders. By uncovering the intricate workings of memory at the genomic level, researchers can develop new approaches and therapies to tackle these disorders.

**Acknowledgments** M.F.-R. is the recipient of a "Formación de Profesorado Universitario (FPU)" fellowship from the Spanish Ministry of Education. A.B. research is supported by grants #202003 from Fundació la Marató de TV3, PID2020-118169RB-100 from AEI co-financed by ERDF, and PCI2021-122087-2A/ERA-Net NEURON NDD-243 from AEI, AC22-00030/JPND2022-115 from ISCIII, PROMETEO/2020/007 from the Generalitat Valenciana, and FCAIXA HR22-00394 from Fundación LaCaixa. The Instituto de Neurociencias is a "Centre of Excellence Severo Ochoa" (CEX2021-001165-S).

## References

Alberini CM, Kandel ER (2015) The regulation of transcription in memory consolidation. Cold Spring Harb Perspect Biol 7:a021741

Asok A, Leroy F, Rayman JB, Kandel ER (2019) Molecular mechanisms of the memory trace. Trends Neurosci 42:14–22

Barondes SH, Jarvik ME (1964) The influence of actinomycin-D on RNA synthesis and on memory. J Neurochem 11:187–195

Bartsch D, Ghirardi M, Skehel PA, Karl KA, Herder SP, Chen M, Bailey CH, Kandel ER (1995) Aplysia CREB2 represses long-term facilitation: relief of repression converts transient facilitation into long-term functional and structural change. Cell 83:979–992

Belgrad J, Fields RD (2018) Epigenome interactions with patterned neuronal activity. Neuroscientist 24:471–485

Benito E, Barco A (2010) CREB's control of intrinsic and synaptic plasticity: implications for CREB-dependent memory models. Trends Neurosci 33:230–240

Benito E, Barco A (2015) The neuronal activity-driven transcriptome. Mol Neurobiol 51:1071–1088

Bourtchuladze R, Frenguelli B, Blendy J, Cioffi D, Schutz G, Silva AJ (1994) Deficient long-term memory in mice with a targeted mutation of the cAMP-responsive element-binding protein. Cell 79:59–68

Buenrostro JD, Wu B, Litzenburger UM, Ruff D, Gonzales ML, Snyder MP, Chang HY, Greenleaf WJ (2015) Single-cell chromatin accessibility reveals principles of regulatory variation. Nature 523:486–490

Cai DJ, Aharoni D, Shuman T, Shobe J, Biane J, Song W, Wei B, Veshkini M, La-Vu M, Lou J et al (2016) A shared neural ensemble links distinct contextual memories encoded close in time. Nature 534:115–118

Chen MB, Jiang X, Quake SR, Südhof TC (2020) Persistent transcriptional programmes are associated with remote memory. Nature 587:437–442

Choi GB, Stettler DD, Kallman BR, Bhaskar ST, Fleischmann A, Axel R (2011) Driving opposing behaviors with ensembles of piriform neurons. Cell 146:1004–1015

Clayton DF, Anreiter I, Aristizabal M, Frankland PW, Binder EB, Citri A (2019) The role of the genome in experience-dependent plasticity: extending the analogy of the genomic action potential. Proc Natl Acad Sci 117:23252–23260

Czajkowski R, Jayaprakash B, Wiltgen B, Rogerson T, Guzman-Karlsson MC, Barth AL, Trachtenberg JT, Silva AJ (2014) Encoding and storage of spatial information in the retrosplenial cortex. Proc Natl Acad Sci USA 111:8661–8666

Day JJJ, Sweatt JDD (2011) Epigenetic mechanisms in cognition. Neuron 70:813–829

DeNardo L, Luo L (2017) Genetic strategies to access activated neurons. Curr Opin Neurobiol 45:121–129

DeNardo LA, Liu CD, Allen WE, Adams EL, Friedmann D, Fu L, Guenthner CJ, Tessier-Lavigne M, Luo L (2019) Temporal evolution of cortical ensembles promoting remote memory retrieval. Nat Neurosci 22:460–469

Dillingham BC, Cameron P, Pieraut S, Cardozo LM, Yoo EJ, Maximov A, Stowers L, Mayford M (2019) Fear learning induces long-lasting changes in gene expression and pathway specific presynaptic growth. BioRxiv

Dong Y, Green T, Saal D, Marie H, Neve R, Nestler EJ, Malenka RC (2006) CREB modulates excitability of nucleus accumbens neurons. Nat Neurosci 9:475–477

Eguchi M, Yamaguchi S (2009) In vivo and in vitro visualization of gene expression dynamics over extensive areas of the brain. NeuroImage 44:1274–1283

Fernandez-Albert J, Lipinski M, Lopez-Cascales MT, Rowley MJ, Martin-Gonzalez AM, del Blanco B, Corces VG, Barco A (2019) Immediate and deferred epigenomic signatures of in vivo neuronal activation in mouse hippocampus. Nat Neurosci 22:1718–1730

Frankland PW, Bontempi B, Talton LE, Kaczmarek L, Silva AJ (2004) The involvement of the anterior cingulate cortex in remote contextual fear memory. Science (80-. ) 304:881–883

Fuentes-Ramos M, Alaiz-Noya M, Barco A (2021) Transcriptome and epigenome analysis of engram cells: next-generation sequencing technologies in memory research. Neurosci Biobehav Rev 127:865–875
Garner AR, Rowland DC, Hwang SY, Baumgaertel K, Roth BL, Kentros C, Mayford M (2012) Generation of a synthetic memory trace. Science (80-. ) 335:1513–1516
Gouty-Colomer LA, Hosseini B, Marcelo IM, Schreiber J, Slump DE, Yamaguchi S, Houweling AR, Jaarsma D, Elgersma Y, Kushner SA (2015) Arc expression identifies the lateral amygdala fear memory trace. Mol Psychiatry 213(21):364–375
Gräff J, Tsai L-H (2013) Histone acetylation: molecular mnemonics on the chromatin. Nat Rev Neurosci 14:97–111
Gräff J, Kim D, Dobbin MM, Tsai LH (2011) Epigenetic regulation of gene expression in physiological and pathological brain processes. Physiol Rev 91:603–649
Gräff J, Joseph NF, Horn ME, Samiei A, Meng J, Seo J, Rei D, Bero AW, Phan TX, Wagner F et al (2014) Epigenetic priming of memory updating during reconsolidation to attenuate remote fear memories. Cell 156:261–276
Guenthner CJ, Miyamichi K, Yang HH, Heller HC, Luo L (2013) Permanent genetic access to transiently active neurons via TRAP: targeted recombination in active populations. Neuron 78:773–784
Gulmez Karaca K, Kupke J, Brito DVC, Zeuch B, Thome C, Weichenhan D, Lutsik P, Plass C, Oliveira AMM (2020) Neuronal ensemble-specific DNA methylation strengthens engram stability. Nat Commun 11:639
Guzowski JF, McNaughton BL, Barnes CA, Worley PF (1999) Environment-specific expression of the immediate-early gene Arc in hippocampal neuronal ensembles. Nat Neurosci 2:1120–1124
Guzowski JF, Timlin JA, Roysam B, McNaughton BL, Worley PF, Barnes CA (2005) Mapping behaviorally relevant neural circuits with immediate-early gene expression. Curr Opin Neurobiol 15:599–606
Halder R, Hennion M, Vidal RO, Shomroni O, Rahman R-U, Rajput A, Centeno TP, van Bebber F, Capece V, Vizcaino JCG et al (2016) DNA methylation changes in plasticity genes accompany the formation and maintenance of memory. Nat Neurosci 19:102–110
Han J-H, Kushner SA, Yiu AP, Cole CJ, Matynia A, Brown RA, Neve RL, Guzowski JF, Silva AJ, Josselyn SA (2007) Neuronal competition and selection during memory formation. Science 316:457–460
Han JH, Kushner SA, Yiu AP, Hsiang HL, Buch T, Waisman A, Bontempi B, Neve RL, Frankland PW, Josselyn SA (2009) Selective erasure of a fear memory. Science (80-. ) 323:1492–1496
Han DH, Park P, Choi DI, Bliss TVP, Kaang BK (2022) The essence of the engram: cellular or synaptic? Semin Cell Dev Biol 125:122–135
Hrvatin S, Hochbaum DR, Nagy MA, Cicconet M, Robertson K, Cheadle L, Zilionis R, Ratner A, Borges-Monroy R, Klein AM et al (2017) Single-cell analysis of experience-dependent transcriptomic states in the mouse visual cortex. Nat Neurosci 211:120–129
Jacob AD, Mocle AJ, Frankland PW, Josselyn SA (2020) Neuronal excitability in memory allocation: mechanisms and consequences. In: Carew TJ (ed) The Oxford handbook of the neurobiology of learning and memory. Oxford University Press
Jaeger BN, Linker SB, Parylak SL, Barron JJ, Gallina IS, Saavedra CD, Fitzpatrick C, Lim CK, Schafer ST, Lacar B et al (2018) A novel environment-evoked transcriptional signature predicts reactivity in single dentate granule neurons. Nat Commun 9:3084
Josselyn SA, Frankland PW (2018) Memory allocation: mechanisms and function. Annu Rev Neurosci 41:389–413
Josselyn SA, Tonegawa S (2020) Memory engrams: recalling the past and imagining the future. Science 367:eaaw4325
Khalaf O, Resch S, Dixsaut L, Gorden V, Glauser L, Gräff J (2018) Reactivation of recall-induced neurons contributes to remote fear memory attenuation. Science (80- ) 360:1239–1242

Kitamura T, Ogawa SK, Roy DS, Okuyama T, Morrissey MD, Smith LM, Redondo RL, Tonegawa S (2017) Engrams and circuits crucial for systems consolidation of a memory. Science 356:73–78

Kyrke-Smith M, Williams JM (2018) Bridging synaptic and epigenetic maintenance mechanisms of the engram. Front Mol Neurosci 11:369

Lacagnina AF, Brockway ET, Crovetti CR, Shue F, McCarty MJ, Sattler KP, Lim SC, Santos SL, Denny CA, Drew MR (2019) Distinct hippocampal engrams control extinction and relapse of fear memory. Nat Neurosci 22:753–761

Lacar B, Linker SB, Jaeger BN, Krishnaswami SR, Barron JJ, Kelder MJE, Parylak SL, Paquola ACM, Venepally P, Novotny M et al (2016) Nuclear RNA-seq of single neurons reveals molecular signatures of activation. Nat Commun 7:11022

Lamothe-Molina PJ, Franzelin A, Beck L, Li D, Auksutat L, Fieblinger T, Laprell L, Alhbeck J, Gee CE, Kneussel M et al (2022) ΔFosB accumulation in hippocampal granule cells drives cFos pattern separation during spatial learning. Nat Commun 131(13):1–16

Lee H, Kaang B-K (2023) How engram mediates learning, extinction, and relapse. Curr Opin Neurobiol 81:102723

Lee JLC, Nader K, Schiller D (2017) An update on memory reconsolidation updating. Trends Cogn Sci 21:531–545

Lee JH, Kim W, Bin Park EH, Cho JH (2022) Neocortical synaptic engrams for remote contextual memories. Nat Neurosci 262(26):259–273

Liu X, Ramirez S, Pang PT, Puryear CB, Govindarajan A, Deisseroth K, Tonegawa S (2012) Optogenetic stimulation of a hippocampal engram activates fear memory recall. Nature 484:381–385

Lopez-Atalaya JP, Barco A (2014) Can changes in histone acetylation contribute to memory formation? Trends Genet 30:529–539

Lopez de Armentia M, Jancic D, Olivares R, Alarcon JM, Kandel ER, Barco A (2007) cAMP response element-binding protein-mediated gene expression increases the intrinsic excitability of CA1 pyramidal neurons. J Neurosci 27:13909–13918

Lv L, Liu Y, Xie J, Wu Y, Zhao J, Li Q, Zhong Y (2019) Interplay between α2-chimaerin and Rac1 activity determines dynamic maintenance of long-term memory. Nat Commun 101(10):1–12

Mamiya N, Fukushima H, Suzuki A, Matsuyama Z, Homma S, Frankland PW, Kida S (2009) Brain region-specific gene expression activation required for reconsolidation and extinction of contextual fear memory. J Neurosci 29:402–413

Marco A (2022) Activity-dependent remodeling of genome architecture in engram cells facilitates memory formation and recall. Neural Regen Res 17:991

Marco A, Meharena HS, Dileep V, Raju RM, Davila-Velderrain J, Zhang AL, Adaikkan C, Young JZ, Gao F, Kellis M et al (2020) Mapping the epigenomic and transcriptomic interplay during memory formation and recall in the hippocampal engram ensemble. Nat Neurosci 23:1606–1617

Matos MR, Visser E, Kramvis I, van der Loo RJ, Gebuis T, Zalm R, Rao-Ruiz P, Mansvelder HD, Smit AB, van den Oever MC (2019) Memory strength gates the involvement of a CREB-dependent cortical fear engram in remote memory. Nat Commun 10:2315

McGaugh JL (2000) Memory--a century of consolidation. Science (80-. ) 287:248–251

Nader K (2003) Memory traces unbound. Trends Neurosci 26:65–72

O'Leary JD, Bruckner R, Autore L, Ryan TJ (2023) Natural forgetting reversibly modulates engram expression. eLife 12:RP92860

Odell SC, Taki F, Klein SL, Chen RJ, Levine OB, Skelly MJ, Nabila A, Brindley E, Gal Toth J, Dündar F et al (2020) Epigenomically Bistable regions across neuron-specific genes govern neuron eligibility to a coding ensemble in the hippocampus. Cell Rep 31:107789

Ortega-de San Luis C, Ryan TJ (2022) Understanding the physical basis of memory: molecular mechanisms of the engram. J Biol Chem 298:101866

Park S, Kramer EE, Mercaldo V, Rashid AJ, Insel N, Frankland PW, Josselyn SA (2016) Neuronal allocation to a hippocampal engram. Neuropsychopharmacology 4113(41):2987–2993

Park A, Jacob AD, Walters BJ, Park S, Rashid AJ, Jung JH, Lau J, Woolley GA, Frankland PW, Josselyn SA (2020) A time-dependent role for the transcription factor CREB in neuronal allocation to an engram underlying a fear memory revealed using a novel in vivo optogenetic tool to modulate CREB function. Neuropsychopharmacology 45:916–924
Pettit NL, Yap EL, Greenberg ME, Harvey CD (2022) Fos ensembles encode and shape stable spatial maps in the hippocampus. Nature 6097926(609):327–334
Rao-Ruiz P, Couey JJ, Marcelo IM, Bouwkamp CG, Slump DE, Matos MR, van der Loo RJ, Martins GJ, van den Hout M, van IJcken WF et al (2019) Engram-specific transcriptome profiling of contextual memory consolidation. Nat Commun 10:2232
Rashid AJ, Yan C, Mercaldo V, Hsiang H-L(L), Park S, Cole CJ, De Cristofaro A, Yu J, Ramakrishnan C, Lee SY et al (2016) Competition between engrams influences fear memory formation and recall. Science 80-. ). 353:383–387
Roy DS, Park Y-GG, Kim ME, Zhang Y, Ogawa SK, DiNapoli N, Gu X, Cho JH, Choi H, Kamentsky L et al (2022) Brain-wide mapping reveals that engrams for a single memory are distributed across multiple brain regions. Nat Commun 13:1799
Ryan TJ, Frankland PW (2022) Forgetting as a form of adaptive engram cell plasticity. Nat Rev Neurosci 23:173–186
Ryan TJ, Roy DS, Pignatelli M, Arons A, Tonegawa S (2015) Engram cells retain memory under retrograde amnesia. Science (80-. ) 348:1007–1013
Sano Y, Shobe JL, Zhou M, Huang S, Shuman T, Cai DJ, Golshani P, Kamata M, Silva AJ (2014) CREB regulates memory allocation in the insular cortex. Curr Biol 24:2833–2837
Shen Y, Zhou M, Cai D, Filho DA, Fernandes G, Cai Y, de Sousa AF, Tian M, Kim N, Lee J et al (2022) CCR5 closes the temporal window for memory linking. Nature 6067912(606):146–152
Shpokayte M, McKissick O, Guan X, Yuan B, Rahsepar B, Fernandez FR, Ruesch E, Grella SL, White JA, Liu XS et al (2022) Hippocampal cells segregate positive and negative engrams. Commun Biol 5:1009
Squire LR (1986) Mechanisms of memory. Science (80-. ) 232:1612–1619
Squire LR, Barondes SH (1970) Actinomycin-D: effects on memory at different times after training. Nature 225:649–650
Sullivan KE, Kendrick RM, Cembrowski MS (2021) Elucidating memory in the brain via single-cell transcriptomics. J Neurochem 157:982–992
Sun X, Bernstein MJ, Meng M, Rao S, Sørensen AT, Yao L, Zhang X, Anikeeva PO, Lin Y (2020a) Functionally distinct neuronal ensembles within the memory engram. Cell 181:410–423.e17
Sun H, Fu S, Cui S, Yin X, Sun X, Qi X, Cui K, Wang J, Ma L, Liu FY et al (2020b) Development of a CRISPR-SaCas9 system for projection- and function-specific gene editing in the rat brain. Sci Adv 6:6687–6705
Sun W, Liu Z, Jiang X, Chen MB, Dong H, Liu J, Südhof TC, Quake SR (2024) Spatial transcriptomics reveal neuron-astrocyte synergy in long-term memory. Nature 627(8003):374–381
Sweatt JD (2013) The emerging field of neuroepigenetics. Neuron 80:624–632
Tonegawa S, Morrissey MD, Kitamura T (2018) The role of engram cells in the systems consolidation of memory. Nat Rev Neurosci 19:485–498
Tyssowski KM, DeStefino NR, Cho JH, Dunn CJ, Poston RG, Carty CE, Jones RD, Chang SM, Romeo P, Wurzelmann MK et al (2018) Different neuronal activity patterns induce different gene expression programs. Neuron 98:530–546.e11
Yap EL, Greenberg ME (2018) Activity-regulated transcription: bridging the gap between neural activity and behavior. Neuron
Yap EL, Pettit NL, Davis CP, Nagy MA, Harmin DA, Golden E, Dagliyan O, Lin C, Rudolph S, Sharma N et al (2021) Bidirectional perisomatic inhibitory plasticity of a Fos neuronal network. Nature 590:115–121
Ye L, Allen WE, Thompson KR, Tian Q, Hsueh B, Ramakrishnan C, Wang AC, Jennings JH, Adhikari A, Halpern CH et al (2016) Wiring and molecular features of prefrontal ensembles representing distinct experiences. Cell 165:1776–1788

Yin JC, Wallach JS, Del Vecchio M, Wilder EL, Zhou H, Quinn WG, Tully T (1994) Induction of a dominant negative CREB transgene specifically blocks long-term memory in Drosophila. Cell 79:49–58

Yiu AP, Mercaldo V, Yan C, Richards B, Rashid AJ, Hsiang H-LL, Pressey J, Mahadevan V, Tran MM, Kushner SA et al (2014) Neurons are recruited to a memory trace based on relative neuronal excitability immediately before training. Neuron 83:722–735

Zhou Y, Won J, Karlsson MG, Zhou M, Rogerson T, Balaji J, Neve R, Poirazi P, Silva AJ (2009) CREB regulates excitability and the allocation of memory to subsets of neurons in the amygdala. Nat Neurosci 1211(12):1438–1443

# Synaptic Engram

**Hyunsu Jung, Daehee Han, Chaery Lee, and Bong-Kiun Kaang**

**Abstract** The concept of the engram refers to structural and/or physiological changes that underlie memory associations during learning. However, the precise biological basis of the engram remains elusive, with ongoing controversy regarding whether it resides at the cellular level or within the synaptic connections between activated cells. Here, we briefly review the studies investigating the cellular engram and the challenges they encounter. Subsequently, we delve into the biological basis of the engram within synaptic connections. In this regard, we introduce the history of synaptic engrams and discuss recent findings suggesting that synaptic plasticity serves as a substrate for memory. Additionally, we provide an overview of key technologies utilized in the study of synaptic plasticity.

**Keywords** Synaptic engram · Cellular engram · Plasticity · Circuits · Connectivity

---

H. Jung · D. Han
Center for Cognition and Sociality, Life Science Institute, Institute for Basic Science (IBS), Daejeon, South Korea

Interdisciplinary Program in Neuroscience, College of Natural Sciences, Seoul National University, Seoul, South Korea

C. Lee
Center for Cognition and Sociality, Life Science Institute, Institute for Basic Science (IBS), Daejeon, South Korea

Department of Biological Sciences, College of Natural Sciences, Seoul National University, Seoul, South Korea

B.-K. Kaang (✉)
Center for Cognition and Sociality, Life Science Institute, Institute for Basic Science (IBS), Daejeon, South Korea

Interdisciplinary Program in Neuroscience, College of Natural Sciences, Seoul National University, Seoul, South Korea

Department of Biological Sciences, College of Natural Sciences, Seoul National University, Seoul, South Korea
e-mail: kaang@ibs.re.kr

 
J. Gräff, S. Ramirez (eds.), *Engrams*, Advances in Neurobiology 38,
https://doi.org/10.1007/978-3-031-62983-9_8

## Overview

The engram was first proposed by Richard Semon in 1904. Semon introduced the term "engram" as a memory representation in which memory persists after an experience (Semon 1904). Recently, engrams have been referred to as physiological and/or chemical changes induced by memory associations newly formed through learning (Josselyn and Tonegawa 2020). However, the biological basis of this engram has not yet been fully elucidated. Whether engrams exist at the cellular level or whether the connections between cells have the properties of an engram is controversial. In this chapter, we briefly review cellular engram studies and discuss their limitations. Finally, we interpret the engrams based on synaptic plasticity and introduce technologies to study synaptic engrams.

## Hebbian Plasticity

In 1949, Donald Hebb asserted that if any two cells are activated repeatedly and simultaneously, they tend to associate so that activity in one facilitates activity in the other (Hebb 1949). This cell assembly theory, also known as Hebbian plasticity, has been expanded to include any long-lasting form of synaptic strengthening or weakening (Abbott and Nelson 2000; Josselyn and Tonegawa 2020). Correlated synaptic inputs and action potential outputs induce long-term potentiation (LTP) and increase synaptic strength (Magee and Johnston 1997), whereas long-term depression (LTD) promotes synaptic weakening (Abbott and Nelson 2000). Recent studies demonstrated that LTP and LTD are associated with learning and memory (Andersen et al. 2017; Nabavi et al. 2014).

If learning is achieved through LTP, causing synaptic enhancement, and this process is saturable, the subsequent induction of electrical LTP should produce less LTP. Rioult-Pedotti et al. (2000) used this limitation of LTP induction to identify the synaptic modification ranges (Rioult-Pedotti et al. 2000). In the rat motor cortex (MI), repeated learning to grasp food pellets using the forelimb strengthened layer II/III horizontal connections. After training, repeated theta burst stimulation (TBS) induced less LTP in trained MI than in untrained MI. Repeated LTD induction produced a larger number of LTD to reach the minimum synaptic strength in the trained MI than in the untrained MI. The reduction in LTP and increase in LTD in the MI layer II/III horizontal connections after learning suggest that LTP is associated with synaptic strengthening during skill learning.

An optogenetic study elucidated a causal relationship between memory encoding and synaptic plasticity in associative learning using auditory fear conditioning. The lateral amygdala (LA) receives inputs from the medial geniculate nucleus (MGN) and auditory cortex (AC). These neural circuits are involved in the encoding of auditory fear memory. To optogenetically induce LTP or LTD in the MGN-to-LA and AC-to-LA circuits, oChIEF, a variant of the light-activated channel ChR2, was expressed in the MGN and AC (Nabavi et al. 2014). An optical fiber for light delivery was placed in the LA to stimulate the axon terminals. oChIEF is suitable for axon terminal stimulation because it produces a reliable signal in response to high-frequency stimulation (Hass and Glickfeld 2016). After fear memory formation, optogenetically induced LTD attenuates the fear response and LTP reactivates memory (Nabavi et al. 2014). Therefore, LTP and LTD are associated with Pavlovian conditioning as well as skill learning. Hebbian plasticity is likely to underlie physiological alterations of synaptic strengthening during learning and may also contribute to the physiological and structural changes of the engram.

## History of Synaptic Engrams

### Cellular Engram and Memory Allocation

The molecular mechanisms of an engram are complex and have not yet been fully elucidated. This topic is further covered in Chapter "Unveiling Transcriptional and Epigenetic Mechanisms Within Engram Cells: Insights into Memory Formation and Stability"; thus, here we only briefly review it. Because the engram stores memories formed during a particular experience, genetic markers that reflect neuronal activity are used in the engram field (DeNardo et al. 2019; Guenthner et al. 2013; Sun et al. 2020). Immediate early genes (IEGs) such as c-Fos and Arc are markers for engram tagging and are used to observe cell reactivation during memory recall (Denny et al. 2014; Sorensen et al. 2016). It is thought that the expression of IEGs is related to the mechanism by which engrams are formed (Miyashita et al. 2018). Neuronal excitation increases the intracellular calcium concentration (Smetters et al. 1999) and evokes ERK/MAPK signaling (Bozon et al. 2003). In turn, cyclic adenosine 3′,5′-monophosphate response element binding protein (CREB) is phosphorylated by ELK1 (also known as ternary complex factor [TCF]) -SRF signaling or Rsk-2 kinase (De Cesare et al. 1998). IEGs are the target genes of CREB and are upregulated by neuronal activation (Kawashima et al. 2009; Sheng et al. 1991).

An ongoing hypothesis in contemporary neuroscience is that memory is allocated to neurons, which is accompanied by global gene expression—indeed, engrams can be interpreted to exist at the cellular level. From this perspective, neurons are places where memories are stored, and physiological and/or chemical changes occur following training. In the hippocampus, cAMP response element (CRE)-mediated gene expression increases after long-lasting long-term potentiation (L-LTP) (Impey et al. 1996). In particular, the cAMP/cAMP-dependent protein kinase (PKA) signaling pathway mediates L-LTP and protein synthesis-dependent memory formation (Bernabeu et al. 1997). CREB plays a central role in memory allocation (Park et al. 2016). A random population of dentate gyrus neurons that overexpress CREB during contextual fear learning induces increased excitability and is allocated to the engram. CREB increases neuronal excitability in the lateral amygdala and modulates memory allocation to specific neurons during auditory fear conditioning (Zhou et al. 2009). The function of CREB can be modulated in freely behaving animals using opto-DN-CREB, an optogenetic tool for regulating CREB expression with high spatiotemporal resolution (Park et al. 2020). Increased CREB expression in a random population of LA neurons an hour before auditory fear conditioning enhanced learning. If CREB expression is required for memory formation, the deletion of CREB function would impair the learning process. The ablation of CREB-expressing LA neurons impairs memory retrieval during auditory fear conditioning (Han et al. 2009).

Interestingly, neuronal populations interact to determine memory allocation. If the two events are independent, encoding them in distinct neuronal populations would efficiently minimize interference. By contrast, if the two experiences are related, it would be productive to allocate them to overlapping populations and functionally link these memories. This hypothesis was verified using auditory fear conditioning in mice (Rashid et al. 2016). Mice received two distinct tones, followed by foot shocks at different intervals. When two events occurred at 6-h intervals, the ratio of overlapping engram cells in the LA was higher than when events occurred at 24-h intervals. Furthermore, optogenetic inhibition of overlapping LA engram cells for two events disrupted the fear response to the second event at 6-h intervals but not at 24-h intervals. Optogenetic activation of LA neurons immediately before the events can also induce artificial linking between two events, even at 24-h intervals. These results suggest that the co-allocation of the engram is caused by neuronal excitability between temporally close events.

**Emergent Properties of Cellular Engram: Interregional Connectivity, Spatiotemporal Dynamics and Reciprocal Inhibition**

Although memories are likely to be encoded in engram cells through allocation in various brain regions, fear memory is encoded by complex patterns of interregional synaptic connections. For example, engram cells encoding recent contextual fear memories are found in the hippocampal tripartite synapses (Choi et al. 2018), ventral hippocampus to the basal amygdala (Kim and Cho 2020) and medial prefrontal cortex (mPFC) to basal amygdala circuit (Kitamura et al. 2017), which are regions previously implicated in contextual fear memory. Likewise, auditory fear consists of the auditory cortex (AC)/auditory thalamus (AT) to lateral-amygdala (LA) synapse (Nabavi et al. 2014), which are subsequently connected to the central amygdala through the basal amygdala (Ciocchi et al. 2010). As an extension of the first discovery of contextual fear memory engram cells in the dentate gyrus of the hippocampus (Liu et al. 2012), a series of studies demonstrated the existence of engram cells in the AC/AT-LA circuits, which encode auditory fear memories (Abdou et al. 2018; Choi et al. 2021; Jeong et al. 2021; Kim and Cho 2017; Ryan et al. 2015). Finally, a recent study reported that engram ensembles of contextual fear memory are widely distributed in the brain, emphasizing the need to understand the functions of interregional interactions of engram ensembles (Roy et al. 2022).

Furthermore, several studies pointed out the dynamic changes in the location of memory as a recent memory develops into a remote memory, which entails a gradual shift in the dominance of memory encoding from recent memory engram cells to remote memory engram cells. Even though engram cells for remote memory emerge at the early stage of memory formation, they become responsible for the recall of the memory only after a certain amount of time passes (DeNardo et al. 2019; Kitamura et al. 2017; Lee et al. 2023b). These changes in the brain regions that are actively engaged in remote memory and the concomitant maturation of remote memory engram cells in those regions are believed to occur through a process called systems consolidation, which requires sleep and signals from recent memory engram cells (de Sousa et al. 2019; Kitamura et al. 2017).

Repeated exposure to conditioned stimuli could extinguish the learned memory, a process called extinction. Another line of studies showed that the extinction of memory generates extinction engram ensembles in the same brain region where fear memory engram ensembles reside (Lacagnina et al. 2019; Teng et al. 2023; Zhang et al. 2020). Extinction engram cells are spatially intermingled with fear engram cells in the basal amygdala, medial prefrontal cortex, and dentate gyrus of the hippocampus, suppressing fear engram cells (Fig. 1a, b). Extinction engram cells and fear engram cells reciprocally inhibit each other, where extinction engram cells inhibit fear engram cells when fear is suppressed, while the opposite occurs when fear is heightened by spontaneous recovery (Fig. 1c). Other lines of research that suggest the activation of fear engram cells during extinction learning complicate the view of the extinction mechanism, however. The activation of the fear engram

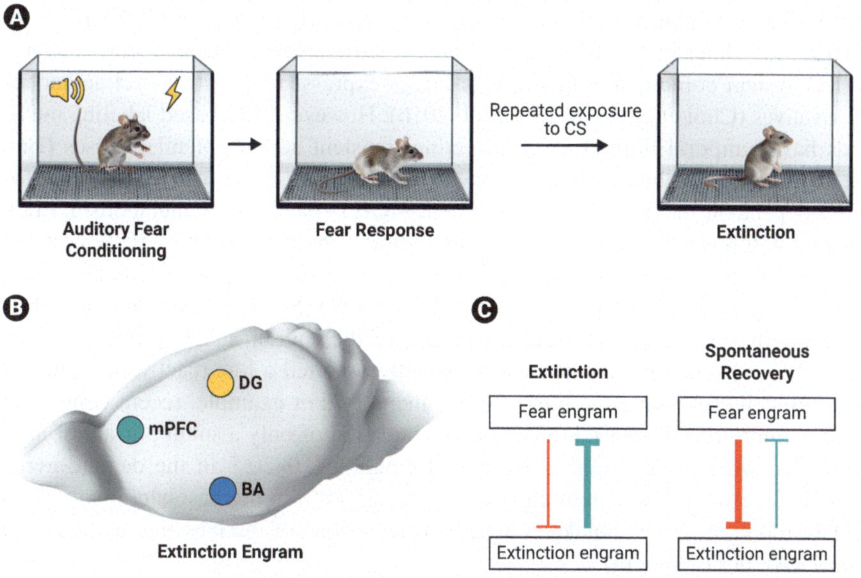


**Fig. 1** Extinction of fear memory. (**a**) The schematic illustration of auditory fear extinction. The rodents freeze after experiencing auditory fear conditioning. However, repeated exposure to the conditioned stimulus, such as tone, reduces the fear response. (**b**) Extinction engrams are known to exist in mPFC, DG, and BA. mPFC, medial prefrontal cortex; DG, dentate gyrus; BA, basal amygdala. (**c**) Fear engram and extinction engram reciprocally suppress each other. After fear extinction, the extinction engram inhibits fear engram. However, spontaneous recovery of fear memory occurs as the fear engram's oppression of the extinction engram becomes fortified

ensemble may mediate the updating of the information encoded by fear engram cells through synaptic remodeling (An et al. 2017; Choi et al. 2021; Teng et al. 2023). Together, the extinction process appears to encompass both the emergence of inhibitory extinction engram cells and the reconstruction of fear memory engram cells.

To summarize, engram cells are distributed throughout the brain, and this brain-wide distribution pattern of engram cells changes with time. Also, the reversal of valence that is associated with fear memory through extinction learning induces a new group of engram cells that inhibit previously formed fear memory engram cells. Altogether, engram cells function as a complicated network that gives rise to emergent properties, and further examination of the inter/intraregional synaptic connections of engram cells is needed.

**A Remaining Question on Cellular Engram: What Are the Dynamics of Cellular Engrams?**

To study engrams, it is essential to mark the cells that store memories of specific experiences and observe their reactivation ratios when recalling them. Several studies have used IEGs such as c-Fos and Arc to label engrams and analyze the reactivation ratio by immunohistochemistry after retrieval (Denny et al. 2014; Sorensen et al. 2016). For example, targeted recombination in the active population (TRAP) system is induced in an activity-dependent manner from Arc or c-Fos using CreERT2, in which recombination occurs by tamoxifen injection (DeNardo et al. 2019; Guenthner et al. 2013). In addition, a tetracycline-controlled transactivator (tTA) system combined with Arc or c-Fos is expressed using tetracycline and its derivatives (Choi et al. 2018; Roy et al. 2016). However, IEG-based labeling methods have temporal limitations in detecting transient action potential pulses (Han et al. 2022). The labeling time window is ambiguous because many drugs used for IEG-dependent induction (i.e., Tamoxifen, 4-OHT) have a slow metabolism, often lasting 4–6 h (DeNardo et al. 2019). In addition, the expression of IEGs requires several upstream signaling pathways following an intracellular calcium increase during neuronal activation (Kiessling and Gass 1993; Sheng and Greenberg 1990).

Partially due to the technical limitations of the engram labeling method mentioned above, engram cell studies consistently reported a low reactivation ratio of engram cells following the retrieval of a memory. For example, recent contextual fear memory recall induced the neuronal activation of only a small portion of neurons that were marked during memory formation (10–20% in the dentate gyrus (DeNardo et al. 2019; Kitamura et al. 2017), 20–40% in the basolateral amygdala (DeNardo et al. 2019; Tanaka et al. 2014)) regardless of the labeling method (i.e., the TRAP or tTa system).

At a glance, it seems paradoxical because engram cells are necessary and sufficient for memory recall. One possible explanation would be that the characteristic temporal dynamics of IEGs (i.e., c-Fos, Arc) are relatively slow compared to neuronal activity, resulting in imprecise labeling of neuronal ensembles that are activated during memory formation. Another possibility is that the size of the engram cells ensemble naturally tends to shrink with time. In-vivo imaging of neuronal ensembles in the retrosplenial cortex (RSC) showed that only 4% of neurons consistently expressed Arc upon the repeated recall of a contextual fear memory. In contrast, almost none of the neurons in the CA1 region stably expressed Arc (Lee et al. 2022). This result may not be restricted to the expression of Arc, which was previously shown to be directly related to synaptic plasticity, because imaging of the in-vivo dynamics of calcium influx as a readout for neuronal activity also revealed similar trends of a high turnover rate of active neuronal ensembles in the basolateral amygdala (BLA) (Grewe et al. 2017). Further studies are needed to understand how engram cells maintain the integrity of memory despite the high turnover of this component.

### The Rationale of Synaptic Engrams

Memory allocation is caused by neuronal excitability and two populations of engram cells compete to link or separate distinct events (Rashid et al. 2016). Engram cells form neuronal networks in various regions of the central nervous system and are involved in memory formation, retrieval, and systems consolidation (Kitamura et al. 2017; Long and Kahana 2015). In other words, the engram cells encode memory by interacting through synaptic connections. Morris et al. proposed the synaptic plasticity and memory (SPM) theory, which states that changes in the strength of connections between neurons store memory (Martin et al. 2000). The SPM theory suggests that activity-dependent synaptic plasticity is a substrate for memory and is both necessary and sufficient for storing information.

## Study of Synaptic Engrams

### Technologies for Studying Synaptic Engrams

Morris et al. proposed four assessment criteria to evaluate the SPM hypothesis (Martin et al. 2000; Martin and Morris 2002). According to these criteria, detectability, mimicry, anterograde alteration, and retrograde alteration must be satisfied. If an animal encodes the behavioral memory of an experience, it should be accompanied by detectable changes in synaptic efficacy. A mimicry criterion is required to verify whether synaptic potentiation is sufficient for memory formation. Under these conditions, artificial strengthening of synaptic function induces behavioral changes in animals. While mimicry studies have focused on the sufficiency of synaptic potentiation for memory encoding, the alteration of synaptic plasticity is related to the necessity of synaptic plasticity in memory traces. Anterograde alteration states that the inhibition of synaptic function during learning interrupts the memory of the experience. Unlike inhibition, retrograde alteration suggests that interference with the pattern of synaptic weights induced by learning causes memory impairment.

The first criterion, detectability, can be proven by research on synaptic properties such as the AMPA/NMDA ratio and AMPAR incorporation (Diering and Huganir 2018; Han et al. 2022; Makino and Malinow 2009). However, visualizing synaptic

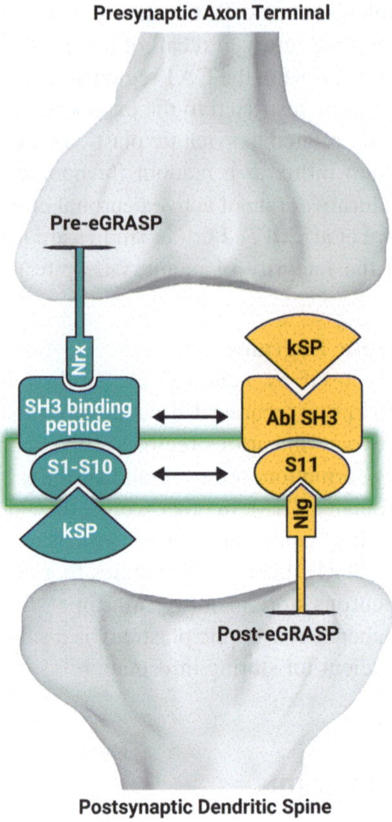


**Fig. 2** The constructs of the dual-eGRASP system. The dual-eGRASP system consists of two constructs: pre-eGRASP and post-eGRASP. Each construct includes four domains. The domain common to both constructs is IgG kappa signal peptide (kSP). kSP secretes eGRASP protein into the extracellular space. S1–10 from pre-eGRASP and S11 from post-eGRASP are split parts of GFP, emitting fluorescence only when reconstructing. SH3 binding peptide and Abl SH3 enable interaction between the two constructs and can modify the intensity of fluorescence by using variants of SH3 binding peptide. Neurexin (Nrx) and neuroligin (Nlg) allow the constructs to be located in the synaptic cleft and to anchor the transmembrane. (Based on Choi and Kaang 2022)

formation and elimination can more intuitively reveal the synaptic dynamics among memory traces. In 2008, green fluorescent protein reconstitution across synaptic partners (GRASP) was developed to materialize specific synapses in *Caenorhabditis elegans* (Feinberg et al. 2008) and was genetically modified for application in the mammalian brain, called mGRASP (Kim et al. 2012). The GRASP system consists of two components divided by a green fluorescent protein (GFP) that only expresses fluorescence at the synapse formation site. Recently, the dual-enhanced GFP reconstitution across synaptic partners (dual-eGRASP) technique, a multicolor variant of the mGRASP system, has evolved (Choi et al. 2018). The dual-eGRASP technique marks synapses between engram cells (E-E) and between engram and non-engram cells (E-N or N-E) in different colors to distinguish them. The dual-eGRASP system consists of two constructs (Choi and Kaang 2022) to separately label presynaptic and postsynaptic cells. The pre-eGRASP component includes four domains: IgG kappa signal peptide for secreting eGRASP protein into the extracellular space, s1–10 strands among 11 strands of GFP, SH3 binding peptides that interact with Abl-SH3 (p30, 32, 40), and neurexin lacking a neuroligin-binding domain for both targeting presynaptic axon terminals and transmembrane anchors. Post-eGRASP also consists of four domains: IgG kappa signal peptide, Abl-SH3 for interaction with pre-eGRASP, the s11 strand of GFP, and neuroligin lacking the neurexin-binding domain for postsynaptic positioning and transmembrane anchoring (Fig. 2).

**Alteration of Synaptic Connection and Spine Morphology Within a Synaptic Engram**

Recent studies have supported the idea that synaptic plasticity is the underlying mechanism of engram formation. By applying the dual-eGRASP technique to the hippocampal CA3 to CA1 circuit, Choi et al. demonstrated how synaptic engrams were specifically enhanced after fear memory formation (Choi et al. 2018). The synapses between CA3 and CA1 engram cells (E-E) exhibited significantly higher relative spine density and strengthened spine morphology, such as head volume and diameter. Moreover, this interregional synaptic connectivity was directly correlated with memory strength. As the researchers increased the intensity of footshock during fear conditioning, both spine density and morphology were enhanced only in E-E synapses (Fig. 3). While these data illustrated the increased structural connectivity of the synaptic engram, synaptic strengths were examined further in hippocampal slices. Interestingly, presynaptic inputs from CA3 engram neurons showed significantly increased release probability, as the paired-pulse-ratio (PPR) was particularly decreased in E-E synapses. A significant increase in postsynaptic AMPA receptors measured in evoked miniature excitatory postsynaptic currents (mEPSCs) also indicated a strong potentiation of CA1 engram neurons due to memory formation. Furthermore, the long-term potentiation of synapses between engrams during memory formation was proved by their LTP occlusion during pairing LTP. These overall findings strongly supported the idea that synaptic engrams in neural circuits could be the substrate for memory encoding.

Additional studies have revealed how these neural substrates would be altered according to memory extinction. How synapses were modified in the auditory cortex (AC) to the lateral amygdala (LA) circuit after auditory fear conditioning, extinction, or re-conditioning was investigated with dual-eGRASP (Choi et al. 2021) (Fig. 4a, b). The enhanced spine morphology of E-E synapses between AC and LA was observed in accordance with a previous study (Choi et al. 2018) after auditory fear conditioning. Such enhancement of E-E synapses was selectively reversed by 3 days of extinction training, during which fear memory is assumed to be extinguished as mice are repeatedly exposed to a conditioned auditory cue without any footshock. Intriguingly, re-conditioning of the extinction group restored the fear memory and increased the spine size of E-E synapses. Thus, the altered spine morphology of synaptic engrams according to different states of associative fear memory implicated their role in encoding specific memories (Fig. 4c).

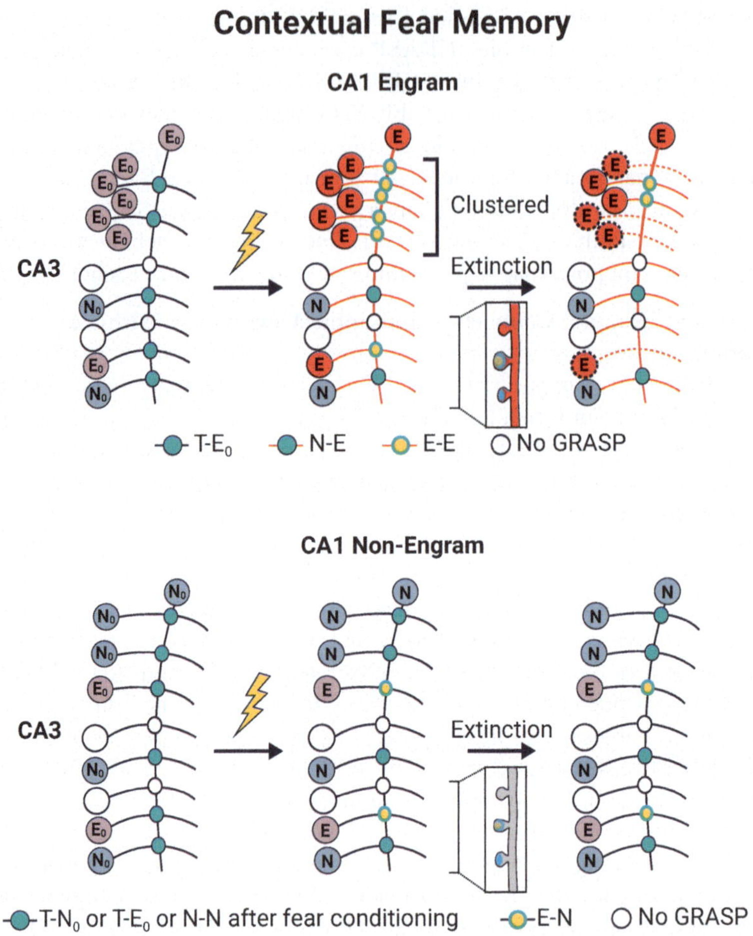


**Fig. 3** Synaptic plasticity during contextual fear memory formation and extinction. The schematic illustration of synaptic dynamics of a contextual fear memory trace in the hippocampal CA3 to CA1 circuit. The relative spine density of engram to engram connection (E-E) increased after fear conditioning. Moreover, spine head volume and diameter are also enhanced during learning. The E-E synapses are clustered by synaptogenesis after contextual fear conditioning (CFC). After contextual fear extinction, some proportion of the E-E connections specifically disappeared whereas almost none of the E-N connections were affected by extinction. (Based on Choi et al. 2018; Lee et al. 2023a)

How these synaptic connections are altered in vivo in the dorsal hippocampus was elucidated by a study that combined in vivo two-photon imaging with the dual-eGRASP technique (Lee et al. 2023a). This approach enabled the comparison of identical synapses in multiple memory states, such as before and after memory formation or after memory extinction (Fig. 3). Fear conditioning induced the formation of new synapses, primarily in the connections between the CA3 engram and CA1 engram neurons, in which about 20% of the analyzed E-E synapses were newly

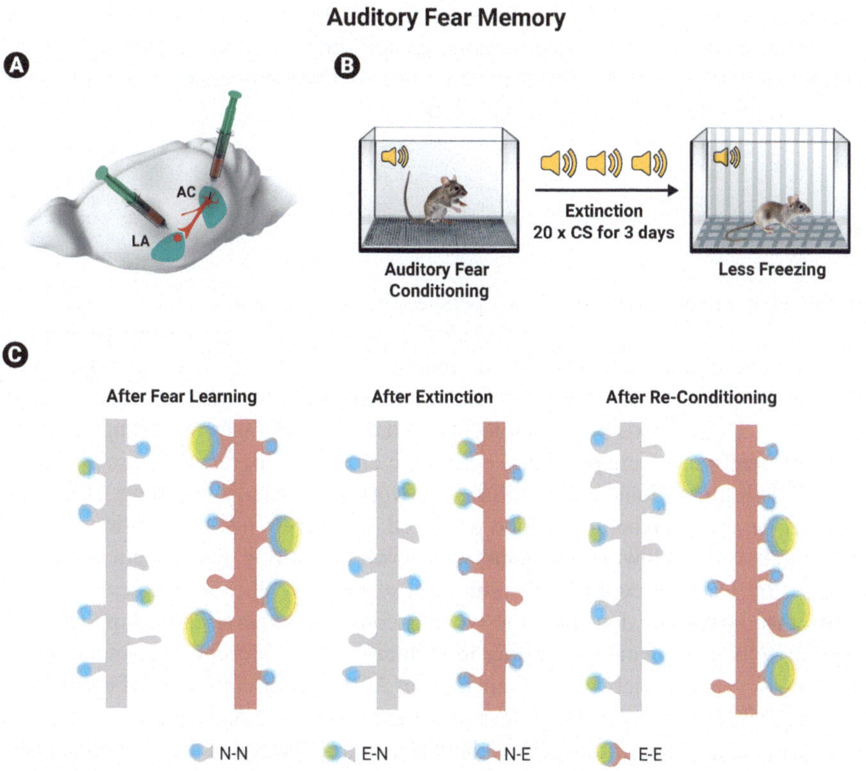


**Fig. 4** Alteration of spine morphology after auditory fear memory formation and extinction. (**a**) An adeno-associated virus (AAV) expressing the dual-eGRASP system is injected into the auditory cortex (AC) and lateral amygdala (LA). (**b**) Schematic illustration of auditory fear extinction. (**c**) The alteration of spine morphology of AC to LA circuits is correlated with auditory fear memory traces. After fear learning, spine head diameter, head volume, and spine volume are increased in engram-to-engram connections (E-E) but not in the nonengram-to-engram connections (N-E). Fear extinction reduces the fear-learning-induced strengthening of spine morphology in E-E synapses. Furthermore, re-conditioning after the extinction restores the extinction-induced weakening of spine morphology in E-E connections

generated. Surprisingly, the CA1 dendrites that developed into engrams after memory formation exhibited lower spine density. Such data implied that neurons with a higher capacity for synaptogenesis could evolve into engrams. Memory attenuation eliminated pre-existing synapses, specifically E-E synapses, while almost none of the connections with non-engram neurons were affected. Since the synaptic dynamics of mice that only underwent repeated context exposure were intact, the modification of E-E synapses was highly correlated to the state of their encoding memory. Moreover, the particular clustering pattern of E-E synapses between CA3 and CA1 was also revealed from the measured distance between each type of synapse. New synaptic engrams were closely formed in dendritic areas with relatively lower spine density, which indicated higher capacity. While pre-existing E-E synapses were evenly dispersed across CA1 dendrites, adding new E-E synapses nearby resulted in the clustering of synaptic engrams. Thus, synaptic engrams could mediate clustered plasticity, which is well known to affect biophysical properties such as synaptic tagging or capturing (Govindarajan et al. 2006).

## Conclusion

In this chapter, we overviewed two perspectives of engrams: cellular engram and synaptic engram. Over the past decades, most engram studies were performed focusing on engram cells. The roles of engram cells have also been investigated in various memory stages: memory formation, retrieval, systems consolidation and extinction. They found specific molecular and cellular changes in the engram cells such as membrane excitability and epigenetic changes. Engram cells are allocated in various brain regions and establish a network for memory encoding. However, the IEG-dependent labeling method produced a relatively low reactivation ratio of the engram cells following the memory retrieval. In addition, the synaptic plasticity induced by learning could be the substrate of memory encoding. Recent studies reveal that the relative density of the synaptic connections between the hippocampal engram cells may change according to memory stages. Moreover, the spine morphology of synapses between engram cells of AC to LA circuits showed correlative structural changes after memory formation and extinction. As proposed by Hebb's theory, connections are strengthened between engram cells that are simultaneously stimulated by experience. However, it requires more studies in the future to understand how much the synaptic engram contributes structurally and functionally to the formation of memory traces.

**Acknowledgments** This work was supported by the Institute for Basic Science (IBS-R001-D3).

## References

Abbott LF, Nelson SB (2000) Synaptic plasticity: taming the beast. Nat Neurosci 3(Suppl):1178–1183. https://doi.org/10.1038/81453

Abdou K, Shehata M, Choko K, Nishizono H, Matsuo M, Muramatsu SI, Inokuchi K (2018) Synapse-specific representation of the identity of overlapping memory engrams. Science 360:1227–1231. https://doi.org/10.1126/science.aat3810

An B, Kim J, Park K, Lee S, Song S, Choi S (2017) Amount of fear extinction changes its underlying mechanisms. elife 6. https://doi.org/10.7554/eLife.25224

Andersen N, Krauth N, Nabavi S (2017) Hebbian plasticity in vivo: relevance and induction. Curr Opin Neurobiol 45:188–192. https://doi.org/10.1016/j.conb.2017.06.001

Bernabeu R, Bevilaqua L, Ardenghi P, Bromberg E, Schmitz P, Bianchin M, Izquierdo I, Medina JH (1997) Involvement of hippocampal cAMP/cAMP-dependent protein kinase signaling pathways in a late memory consolidation phase of aversively motivated learning in rats. Proc Natl Acad Sci USA 94:7041–7046. https://doi.org/10.1073/pnas.94.13.7041

Bozon B, Kelly A, Josselyn SA, Silva AJ, Davis S, Laroche S (2003) MAPK, CREB and zif268 are all required for the consolidation of recognition memory. Philos T R Soc B 358:805–814. https://doi.org/10.1098/rstb.2002.1224

Choi DI, Kaang BK (2022) Interrogating structural plasticity among synaptic engrams. Curr Opin Neurobiol 75:102552. https://doi.org/10.1016/j.conb.2022.102552

Choi JH, Sim SE, Kim JI, Choi DI, Oh J, Ye S, Lee J, Kim T, Ko HG, Lim CS, Kaang BK (2018) Interregional synaptic maps among engram cells underlie memory formation. Science 360:430–435. https://doi.org/10.1126/science.aas9204

Choi DI, Kim J, Lee H, Kim JI, Sung Y, Choi JE, Venkat SJ, Park P, Jung H, Kaang BK (2021) Synaptic correlates of associative fear memory in the lateral amygdala. Neuron 109:2717–2726 e2713. https://doi.org/10.1016/j.neuron.2021.07.003

Ciocchi S, Herry C, Grenier F, Wolff SBE, Letzkus JJ, Vlachos I, Ehrlich I, Sprengel R, Deisseroth K, Stadler MB et al (2010) Encoding of conditioned fear in central amygdala inhibitory circuits. Nature 468:277–U239. https://doi.org/10.1038/nature09559

De Cesare D, Jacquot S, Hanauer A, Sassone-Corsi P (1998) Rsk-2 activity is necessary for epidermal growth factor-induced phosphorylation of CREB protein and transcription of c-fos gene. Proc Natl Acad Sci USA 95:12202–12207. https://doi.org/10.1073/pnas.95.21.12202

de Sousa AF, Cowansage KK, Zutshi I, Cardozo LM, Yoo EJ, Leutgeb S, Mayford M (2019) Optogenetic reactivation of memory ensembles in the retrosplenial cortex induces systems consolidation. Proc Natl Acad Sci USA 116:8576–8581. https://doi.org/10.1073/pnas.1818432116

DeNardo LA, Liu CD, Allen WE, Adams EL, Friedmann D, Fu L, Guenthner CJ, Tessier-Lavigne M, Luo L (2019) Temporal evolution of cortical ensembles promoting remote memory retrieval. Nat Neurosci 22:460–469. https://doi.org/10.1038/s41593-018-0318-7

Denny CA, Kheirbek MA, Alba EL, Tanaka KF, Brachman RA, Laughman KB, Tomm NK, Turi GF, Losonczy A, Hen R (2014) Hippocampal memory traces are differentially modulated by experience, time, and adult neurogenesis. Neuron 83:189–201. https://doi.org/10.1016/j.neuron.2014.05.018

Diering GH, Huganir RL (2018) The AMPA receptor code of synaptic plasticity. Neuron 100:314–329. https://doi.org/10.1016/j.neuron.2018.10.018

Feinberg EH, VanHoven MK, Bendesky A, Wang G, Fetter RD, Shen K, Bargmannl CI (2008) GFP reconstitution across synaptic partners (GRASP) defines cell contacts and synapses in living nervous systems. Neuron 57:353–363. https://doi.org/10.1016/j.neuron.2007.11.030

Govindarajan A, Kelleher RJ, Tonegawa S (2006) A clustered plasticity model of long-term memory engrams. Nat Rev Neurosci 7:575–583. https://doi.org/10.1038/nrn1937

Grewe BF, Grundemann J, Kitch LJ, Lecoq JA, Parker JG, Marshall JD, Larkin MC, Jercog PE, Grenier F, Li JZ et al (2017) Neural ensemble dynamics underlying a long-term associative memory. Nature 543:670–675. https://doi.org/10.1038/nature21682

Guenthner CJ, Miyamichi K, Yang HH, Heller HC, Luo L (2013) Permanent genetic access to transiently active neurons via TRAP: targeted recombination in active populations. Neuron 78:773–784. https://doi.org/10.1016/j.neuron.2013.03.025

Han JH, Kushner SA, Yiu AP, Hsiang HL, Buch T, Waisman A, Bontempi B, Neve RL, Frankland PW, Josselyn SA (2009) Selective erasure of a fear memory. Science 323:1492–1496. https://doi.org/10.1126/science.1164139

Han DH, Park P, Choi DI, Bliss TVP, Kaang BK (2022) The essence of the engram: cellular or synaptic? Semin Cell Dev Biol 125:122–135. https://doi.org/10.1016/j.semcdb.2021.05.033

Hass CA, Glickfeld LL (2016) High-fidelity optical excitation of cortico-cortical projections at physiological frequencies. J Neurophysiol 116:2056–2066. https://doi.org/10.1152/jn.00456.2016

Hebb DO (1949) The organization of behavior; a neuropsychological theory. Wiley

Impey S, Mark M, Villacres EC, Poser S, Chavkin C, Storm DR (1996) Induction of CRE-mediated gene expression by stimuli that generate long-lasting LTP in area CA1 of the hippocampus. Neuron 16:973–982. https://doi.org/10.1016/S0896-6273(00)80120-8

Jeong Y, Cho HY, Kim M, Oh JP, Kang MS, Yoo M, Lee HS, Han JH (2021) Synaptic plasticity-dependent competition rule influences memory formation. Nat Commun 12:3915. https://doi.org/10.1038/s41467-021-24269-4

Josselyn SA, Tonegawa S (2020) Memory engrams: recalling the past and imagining the future. Science 367. https://doi.org/10.1126/science.aaw4325

Kawashima T, Okuno H, Nonaka M, Adachi-Morishima A, Kyo N, Okamura M, Takemoto-Kimura S, Worley PF, Bito H (2009) Synaptic activity-responsive element in the Arc/Arg3.1 promoter essential for synapse-to-nucleus signaling in activated neurons. Proc Natl Acad Sci USA 106:316–321. https://doi.org/10.1073/pnas.0806518106

Kiessling M, Gass P (1993) Immediate early gene expression in experimental epilepsy. Brain Pathol 3:381–393. https://doi.org/10.1111/j.1750-3639.1993.tb00766.x

Kim WB, Cho JH (2017) Encoding of discriminative fear memory by input-specific LTP in the amygdala. Neuron 95:1129–1146 e1125. https://doi.org/10.1016/j.neuron.2017.08.004

Kim WB, Cho JH (2020) Encoding of contextual fear memory in hippocampal-amygdala circuit. Nat Commun 11:1382. https://doi.org/10.1038/s41467-020-15121-2

Kim J, Zhao T, Petralia RS, Yu Y, Peng HC, Myers E, Magee JC (2012) mGRASP enables mapping mammalian synaptic connectivity with light microscopy. Nat Methods 9:96–U139. https://doi.org/10.1038/Nmeth.1784

Kitamura T, Ogawa SK, Roy DS, Okuyama T, Morrissey MD, Smith LM, Redondo RL, Tonegawa S (2017) Engrams and circuits crucial for systems consolidation of a memory. Science 356:73–78. https://doi.org/10.1126/science.aam6808

Lacagnina AF, Brockway ET, Crovetti CR, Shue F, McCarty MJ, Sattler KP, Lim SC, Santos SL, Denny CA, Drew MR (2019) Distinct hippocampal engrams control extinction and relapse of fear memory. Nat Neurosci 22:753–761. https://doi.org/10.1038/s41593-019-0361-z

Lee BH, Shim JY, Moon HC, Kim DW, Kim J, Yook JS, Kim J, Park HY (2022) Real-time visualization of mRNA synthesis during memory formation in live mice. Proc Natl Acad Sci USA 119:e2117076119. https://doi.org/10.1073/pnas.2117076119

Lee C, Lee BH, Jung H, Lee C, Sung Y, Kim H, Kim J, Shim JY, Kim JI, Choi DI et al (2023a) Hippocampal engram networks for fear memory recruit new synapses and modify pre-existing synapses in vivo. Curr Biol 33:507–516 e503. https://doi.org/10.1016/j.cub.2022.12.038

Lee JH, Kim WB, Park EH, Cho JH (2023b) Neocortical synaptic engrams for remote contextual memories. Nat Neurosci 26:259–273. https://doi.org/10.1038/s41593-022-01223-1

Liu X, Ramirez S, Pang PT, Puryear CB, Govindarajan A, Deisseroth K, Tonegawa S (2012) Optogenetic stimulation of a hippocampal engram activates fear memory recall. Nature 484:381–385. https://doi.org/10.1038/nature11028

Long NM, Kahana MJ (2015) Successful memory formation is driven by contextual encoding in the core memory network. NeuroImage 119:332–337. https://doi.org/10.1016/j.neuroimage.2015.06.073

Magee JC, Johnston D (1997) A synaptically controlled, associative signal for Hebbian plasticity in hippocampal neurons. Science 275:209–213. https://doi.org/10.1126/science.275.5297.209

Makino H, Malinow R (2009) AMPA receptor incorporation into synapses during LTP: the role of lateral movement and exocytosis. Neuron 64:381–390. https://doi.org/10.1016/j.neuron.2009.08.035

Martin SJ, Morris RG (2002) New life in an old idea: the synaptic plasticity and memory hypothesis revisited. Hippocampus 12:609–636. https://doi.org/10.1002/hipo.10107

Martin SJ, Grimwood PD, Morris RG (2000) Synaptic plasticity and memory: an evaluation of the hypothesis. Annu Rev Neurosci 23:649–711. https://doi.org/10.1146/annurev.neuro.23.1.649

Miyashita T, Kikuchi E, Horiuchi J, Saitoe M (2018) Long-term memory engram cells are established by c-Fos/CREB transcriptional cycling. Cell Rep 25:2716–2728 e2713. https://doi.org/10.1016/j.celrep.2018.11.022

Nabavi S, Fox R, Proulx CD, Lin JY, Tsien RY, Malinow R (2014) Engineering a memory with LTD and LTP. Nature 511:348–352. https://doi.org/10.1038/nature13294

Park S, Kramer EE, Mercaldo V, Rashid AJ, Insel N, Frankland PW, Josselyn SA (2016) Neuronal allocation to a hippocampal engram. Neuropsychopharmacology 41:2987–2993. https://doi.org/10.1038/npp.2016.73

Park A, Jacob AD, Walters BJ, Park S, Rashid AJ, Jung JH, Lau J, Woolley GA, Frankland PW, Josselyn SA (2020) A time-dependent role for the transcription factor CREB in neuronal allocation to an engram underlying a fear memory revealed using a novel in vivo optogenetic tool to modulate CREB function. Neuropsychopharmacology 45:916–924. https://doi.org/10.1038/s41386-019-0588-0

Rashid AJ, Yan C, Mercaldo V, Hsiang HLL, Park S, Cole CJ, De Cristofaro A, Yu J, Ramakrishnan C, Lee SY et al (2016) Competition between engrams influences fear memory formation and recall. Science 353:383–387. https://doi.org/10.1126/science.aaf0594

Rioult-Pedotti MS, Friedman D, Donoghue JP (2000) Learning-induced LTP in neocortex. Science 290:533–536. https://doi.org/10.1126/science.290.5491.533

Roy DS, Arons A, Mitchell TI, Pignatelli M, Ryan TJ, Tonegawa S (2016) Memory retrieval by activating engram cells in mouse models of early Alzheimer's disease. Nature 531:508–512. https://doi.org/10.1038/nature17172

Roy DS, Park YG, Kim ME, Zhang Y, Ogawa SK, DiNapoli N, Gu XY, Cho JH, Choi H, Kamentsky L et al (2022) Brain-wide mapping reveals that engrams for a single memory are distributed across multiple brain regions. Nat Commun 13:ARTN 1799. https://doi.org/10.1038/s41467-022-29384-4

Ryan TJ, Roy DS, Pignatelli M, Arons A, Tonegawa S (2015) Engram cells retain memory under retrograde amnesia. Science 348:1007–1013. https://doi.org/10.1126/science.aaa5542

Semon R (1904) Die Mneme : als erhaltendes Prinzip im Wechsel des organischen Geschehens (Engelmann)

Sheng M, Greenberg ME (1990) The regulation and function of c-fos and other immediate early genes in the nervous system. Neuron 4:477–485. https://doi.org/10.1016/0896-6273(90)90106-p

Sheng M, Thompson MA, Greenberg ME (1991) Creb - a Ca2+-regulated transcription factor phosphorylated by Calmodulin-dependent kinases. Science 252:1427–1430. https://doi.org/10.1126/science.1646483

Smetters D, Majewska A, Yuste R (1999) Detecting action potentials in neuronal populations with calcium imaging. Methods 18:215–221. https://doi.org/10.1006/meth.1999.0774

Sorensen AT, Cooper YA, Baratta MV, Weng FJ, Zhang Y, Ramamoorthi K, Fropf R, LaVerriere E, Xue J, Young A et al (2016) A robust activity marking system for exploring active neuronal ensembles. elife 5. https://doi.org/10.7554/eLife.13918

Sun XC, Bernstein MJ, Meng MZ, Rao SY, Sorensen AT, Yao L, Zhang XH, Anikeeva PO, Lin YX (2020) Functionally distinct neuronal ensembles within the memory engram. Cell 181:410-+. https://doi.org/10.1016/j.cell.2020.02.055

Tanaka KZ, Pevzner A, Hamidi AB, Nakazawa Y, Graham J, Wiltgen BJ (2014) Cortical representations are reinstated by the hippocampus during memory retrieval. Neuron 84:347–354. https://doi.org/10.1016/j.neuron.2014.09.037

Teng SW, Wang XR, Du BW, Chen XL, Fu GZ, Liu YF, Xu SQ, Shuai JC, Chen ZY (2023) Altered fear engram encoding underlying conditioned versus unconditioned stimulus-initiated memory updating. Sci Adv 9:eadf0284. https://doi.org/10.1126/sciadv.adf0284

Zhang X, Kim J, Tonegawa S (2020) Amygdala reward neurons form and store fear extinction memory. Neuron 105(1077–1093):e1077. https://doi.org/10.1016/j.neuron.2019.12.025

Zhou Y, Won J, Karlsson MG, Zhou M, Rogerson T, Balaji J, Neve R, Poirazi P, Silva AJ (2009) CREB regulates excitability and the allocation of memory to subsets of neurons in the amygdala. Nat Neurosci 12:1438–1443. https://doi.org/10.1038/nn.2405

# Part IV
# Beyond Associative Learning: Engrams for Other Memory Processes

# Engrams of Fear Memory Attenuation

**Johannes Gräff**

**Abstract** Fear attenuation is an etiologically relevant process for animal survival, since once acquired information needs to be continuously updated in the face of changing environmental contingencies. Thus, when situations are encountered that were originally perceived as fearful but are no longer so, fear must be attenuated, otherwise, it risks becoming maladaptive. But what happens to the original memory trace of fear during fear attenuation? In this chapter, we review the studies that have started to approach this question from an engram perspective. We find evidence pointing to both the original memory trace of fear being suppressed, as well as it being updated towards safety. These seemingly conflicting results reflect a well-established dichotomy in the field of fear memory attenuation, namely whether fear attenuation is mediated by an inhibitory mechanism that suppresses fear expression, called extinction, or by an updating mechanism that allows the fear memory to reconsolidate in a different form, called reconsolidation-updating. Which of these scenarios takes the upper hand is ultimately influenced by the behavioral paradigms used to induce fear attenuation, but is an important area for further study as the precise cell populations underlying fear attenuation and the molecular mechanisms therein can now be understood at unprecedented resolution.

**Keywords** Attenuation · Extinction · Reconsolidation-updating · Trace · Persistence

---

J. Gräff (✉)
Brain Mind Institute, School of Life Sciences, EPFL, Lausanne, Switzerland
e-mail: johannes.graeff@epfl.ch



J. Gräff, S. Ramirez (eds.), *Engrams*, Advances in Neurobiology 38,
https://doi.org/10.1007/978-3-031-62983-9_9

## Introduction

Although memories with a strong emotional valence, such as those associated with fear or traumatic experiences, are potentially very long-lasting, they are nevertheless not written in stone and can be attenuated by different means. One such means is termed extinction learning (see Glossary), during which a once fearful stimulus (in the case of an associative conditioning paradigm, the conditioned stimulus, CS) is repeatedly presented in a non-fearful environment, which eventually leads to the attenuation of the fear response (Fig. 1a). In theory, extinction learning can be mediated by two processes in the brain. On the one hand, extinction learning can create a new memory trace of safety that competes with and suppresses the original memory trace of fear. This idea originated with Ivan Pavlov (1849–1936), who saw himself "... forced to the conclusion that experimental extinction is based on inhibition" (Pavlov 1927), and has since been met with experimental evidence, for example, from a neurocircuit perspective showing that during extinction learning, different neuronal subpopulations are active than during fear recall (Herry et al. 2008). On the other hand, extinction learning can lead to an updating of the original memory trace of fear, which dates back to two seminal studies in rodents and humans by the laboratories of Joe LeDoux and Elizabeth Phelps, respectively, who showed that when applied during the reconsolidation window upon recalling a fearful memory, extinction training leads a permanent attenuation of the fear response, as evidenced by a complete absence of spontaneous recovery of fear, its reinstatement or its renewal (Monfils et al. 2009; Schiller et al. 2010).

Another means to attenuate fear is termed reconsolidation blockade. Each time a memory is recalled, it becomes temporally vulnerable to change before it reconsolidates (Alberini 2011; Nader and Hardt 2009; Tronson and Taylor 2007). Such malleability is an integral component for lifelong learning, as only that way can memories adapt to changing environmental contingencies: Memory malleability helps the memory to be either strengthened—when a higher valence is encountered at recall, maintained—when similar valences are present at learning and recall, or weakened—when a lower valence is encountered at recall. To reduce fear memories, this memory malleability can be capitalized upon by so-called reconsolidation blockers such as the β-adrenergic receptor antagonist propranolol (Beckers and Kindt 2017), which aims at preventing the reinforcement of the fear memory by blocking fear memory recall.

Whether and the extent to which these scenarios—reconsolidation-updating, extinction or reconsolidation blockade—are at play at the level of individual cells has only recently started to be investigated. Given its conceptual framework—a query about the identity and interaction of cellular ensembles—this question is ideally suited to be investigated with engram tagging technologies. Engram tagging relies on either a reporter gene or a chemo-/opto-genetic effector gene coupled to an inducible, neuronal activity-dependent promoter such as an immediate early gene (IEG), the expression of which can be temporally controlled by providing the chemical inducer during a restricted period in time. This way, fear encoding, fear recall

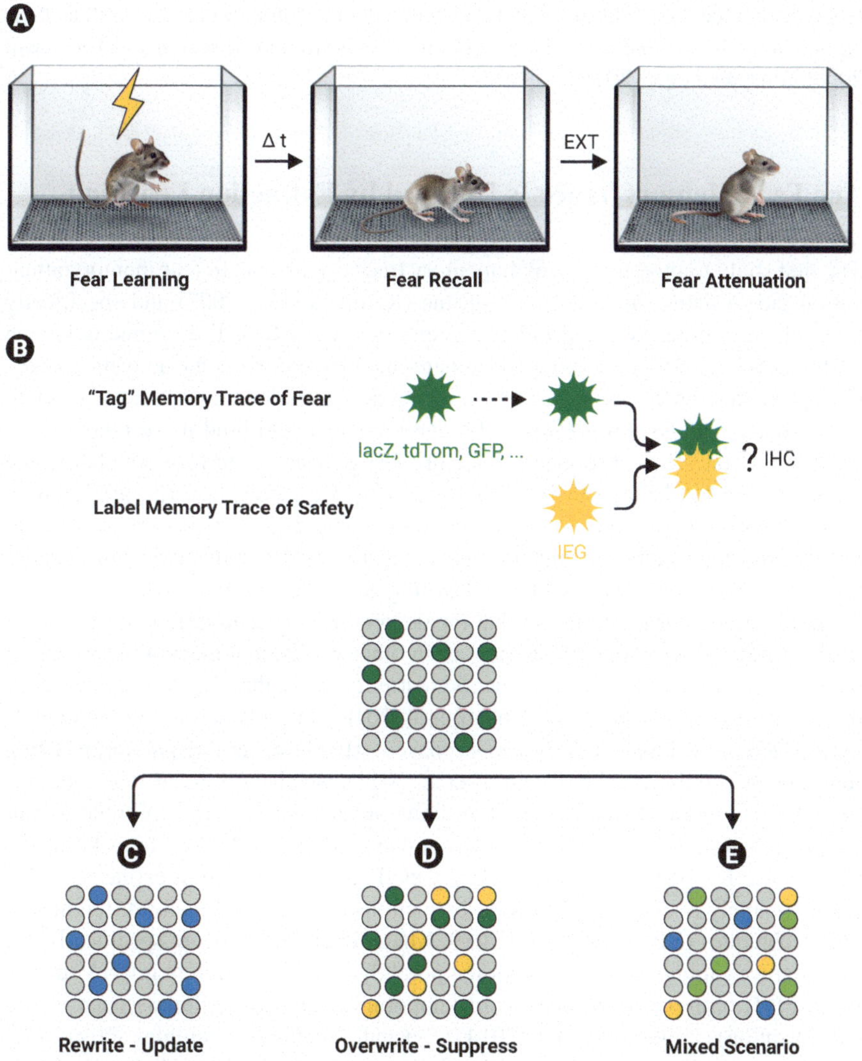


**Fig. 1** Fear memory attenuation: Learning anew or new learning? (**a**) Schematic representation of behavioral fear attenuation, which is brought about by extinction learning (EXT) following memory recall. (**b**) In order to identify the cellular populations responsible for fear memory attenuation, the memory trace of fear (symbolized by green "cells") can be identified and tagged by an exogenous tagging system (such as lacZ, tdTom, or GFP) regulated by neuronal activity-responsive promoters, which can then be studied for its reactivation following fear attenuation by endogenously occurring immediate early gene (IEG) expression (symbolized by red "cells"), such as cFos, Homer 1a and others. (**c**) In reconsolidation-updating, the original memory trace of fear is reactivated after fear attenuation (symbolized by blue "cells"). The key to successful fear attenuation lies within the original memory trace of fear. (**d**) In extinction, the original memory trace of fear is suppressed by the creation of a new memory trace of safety (yellow "cells"). The key to successful fear attenuation lies within a newly learned memory trace. (**e**) In the most parsimonious interpretation of the data available so far, fear attenuation is mediated by both updating and suppression mechanisms

or extinction learning engrams can be visualized and/or manipulated as well as their co-activation by the endogenous expression of IEGs studied (Josselyn and Tonegawa 2020; Tonegawa et al. 2015).

## The Fear Memory Trace Is Updated by Extinction Learning

The first study to investigate which memory traces contribute to fear memory attenuation did so using the Tettag mouse line (Reijmers et al. 2007) and specifically focused on remote, i.e., 28d-old fear memories, for which little remains known (Albo and Gräff 2018). In Tettag mice, neuronal activity drives the activation of the cFos promoter, which is coupled to a tetracycline-controlled transactivator (tTA). In the absence of Doxyclince (Dox), tTA can recognize and bind its responsive element, the tetracycline responsive element (TRE), which is in turn coupled to the reporter gene lacZ that can be visualized with β-galactosidase (β-gal) staining. Using this system, neurons activated by remote contextual fear memory recall were tagged with β-gal, and subsequently analyzed for their reactivation with endogenous cFos expression after extinction learning had occurred (Fig. 1b).

In this study, our group found that the degree of fear memory reactivation positively correlated with fear attenuation, or in other words, that the amount of recall-induced cells reactivated after extinction learning was higher the more pronounced the attenuation of the fear was (Khalaf et al. 2018). Initially studied in the dentate gyrus (DG) of the hippocampus using a spaced extinction learning paradigm lasting four days (Costanzi et al. 2011; Gräff et al. 2014), we also observed such reactivation of recall-induced engrams in other brain areas, namely the infralimbic part of the prefrontal cortex (IL) and the basolateral amygdala (BLA), key areas known to be activated by extinction learning (Herry et al. 2010), in different extinction paradigms (namely a spaced and massed extinction training), and using different tagging techniques (Khalaf and Gräff 2019; Khalaf et al. 2018). More specifically, for the latter we capitalized on the spatiotemporal dynamics of different mRNA species by catFISH (compartment analysis of temporal activity using fluorescence in situ hybridization) (Zelikowsky et al. 2014) to visualize cellular ensembles activated by remote memory recall—evidenced by the transcript of the immediate early gene Homer1a—and by massed extinction learning—consisting of six consecutive 3 min epochs of CS exposure (Monfils et al. 2009)—which was evidenced by the nuclear abundance of cFos mRNA species. Also here, the extent of reactivation of recall-induced neurons following extinction training correlated positively with the degree of fear attenuation.

What about the functional relevance of such reactivation for fear attenuation? For this, we bidirectionally altered the activity of recall-induced neurons during the extinction paradigm using chemogenetics. When remote fear recall-induced neurons were chemogenetically activated using the activatory DREADD (designer receptor exclusively activated by designer drug) hM3Dq (Armbruster et al. 2007), recall-induced neurons were preferentially reactivated following extinction learning and fear attenuation was improved; conversely, when remote fear recall-induced neurons were chemogenetically silenced using the Daun02 inactivation method (Koya et al. 2009), the reactivation of recall-induced neurons was reduced and fear attenuation prevented (Khalaf et al. 2018). This at first sight paradoxical finding can be readily explained by reconsolidation-updating mechanisms: The key for successful attenuation of remote fear memories therefore lies in the same cells that were used to store the memory in the first place, i.e., in an updating of the original memory trace of fear towards safety (Frankland and Josselyn 2018).

Importantly, these findings, i.e., the evidence of reconsolidation-updating at the engram level (Fig. 1c), have since been replicated. Directly, insofar as optogenetic and chemogenetic stimulation of the original hippocampal fear engram facilitated recent fear attenuation (Cincotta et al. 2021; Zhang et al. 2021); and indirectly, because in extinction protocols that failed to permanently alter fear memory expression, part of the original fear engram continued to exist (Cincotta et al. 2021; Yoshii et al. 2017), which can be explained by the original memory trace of fear having been merely inhibited, rather than permanently changed.

Taking manipulating the original fear engram one step further, a recent study investigated whether fear attenuation can be artificially achieved without extinction learning, i.e., by solely altering the activity of fear memory engram cells in the absence of any behavioral intervention. To this end, the authors subjected mice to auditory fear conditioning in context A, and then repeatedly reactivated the original memory engram of fear (tagged during conditioning) by the optogenetic activator channelrhodopsin (ChR2) in context B, before re-assessing the animals' memory performance to a tone exposure in context A again. Optogenetically reactivating the LA fear engram per se, i.e., without tone re-exposure in context B, efficiently reduced the animals' fear response, even when subsequently tested for fear renewal (Park et al. 2022). Thus, repeated optogenetic reactivation of the original fear engram was sufficient to lead to behavioral fear attenuation.

Beyond fear attenuation, similar engram reactivation mechanisms might also be at play for drug-induced memories. For example, a recent study found that optogenetic activation of a locus coeruleus (LC) to DG engram pathway was beneficial for the attenuation of morphine-related memories: Activation of the LC-to-DG pathway prior to drug extinction training significantly facilitated drug memory attenuation (Dai et al. 2023), again speaking to the importance of recalling the (drug) memory engram prior to extinction learning.

These studies highlight that the key to unlocking fear attenuation lies within the original population of fear-bearing neurons. This notion is doubly important. First, for the process of reconsolidation-updating, which has now been met with experimental evidence at the level of memory engrams. And second, for a long-standing tenet in psychotherapy, which had only been empirically observed, but which has now found its physiological underpinning at the engram level, namely the importance of recalling a traumatic memory for it to be changed (Foa and Kozak 1986).

## The Fear Memory Trace Is Suppressed by Extinction Learning

These findings notwithstanding, evidence at the level of engram cells also shows that fear attenuation is mediated by a new memory trace of safety that suppresses the original memory trace of fear. The first study alluding to this fact used a ramification of the original Tettag mouse strain, which contains H2B-GFP as a reporter gene instead of lacZ, as well as a massed extinction paradigm to a cued fear memory. Upon fear memory attenuation, the original fear engram trace (H2B-GFP positive) in the BA was found to be less reactivated (Zif268 positive) compared to fear recall, and this reduced reactivation correlated with a perisomatic sheathing of the original fear engram cells by parvalbumin (PV)-positive interneurons (Trouche et al. 2013). Since PV interneurons are known to silence excitatory neurons, including within the amygdala (Lucas and Clem 2018), this finding stipulates that the expression of the original fear memory trace, and with it fear expression itself, was suppressed by PV inhibition, although no functional manipulation to support this claim was conducted.

A more explicit testimony of the co-existence of a new memory trace of safety being formed and competing with the original memory trace of fear upon behavioral attenuation surfaced a few years later (Lacagnina et al. 2019). Here, a tamoxifen-dependent system to tag memory engram cells was employed, the Arc-Cre$^{ERT2}$ mice, which were crossed to Cre-dependent halorhodopsin or channelrhodopsin-expressing mice. In Arc-Cre$^{ERT2}$ mice, Cre-recombinase is expressed upon neuronal activity, and its action can be temporally controlled by tamoxifen injection. This way, the original engram of fear was tagged at encoding and subsequently visualized and/or manipulated during extinction training of recent (i.e., 5d-old) contextual fear memories. Extinction training was found to recruit a second engram in the DG of these animals, which exhibited high activity when the fear expression was low and suppressed fear memory expression when optogenetically activated. Conversely, when the animals showed signs of spontaneous recovery, the original fear memory engram proved to be more active. As a result, its optogenetic activation was found to be necessary for fear resurfacing. These results indicate that within the DG there are two distinct ensembles present during behavioral fear attenuation, one for fear expression and one for fear suppression (Fig. 1d), with strong experimental evidence for the importance of the latter.

Subsequent studies have confirmed and expanded on these findings. In the mouse BLA, extinction learning was found to stimulate the formation of a new engram that was not only different from the original memory trace of fear, but also activated by natural rewards such as food and water (Zhang et al. 2020). Since optogenetic activation of this extinction engram was capable of altering reward-driven behavior and vice versa, it is possible that a fear extinction memory represents a newly formed reward memory, a reward that would thus be mediated by the absence of the CS. Intriguingly, similar mechanisms might also be at play in the human brain, where counterconditioning, a form of rewarded extinction, was found to outperform standard extinction learning, and this was paralleled by increased functional connectivity between the amygdala and the ventral striatum, a brain area long known to be implicated in reward processing (Burton et al. 2015), as revealed by functional magnetic resonance imaging (fMRI) (Keller et al. 2022).

Beyond a single brain area, several other studies have started to combine neuronal activity-dependent tagging with monosynaptic rabies tracing and revealed that fear extinction engrams in the mouse mPFC are part of an extinction-specific circuit with mPFC-projecting neurons found in the BLA and hippocampus, which is not activated by fear memory recall (Gu et al. 2022). Similarly, using cFos-Cre$^{ERT2}$ mice combined with various virus-based labeling, calcium recording and optogenetic constructs, fear and extinction memories were found to be encoded by distinct subpopulations of projection neurons in the insular cortex (IC), an important cortical area for valence processing (Davidovic et al. 2019), namely IC->CeA for fear recall and IL->nucleus accumbens (NAc) projections for fear attenuation (Wang et al. 2022). Moreover, IC->CeA and IC->NAc projections were found to reciprocally inhibit each other via local interneuron recruitment, thus providing further experimental evidence of extinction training leading to the inhibition of the original memory trace of fear.

Taken together, from both an engram and circuit perspective, there is thus also ample experimental evidence in support of Pavlov's original theory of behavioral extinction being mediated by inhibition of fear expression.

## The Expression of the Fear Memory Trace Is Constrained by Reconsolidation Blockers

Another means to inhibit fear expression can also be achieved by blocking its reemergence after learning. Indeed, over the past two decades, several pharmacological agents have been identified as powerful anxiolytics because they impinge on the neural processes needed to restabilize memory. For example, fear memories can be weakened by NMDA receptor antagonists (Lee et al. 2006), glucocorticoid receptors (Pitman et al. 2011), or noradrenergic blockade by the beta-blocker propranolol (Kindt et al. 2009). But what happens to the fear memory engram upon reconsolidation blockade?

The first study to address this question assessed how propranolol affected fear memory trace reactivation upon memory recall. To this end, ArcCreER$^{T2}$ x eYFP mice were used to tag the original memory trace of fear, injected with propranolol before fear memory recall, and stained for cFos to measure the fear engram's trace reactivation. Propranolol was not only found to decrease the animals' fear response, as expected, but also to reduce reactivated fear memory engram cells in the DG, albeit not in other brain areas (Leal Santos et al. 2021). In a follow-up study by the same group, a similarly reduced reactivation of fear engram cells was observed in the hilus, a polymorphic hippocampal cell layer consisting of neurons, mossy cells, and GABAergic interneurons, of which PV-containing interneurons showed an altered activity profile (Leal Santos et al. 2022).

Together, these studies indicate that reconsolidation blockers such as propranolol are likely to decrease fear expression by weakening the reactivation of the initial traumatic memory trace, although functional experiments to manipulate memory trace activity in the presence of propranolol to substantiate this interpretation have yet to be conducted. In light of the controversy of whether fear attenuation is achieved by processes targeting reconsolidation versus extinction, such studies are highly warranted.

## Discussion

Aside from the scarce literature on the destiny of the fear memory trace under conditions of reconsolidation blockade, the experimental evidence of whether fear memory attenuation is mediated by a relearning of the original memory trace of fear towards safety or, conversely, by the formation of a new memory trace of safety suppressing it, is conflicting. Certainly, methodological differences such as the tagging system (e.g., Tettag vs TRAP), the reporter gene (e.g., cFos vs Arc), the functional tool employed (e.g., chemo- vs optogenetics), or the timepoints studied (e.g., tagging at encoding vs tagging at recall; recent vs remote forms of fear memory) are likely to contribute to this contrasting findings. Notwithstanding, the most parsimonious explanation for these seemingly opposing results is that, in all likelihood, both processes co-exist (Fig. 1e). For example, although the reactivation of the fear memory trace by extinction training clearly correlated with behavioral fear attenuation (Khalaf et al. 2018), thus speaking for a relearning of the original memory trace towards safety, fear attenuation also triggered cFos activation outside recall-induced neurons, which indicates the parallel creation of a new memory trace of fear. Likewise, although fear attenuation was found to decrease the reactivation of the learning-induced ensemble (Lacagnina et al. 2019), speaking for fear neurons being suppressed, such reactivation was not zero, indicating that part of the original memory trace of fear was still reactivated by the extinction paradigm. Furthermore, as fear memory and attenuation engrams can only be functionally studied in one or, at best, two interconnected brain areas, which scenario predominates in other brain areas remains completely unknown. Consequently, from an engram perspective, reconsolidation-updating and extinction, two processes known to mediate behavioral fear attenuation, are not mutually exclusive.

In addition to engram-specific evidence for the co-occurrence of both scenarios, the same dichotomy has also been observed on other levels: On a brain area level, where the same brain region can accommodate both reconsolidation-updating and extinction (Baldi and Bucherelli 2015), on a synaptic level, where both reconsolidation-updating and extinction learning have been shown to alter spine formation and density on the same neuron (Lai et al. 2012; Lee et al. 2012); and on a molecular level, where similar gene expression signatures have been reported for either reconsolidation-updating or extinction (Clem and Schiller 2016; Lee et al. 2017; Mamiya et al. 2009; Suzuki et al. 2004; Tedesco et al. 2014). Thus, reconsolidation-updating, i.e., relearning of the original fearful association, and extinction, i.e., inhibition of the original fearful association, likely co-exist (Clem and Schiller 2016).

Ultimately, the revelation of these scenarios is also heavily influenced by the experimental paradigm. For example, reconsolidation-updating paradigms are known to be triggered by memory recall that is temporally separated from a subsequent extinction paradigm. If this separation does not occur, attenuated fear memories spontaneously resurface (Costanzi et al. 2011; Monfils et al. 2009; Ponnusamy et al. 2016), indicating memory suppression or inhibitory mechanisms to have taken the upper hand, despite the use of an identical extinction paradigm. Likewise, if the extinction paradigm is applied outside the time-limited reconsolidation window of six hours, both recent and remote fear memories have been observed to spontaneously resurface (Gräff et al. 2014; Monfils et al. 2009), highlighting the limited period during which extinction paradigms can be applied to lead to an updating of the original memory trace. Thus, the gap between recall and extinction training is a key factor determining whether updating or inhibiting the original memory trace occurs. In addition to its timing, the length of the extinction paradigm itself (Craske et al. 2014), as well as the age (Milekic and Alberini 2002) and strength (Suzuki et al. 2004) of the memory form so-called boundary conditions that dictate whether reconsolidation-updating or extinction proper are triggered (Elsey and Kindt 2017).

To further disentangle reconsolidation-updating and extinction from an engram perspective, a triple engram labeling system would be highly desirable. This way, two timepoints, such as fear memory encoding and recall, could be labeled with an exogenous engram tagging system (e.g., Tettag for encoding and TRAP for recall). In contrast, a third timepoint, such as extinction learning, could be tagged with an endogenous neuronal activity marker (e.g., cFos, Arc or Homer1a). Importantly, such a triple labelling system could then also be employed to better understand the manifold occurrences of fear resurfacing, such as its reinstatement, renewal and spontaneous recovery. Equally important would be an assessment of engram connectivity across brain areas known to be implicated in fear attenuation. In hippocampal circuits, this has already been achieved by combining engram tagging with the eGRASP system, which revealed that following memory formation engram cells are structurally more connected with each other than non-engram cells (Choi et al. 2018).

Honing in on the precise cell populations mediating fear attenuation is undoubtedly of great translational importance as, for example, gene expression changes, and therewith potential pharmacological targets, can be more precisely identified and functionally studied than at bulk resolution. This is of particular value for long-lasting forms of fear memories, which are often at the basis of fear and stress-related disorders such as post-traumatic stress disorder (PTSD), but for which little remains known.

## Glossary

**Attenuation** In studies involving fear conditioning, the term "attenuation" refers to the observable decrease in the animal's fear response, which is measurable by the reduction in the time spent freezing.

**Extinction learning** The term "extinction" can refer to two processes. On the one hand, it is synonymous with "attenuation" insofar as it refers to the reduction in the animal's fear response. On the other hand, it refers to the behavioral paradigm employed to yield to fear attenuation, i.e., the repetitive exposure of the fear-conditioned animal to the CS. In this chapter, we refer to the latter one as "extinction learning."

**Reconsolidation** The process by which memories become restabilized after memory recall.

**Reconsolidation blockade** The process by which a memory is prevented from reconsolidating, for example, by pharmacological agents.

**Reconsolidation window** A limited period of time during which memories are changeable after memory recall.

**Reinstatement of fear** The re-appearance of an attenuated fear response upon another unexpected presentation of the unconditioned stimulus (US).

**Renewal of fear** The re-appearance of an attenuated fear response when the CS is presented in a different context.

**Spontaneous recovery of fear** The unsolicited resurfacing of an attenuated fear response with time.

**Updating** Often used in conjunction with reconsolidation, "reconsolidation-updating" refers to the process by which, after memory recall, memories are reinterpreted in light of environmental contingencies that are no longer the same as at the time of memory encoding.

## References


Alberini CM (2011) The role of reconsolidation and the dynamic process of long-term memory formation and storage. Front Behav Neurosci 5:12

Albo Z, Gräff J (2018) The mysteries of remote memory. Philos Trans R Soc Lond Ser B Biol Sci:373, 20170029

Armbruster BN, Li X, Pausch MH, Herlitze S, Roth BL (2007) Evolving the lock to fit the key to create a family of G protein-coupled receptors potently activated by an inert ligand. Proc Natl Acad Sci USA 104:5163–5168

Baldi E, Bucherelli C (2015) Brain sites involved in fear memory reconsolidation and extinction of rodents. Neurosci Biobehav Rev 53:160–190

Beckers T, Kindt M (2017) Memory reconsolidation interference as an emerging treatment for emotional disorders: strengths, limitations, challenges, and opportunities. Annu Rev Clin Psychol 13:99–121

Burton AC, Nakamura K, Roesch MR (2015) From ventral-medial to dorsal-lateral striatum: neural correlates of reward-guided decision-making. Neurobiol Learn Mem 117:51–59

Choi JH, Sim SE, Kim JI, Choi DI, Oh J, Ye S, Lee J, Kim T, Ko HG, Lim CS et al (2018) Interregional synaptic maps among engram cells underlie memory formation. Science 360:430–435

Cincotta C, Murawski NJ, Grella SL, McKissick O, Doucette E, Ramirez S (2021) Chronic activation of fear engrams induces extinction-like behavior in ethanol-exposed mice. Hippocampus 31:3–10

Clem RL, Schiller D (2016) New learning and unlearning: strangers or accomplices in threat memory attenuation? Trends Neurosci 39:340–351

Costanzi M, Cannas S, Saraulli D, Rossi-Arnaud C, Cestari V (2011) Extinction after retrieval: effects on the associative and nonassociative components of remote contextual fear memory. Learn Mem 18:508–518

Craske MG, Treanor M, Conway CC, Zbozinek T, Vervliet B (2014) Maximizing exposure therapy: an inhibitory learning approach. Behav Res Ther 58:10–23

Dai Z, Liu Y, Nie L, Chen W, Xu X, Li Y, Zhang J, Shen F, Sui N, Liang J (2023) Locus coeruleus input-modulated reactivation of dentate gyrus opioid-withdrawal engrams promotes extinction. Neuropsychopharmacology 48:327–340

Davidovic M, Starck G, Olausson H (2019) Processing of affective and emotionally neutral tactile stimuli in the insular cortex. Dev Cogn Neurosci 35:94–103

Elsey JWB, Kindt M (2017) Breaking boundaries: optimizing reconsolidation-based interventions for strong and old memories. Learn Mem 24:472–479

Foa EB, Kozak MJ (1986) Emotional processing of fear: exposure to corrective information. Psychol Bull 99:20–35

Frankland PW, Josselyn SA (2018) Facing your fears. Science 360:1186–1187

Gräff J, Joseph NF, Horn ME, Samiei A, Meng J, Seo J, Rei D, Bero AW, Phan TX, Wagner F et al (2014) Epigenetic priming of memory updating during reconsolidation to attenuate remote fear memories. Cell 156:261–276

Gu X, Wu YJ, Zhang Z, Zhu JJ, Wu XR, Wang Q, Yi X, Lin ZJ, Jiao ZH, Xu M et al (2022) Dynamic tripartite construct of interregional engram circuits underlies forgetting of extinction memory. Mol Psychiatry 27:4077–4091

Herry C, Ciocchi S, Senn V, Demmou L, Muller C, Luthi A (2008) Switching on and off fear by distinct neuronal circuits. Nature 454:600–606

Herry C, Ferraguti F, Singewald N, Letzkus JJ, Ehrlich I, Luthi A (2010) Neuronal circuits of fear extinction. Eur J Neurosci 31:599–612

Josselyn SA, Tonegawa S (2020) Memory engrams: recalling the past and imagining the future. Science 367:eaaw4325

Keller NE, Hennings AC, Leiker EK, Lewis-Peacock JA, Dunsmoor JE (2022) Rewarded extinction increases amygdalar connectivity and stabilizes long-term memory traces in the vmPFC. J Neurosci 42:5717–5729

Khalaf O, Gräff J (2019) Reactivation of recall-induced neurons in the infralimbic cortex and the basolateral amygdala after remote fear memory attenuation. Front Mol Neurosci 12:70

Khalaf O, Resch S, Dixsaut L, Gorden V, Glauser L, Gräff J (2018) Reactivation of recall-induced neurons contributes to remote fear memory attenuation. Science 360:1239–1242

Kindt M, Soeter M, Vervliet B (2009) Beyond extinction: erasing human fear responses and preventing the return of fear. Nat Neurosci 12:256–258

Koya E, Golden SA, Harvey BK, Guez-Barber DH, Berkow A, Simmons DE, Bossert JM, Nair SG, Uejima JL, Marin MT et al (2009) Targeted disruption of cocaine-activated nucleus accumbens neurons prevents context-specific sensitization. Nat Neurosci 12:1069–1073

Lacagnina AF, Brockway ET, Crovetti CR, Shue F, McCarty MJ, Sattler KP, Lim SC, Santos SL, Denny CA, Drew MR (2019) Distinct hippocampal engrams control extinction and relapse of fear memory. Nat Neurosci 22:753–761

Lai CS, Franke TF, Gan WB (2012) Opposite effects of fear conditioning and extinction on dendritic spine remodelling. Nature 483:87–91

Leal Santos S, Stackmann M, Munoz Zamora A, Mastrodonato A, De Landri AV, Vaughan N, Chen BK, Lanio M, Denny CA (2021) Propranolol decreases fear expression by modulating fear memory traces. Biol Psychiatry 89:1150–1161

Leal Santos S, Chen BK, Pereira GR, Pham V, Denny CA (2022) Propranolol administration modulates neural activity in the Hippocampal Hilus during fear retrieval. Front Behav Neurosci 16:919831

Lee JL, Milton AL, Everitt BJ (2006) Reconsolidation and extinction of conditioned fear: inhibition and potentiation. J Neurosci 26:10051–10056

Lee SH, Kwak C, Shim J, Kim JE, Choi SL, Kim HF, Jang DJ, Lee JA, Lee K, Lee CH et al (2012) A cellular model of memory reconsolidation involves reactivation-induced destabilization and restabilization at the sensorimotor synapse in Aplysia. Proc Natl Acad Sci USA 109:14200–14205

Lee JLC, Nader K, Schiller D (2017) An update on memory reconsolidation updating. Trends Cogn Sci 21:531–545

Lucas EK, Clem RL (2018) GABAergic interneurons: the orchestra or the conductor in fear learning and memory? Brain Res Bull 141:13–19

Mamiya N, Fukushima H, Suzuki A, Matsuyama Z, Homma S, Frankland PW, Kida S (2009) Brain region-specific gene expression activation required for reconsolidation and extinction of contextual fear memory. J Neurosci 29:402–413

Milekic MH, Alberini CM (2002) Temporally graded requirement for protein synthesis following memory reactivation. Neuron 36:521–525

Monfils MH, Cowansage KK, Klann E, LeDoux JE (2009) Extinction-reconsolidation boundaries: key to persistent attenuation of fear memories. Science 324:951–955

Nader K, Hardt O (2009) A single standard for memory: the case for reconsolidation. Nat Rev Neurosci 10:224–234

Park S, Jung JH, Karimi SA, Jacob AD, Josselyn SA (2022) Opto-extinction of a threat memory in mice. Brain Res Bull 191:61–68

Pavlov I (1927) Conditioned reflexes. Oxford University Press, Oxford, UK

Pitman RK, Milad MR, Igoe SA, Vangel MG, Orr SP, Tsareva A, Gamache K, Nader K (2011) Systemic mifepristone blocks reconsolidation of cue-conditioned fear; propranolol prevents this effect. Behav Neurosci 125:632–638

Ponnusamy R, Zhuravka I, Poulos AM, Shobe J, Merjanian M, Huang J, Wolvek D, O'Neill PK, Fanselow MS (2016) Retrieval and reconsolidation accounts of fear extinction. Front Behav Neurosci 10:89

Reijmers LG, Perkins BL, Matsuo N, Mayford M (2007) Localization of a stable neural correlate of associative memory. Science 317:1230–1233

Schiller D, Monfils MH, Raio CM, Johnson DC, Ledoux JE, Phelps EA (2010) Preventing the return of fear in humans using reconsolidation update mechanisms. Nature 463:49–53

Suzuki A, Josselyn SA, Frankland PW, Masushige S, Silva AJ, Kida S (2004) Memory reconsolidation and extinction have distinct temporal and biochemical signatures. J Neurosci 24:4787–4795

Tedesco V, Roquet RF, DeMis J, Chiamulera C, Monfils MH (2014) Extinction, applied after retrieval of auditory fear memory, selectively increases zinc-finger protein 268 and phosphorylated ribosomal protein S6 expression in prefrontal cortex and lateral amygdala. Neurobiol Learn Mem 115:78–85

Tonegawa S, Liu X, Ramirez S, Redondo R (2015) Memory engram cells have come of age. Neuron 87:918–931

Tronson NC, Taylor JR (2007) Molecular mechanisms of memory reconsolidation. Nat Rev Neurosci 8:262–275

Trouche S, Sasaki JM, Tu T, Reijmers LG (2013) Fear extinction causes target-specific remodeling of perisomatic inhibitory synapses. Neuron 80:1054–1065

Wang Q, Zhu JJ, Wang L, Kan YP, Liu YM, Wu YJ, Gu X, Yi X, Lin ZJ, Wang Q et al (2022) Insular cortical circuits as an executive gateway to decipher threat or extinction memory via distinct subcortical pathways. Nat Commun 13:5540

Yoshii T, Hosokawa H, Matsuo N (2017) Pharmacogenetic reactivation of the original engram evokes an extinguished fear memory. Neuropharmacology 113:1–9

Zelikowsky M, Hersman S, Chawla MK, Barnes CA, Fanselow MS (2014) Neuronal ensembles in amygdala, hippocampus, and prefrontal cortex track differential components of contextual fear. J Neurosci 34:8462–8466

Zhang X, Kim J, Tonegawa S (2020) Amygdala reward neurons form and store fear extinction memory. Neuron 105:1077–1093 e1077

Zhang Y, Wang Z, Ju J, Liao J, Zhou Q (2021) Elevated activity in the dorsal dentate gyrus reduces expression of fear memory after fear extinction training. J Psychiatry Neurosci 46:E390–E401

# Adult Neurogenesis, Context Encoding, and Pattern Separation: A Pathway for Treating Overgeneralization

**Wei-li Chang and Rene Hen**

**Abstract** In mammals, the subgranular zone of the dentate gyrus is one of two brain regions (with the subventricular zone of the olfactory bulb) that continues to generate new neurons throughout adulthood, a phenomenon known as adult hippocampal neurogenesis (AHN) (Eriksson et al., Nat Med 4:1313–1317, 1998; García-Verdugo et al., J Neurobiol 36:234–248, 1998). The integration of these new neurons into the dentate gyrus (DG) has implications for memory encoding, with unique firing and wiring properties of immature neurons that affect how the hippocampal network encodes and stores attributes of memory. In this chapter, we will describe the process of AHN and properties of adult-born cells as they integrate into the hippocampal circuit and mature. Then, we will discuss some methodological considerations before we review evidence for the role of AHN in two major processes supporting memory that are performed by the DG. First, we will discuss encoding of contextual information for episodic memories and how this is facilitated by AHN. Second, will discuss pattern separation, a major role of the DG that reduces interference for the formation of new memories. Finally, we will review clinical and translational considerations, suggesting that stimulation of AHN may help decrease overgeneralization—a common endophenotype of mood, anxiety, trauma-related, and age-related disorders.

**Keywords** Neurogenesis · Contextual encoding · Pattern separation · Dentate gyrus

---

W.-l. Chang · R. Hen (✉)
Departments of Psychiatry and Neuroscience, Columbia University, New York, NY, USA

Division of Systems Neuroscience, New York State Psychiatric Institute, New York, NY, USA
e-mail: rh95@cumc.columbia.edu



J. Gräff, S. Ramirez (eds.), *Engrams*, Advances in Neurobiology 38,
https://doi.org/10.1007/978-3-031-62983-9_10

## Adult Hippocampal Neurogenesis in the Dentate Gyrus

Within the subgranular zone of the adult DG, there is a pool of neural stem cells (type 1 cells) that mostly exist in a quiescent state and divide infrequently. These neural stem cells can become activated to divide either symmetrically or asymmetrically to produce additional neural stem cells, as well as intermediate progenitor cells (type 2 cells) (Fig. 1). Intermediate progenitor cells divide a few times before differentiating into neuroblasts and then immature neurons, which have distinct morphological and physiological properties. As these immature adult-born neurons continue to develop, they gradually become mostly undistinguishable from mature granule cells (mGCs) and integrate into the existing hippocampal circuitry (Drew et al. 2013; Toni et al. 2007, 2008; Van Praag et al. 2002).

In rodents, inputs to immature adult-born granule cells (iGCs) are found as early as 2 weeks after their birth (Van Praag et al. 2002; Schmidt-Hieber et al. 2004; Ge et al. 2006; Espósito et al. 2005; Overstreet et al. 2004) and functional synapses from 2-week-old iGCs can also be observed (Gu et al. 2012). iGCs that are 4–6 weeks old exhibit increased synaptic plasticity, higher intrinsic excitability, higher input resistance, enhanced long-term potentiation, increased activation in response to input stimulation, and distinct wiring properties compared to mature granule cells (mGCs, over-8-week-old in mice) (Schmidt-Hieber et al. 2004; Luna et al. 2019; Bergami et al. 2015; Ge et al. 2007; Snyder et al. 2001; Marín-Burgin et al. 2012; Dieni et al. 2016; Mongiat et al. 2009; Yang et al. 2015). It is believed

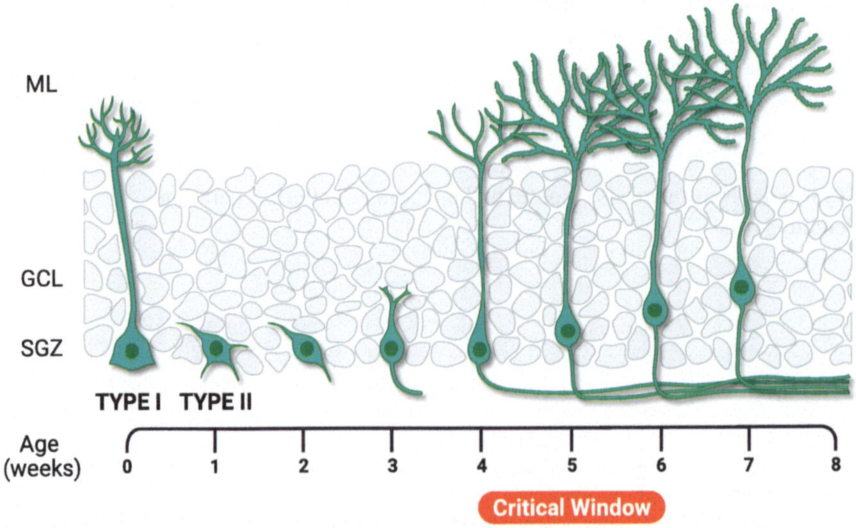


**Fig. 1** Adult hippocampal neurogenesis timeline showing a schematic of neuronal differentiation of adult-born granule cells in the murine dentate gyrus. Cells aged 4–6 weeks are thought to be most important for DG function by modulating mature granule cell activity, both directly and indirectly. After 8 weeks, cells are thought to be fully mature. (Adapted from Denny et al. (2012))

that iGCs at this 4- to 6-week-old stage exert unique effects on hippocampal activity patterns to support functions such as contextual encoding and pattern separation (Fig. 1) (Gu et al. 2012; Sahay et al. 2011; Denny et al. 2012; Nakashiba et al. 2012; Danielson et al. 2016; Huckleberry et al. 2018). There is evidence of inhibitory modulation of mGCs by iGCs, resulting in increased sparseness of DG representations of external stimuli (Drew et al. 2016; Anacker et al. 2018; McHugh et al. 2022; Lacefield et al. 2012; Ikrar et al. 2013). However, there is also recent evidence of direct connections between iGCs and mGCs, with inhibitory or excitatory effects on mGCs depending on the excitation levels of iGCs (Luna et al. 2019). iGCs may therefore support the formation of distinct engrams by modulating the activity of the hippocampal network through varied mechanisms. Furthermore, adult-born granule cells may retain unique properties even after reaching maturity, including increased learning-induced plasticity (Lemaire et al. 2012; Huckleberry and Shansky 2021; Tronel et al. 2010), distinct dendritic arborization structure (Kerloch et al. 2019), and increased spine density and mossy fiber bouton size (Cole et al. 2020). Compared to granule cells born earlier in development, mature adult-born granule cells may also be recruited for different functions (Tronel et al. 2015; Masachs et al. 2021; Abrous et al. 2022).

The total process of AHN involves proliferation, migration, survival, and differentiation of adult-born cells, and so levels of AHN can be increased or decreased by internal and external factors that alter the rates of any one of these steps. In rodents, studies have shown that neurogenesis is increased by enriched environment (Kempermann et al. 1997, 2002; Kobilo et al. 2011; Leiter et al. 2016; Mustroph et al. 2012), exercise (Van Praag et al. 1999; Clark et al. 2009; Garrett et al. 2012; Schoenfeld and Swanson 2021; Speisman et al. 2013), antidepressant treatment (Santarelli et al. 2003; Malberg et al. 2000; Sairanen et al. 2005), and learning (Gould et al. 1999). In contrast, neurogenesis is decreased by aging (Altman and Das 1965; Foster 2006; Olariu et al. 2007; Rai et al. 2007), stress (Gould and Tanapat 1999; Heine et al. 2005; Mirescu and Gould 2006; Surget and Belzung 2022; Veena et al. 2011), impoverished environments (Dranovsky et al. 2011), neurodegenerative disorders (Babcock et al. 2021; Kim et al. 2022; Regensburger et al. 2014), substance use (Castilla-Ortega and Santín 2020; Zhang et al. 2016), and genetic mutations (Li and Zhao 2014; Patzlaff et al. 2018; Ceccarelli et al. 2020). Other events, such as ischemic injuries, traumatic brain injury, and seizures, have a more complicated effect on neurogenesis and depend on the magnitude and duration of the episode (Gao et al. 2023).

In addition, neurogenesis can be directly manipulated to examine effects on behaviors or neural activity. To specifically decrease neurogenesis, neural stem cells can be ablated with X-irradiation applied directly to the hippocampus (Santarelli et al. 2003; Parent et al. 1999; Burghardt et al. 2012). Genetic strategies to decrease neurogenesis include glial fibrillary acidic protein (GFAP)-TK mice or rats (Denny et al. 2012; Lacefield et al. 2012; Campos et al. 2013; Cahill et al. 2019; Kirshenbaum et al. 2014) and constitutive cyclin D2 knockout (Lopes et al. 2023; Kowalczyk et al. 2004; Garthe et al. 2014; Filipkowski and Kaczmarek 2018; Urbach et al. 2013). Immature neurons can also be inhibited or silenced using chemogenetic and optogenetic strategies in transgenic lines such as Nestin$^{CreERT2}$ (Danielson et al. 2016; Dranovsky et al. 2011) and Ascl1$^{CreERT2}$ (Yang et al. 2015; Tuncdemir et al. 2023). One genetic strategy to boost AHN increases the survival of adult-born granule cells in a cell-autonomous manner by deleting the pro-apoptotic gene Bax selectively from neural stem cells and their progeny (Sahay et al. 2011). Other strategies, including enrichment, exercise, and pharmacological treatment, are also used, though these interventions have other effects beyond just increasing AHN (Santarelli et al. 2003; Wu et al. 2015; Dranovsky and Hen 2006; Clemenson et al. 2015; Creer et al. 2010; Van Praag et al. 2005; Drew and Hen 2007; Sahay and Hen 2007).

Whether there is evidence of AHN in humans was thought to be answered in 1998 (Eriksson et al. 1998). Still, there is continued debate and conflicting reports, with many additional groups demonstrating evidence of AHN (Tartt et al. 2022; Spalding et al. 2013; Johansson et al. 1999a, b; Terreros-Roncal et al. 2021; Moreno-Jiménez et al. 2019, 2021; Ernst et al. 2014; Christian et al. 2014; Knoth et al. 2010; Boldrini et al. 2018; Tobin et al. 2019; Ammothumkandy et al. 2022) while others found no evidence of AHN and suggested alternative explanations of AHN evidence in humans (Sorrells et al. 2018, 2021; Cipriani et al. 2018; Sanai et al. 2011). Neurogenesis is usually assessed using immunohistochemical staining for protein markers present in iGCs. In humans, this staining, conducted in postmortem brain tissue, is heavily influenced by factors that affect the quality of the tissue and phenotyping accuracy of the donors (Moreno-Jiménez et al. 2021; Flor-García et al. 2020). Two major factors that affect tissue quality include postmortem delay for tissue collection and prolonged exposure to fixative, which can degrade the visualization of AHN markers (Gallardo-Caballero et al. 2023). Phenotyping accuracy of donors can also affect tissue quality if neuropathological processes such as neurodegeneration, seizures, toxic exposures, or neoplasms are not correctly attributed to or identified in the tissue (Bao and Swaab 2018). Beyond these challenges, interpretation of neuronal birth dates is difficult due to complicated expression patterns of markers in neurons that remain immature for very long periods (Zhou et al. 2022). What is apparent in most studies of human brain tissue is that the number of neural stem cells decreases dramatically with age (Moreno-Jiménez et al. 2019). Still, there is evidence that maturation time also lengthens. Therefore, the total number of neurons born after early childhood that demonstrate properties of immature neurons in human adults is still a significant number (Zhou et al. 2022; Tosoni et al. 2023).

## Methodological Considerations for Examining the Role of AHN in Memory

***Ex Vivo Versus In Vivo Measurements*** The field of engram research has greatly benefited from tools utilizing the detection of immediate early genes like Arc and Fos to label cells that were active during a discrete experience (Denny et al. 2014; Ramirez et al. 2013). These tools allow us to indelibly label relevant cells and test for their behavioral effects. Meanwhile, using in vivo activity recording, either with electrophysiology or calcium imaging, activity patterns and population dynamics can be studied to determine how particular cells are recruited to the engram. Given the complex relationship between neuronal activity and immediate early gene expression, these methods are often complimentary. Special consideration should be taken in studies of the role of AHN, though, since immature GCs within the 4- to 6-week-old range, despite being electrophysiologically hyperexcitable, are less likely to express protein markers of activation compared to mature GCs (Parylak et al. 2023). Immature GCs younger than 3–4 weeks of age have minimal IEG expression (Jessberger and Kempermann 2003; Synder et al. 2009; Jungenitz et al. 2014), even though immature neurons at this age can produce action potentials (Dieni et al. 2016; Li et al. 2017).

Mixed selectivity and population-level coding, which will be discussed more below, also have implications for the utility of IEG studies. There is mounting evidence that cells that do not appear explicitly tuned to specific environmental features still participate in the population coding for these features (Stefanini et al. 2020) but may not be active enough to mount an appreciable level of IEG expression. Moreover, as adult-born granule cells are believed to have a modulatory effect on DG activity patterns, single-cell and population-level decoding strategies, which rely heavily on temporal dynamics at the level of milliseconds, are more likely to detect the impact of altered AHN (Stefanini et al. 2020). Thus, to get a thorough, mechanistic understanding of how episodic memories are encoded and stored, in vivo recording techniques such as single-cell calcium imaging will continue to be an indispensable method for investigation.

***Functional Heterogeneity of the DG Along the Dorsal-Ventral Axis*** As the varied roles of the hippocampus in general, and the DG in particular, have been more finely elucidated, the functional heterogeneity of the hippocampus and its subregions along the dorsal-ventral axis have become apparent (Anacker and Hen 2017). The ventral pole appears to be more involved when information has a positive or negative valence: so-called "mood-related" or emotionally relevant behaviors, while the dorsal pole is thought to mediate cognitive-related behaviors (Kheirbek et al. 2013; Bannerman et al. 2002, 2004; Tannenholz et al. 2014; Fanselow and Dong 2010; Strange et al. 2014; Moser et al. 1995; Ferbinteanu and McDonald 2001; Kjelstrup et al. 2002; Maren and Holt 2004). The functional segregation of the dorsal vs. ventral hippocampus is supported by tract tracing and gene expression data (Fanselow and Dong 2010; Moser and Moser 1998; Leonardo et al. 2006; Dong et al. 2009). This distinction along the homologous posterior-anterior axis in humans seems to be conserved, with the anterior hippocampus more involved in emotional processing and the posterior hippocampus more involved in cognitive tasks (Bannerman et al. 2004; Tannenholz et al. 2014; Fanselow and Dong 2010; Tanti and Belzung 2013). Accordingly, rodents living in an enriched environment have increased AHN in the dorsal DG (Kempermann et al. 1997; Tanti et al. 2012, 2013), and chronic stress models appear to decrease AHN more in the ventral DG (Tanti et al. 2012, 2013; Wu and Hen 2014; Lehmann et al. 2013). Focal irradiation experiments in mice have suggested that AHN in the dorsal DG has a more pronounced effect on behavioral pattern separation in a fear discrimination task than the ventral DG. The converse was true for antidepressant effects on measures of innate anxiety after chronic stress or chronic corticosterone, with the ventral DG irradiation alone able to recapitulate the effects of whole-hippocampus irradiation, while dorsal DG irradiation had a less pronounced effect (Wu and Hen 2014). Methods such as two-photon calcium imaging with head-fixed mice have revealed a lot about how elements of the DG, including iGCs, function, but this technique is largely limited to dorsal structures. As ventral DG studies become more widespread, it will be important to discern which circuit properties are common to all regions of the DG and which are more characteristic of ventral vs. dorsal regions.

***Sex and Species Differences*** Like many areas of biomedical science, the vast majority of preclinical research on AHN has been conducted on sexually naïve adult male mice and rats, so we have less of a sense for sex differences, either before or after mating. There is already some evidence that sex hormones, both androgens and estrogens, can affect AHN differentially in male and female rodents through various mechanisms (Mahmoud et al. 2016; Galea et al. 2013). Additional studies in female subjects, and direct comparisons to evaluate for sex differences, will be critical for a full picture of DG circuits, AHN effects, and how these impact memory.

In this chapter, we also mostly focus on literature from studies in mice, but a considerable amount has been learned about AHN from studies in rats. Some of the references for our current knowledge on the topics discussed in this chapter are from these rat studies. However, there are differences in AHN between mice and rats. Adult-born cells mature, survive, and integrate into the DG at a higher rate in rats than in mice, and timelines for the phases of proliferation, differentiation, and maturation are different as well (Synder et al. 2009). These methodological considerations of measurement techniques, regional heterogeneity, and additional biological variables like sex and species represent possible directions for the field to expand to answer additional open questions. They will not be directly addressed in the work summarized below.

## AHN and Contextual Encoding

***Encoding Stimuli in the External Environment*** Many cells within the DG demonstrate the tendency to fire when the animal is in a specific location or position in their environment, and these neurons are termed "place cells" (Burgess and O'Keefe 1996; O'Keefe and Dostrovsky 1971). Place cells in the DG are thought to integrate sensory and self-motion information for a representation of the spatial environment of the animal (Kesner 2013; Lee and Jung 2017). Granule cells in the DG can also respond directly to sensory cues, including visual (Hainmueller and Bartos 2018; Lee et al. 2009), tactile (Jung et al. 2019) and olfactory modalities (Woods et al.

2020). GCs exhibit firing related to objects and landmarks in the environment at fixed vector relationships to the subject within an arena (same distance and direction) (GoodSmith et al. 2022; Deshmukh and Knierim 2013). These "landmark vector" cells among GCs have diverse responses to landmarks being moved or present in new environments (GoodSmith et al. 2022), further supporting detailed encoding of stimuli from the context of a particular episodic memory. In contrast to place cells, which preferentially fire at fixed locations within the environment, landmark vector cells preferentially fire when objects in the environment are at a fixed distance and direction in relation to the subject (Deshmukh and Knierim 2013; Geiller et al. 2017; Høydal et al. 2019).

Additionally, there are distinct populations of cells in the DG that are either spatially tuned for a specific location within the environment or that respond preferentially to sensory cues in the environment, independent of where they occur in the environment (Tuncdemir et al. 2022). These cue cells seem to be stably distinct from place cells, and while cues could be of various sensory modalities, cue cells for one sensory modality were not responsive to cues of another modality. How these cue cells relate to landmark vector cells is not yet known since they were identified in head-fixed mice traversing a treadmill and are manipulated and sensed differently than landmark objects. Still, both cell types maintain their preferential firing when the cue or landmark is moved to a new location. Thus, the DG appears to be coding for both "what" (non-spatial, sensory cues) and "where" (spatial information tied to discrete locations), sometimes integrating this information in the same cell, as with landmark vectors (objects with fixed spatial relationships to the individual). This is consistent with the information contained in the main inputs to the DG from the lateral entorhinal cortex, which is thought to primarily represent information about sensory cues, and the medial entorhinal cortex, which contains self-motion information (Hargreaves et al. 2005; Knierim et al. 2013). For example, when lateral entorhinal cortex inputs to the DG are silenced, GC representations of odorant stimuli are degraded (Woods et al. 2020).

Place cells in the DG can change their preferred firing location and their firing rate when the environment changes, a process referred to as "remapping" (Latuske et al. 2018; Kubie et al. 2020), which likely supports distinct neural representations of different environments (Colgin et al. 2008). There is preliminary evidence that iGCs preferentially promote the remapping of place cells over cue cells. When AHN is ablated with targeted hippocampal irradiation, or when 4- to 6-week-old iGCs are selectively inhibited with chemogenetics, place cell remapping between different contexts is impaired, and cue cells exhibit even less remapping (Tuncdemir et al. 2023). Calcium imaging of iGCs themselves reveals that iGCs are both more active and more diffusely tuned than mGCs (Danielson et al. 2016; McHugh et al. 2022). These properties, along with observed effects in mGC activity patterns when iGCs are suppressed, support the idea that, rather than acting as additional encoding units, iGCs perform a modulatory role within the DG.

In addition to the cells that are tightly linked to place, cues, or other features in the environment, there is also increasing evidence that many brain areas, including the hippocampus, employ a more distributed code for environmental encoding with individual neurons showing mixed selectivity (Stefanini et al. 2020; Rigotti et al. 2013; Eichenbaum 2018; Fusi et al. 2016) (Fig. 2). Stefanini et al. used calcium imaging data from the DG to decode the animal's position, as well as current direction of motion and speed, based on the population-level activity. Interestingly, they showed that non-place cells, i.e., DG neurons that did not meet the criteria for spatial tuning by standard metrics, still contributed to the population code for the mouse's position. In fact, when the decoder was trained exclusively on non-place cell firing patterns, it could still accurately predict the location of the mouse in the arena. The importance of non-spatially tuned cells and how they contribute to decoding was also shown using simulated neural data (Fig. 2) (Stefanini et al. 2020). Cell populations can also rely on correlated activity involving seemingly non-selective neurons to encode memory and context information (Stefanini et al. 2020; Jimenez et al. 2020). Mixed selectivity is thought to yield higher dimensionality to representations in the neural code, which enables more complex cognitive processes (Rigotti et al. 2013; Fusi et al. 2016). Importantly, there is recent evidence that AHN supports this dimensionality and helps decorrelate neural representations: a decoder trained on DG data from mice lacking iGC activity (due to irradiation or chemogenetic inhibition), re-exposed to a context after several days' delay, is less able to differentiate between the two time points. In other words, without AHN, the population code for spatial position becomes more similar between two experiences that are separated in time (Tuncdemir et al. 2023).

It appears, then, that the DG is encoding several aspects of the environmental context of an animal, both at the single-cell level, with neurons tuned to place, self-motion, landmarks, or sensory cues, and at the population level, with a majority of neurons exhibiting mixed selectivity. This summated encoding of stimuli in particular patterns can then allow for the encoding of specific episodic memories.

***Contextual Fear Conditioning*** The hippocampus is known to be required for the encoding of memories and retrieval of recent memories. In rodents, much has been learned about memory circuits using the contextual fear conditioning (CFC) paradigm. Mice or rats are exposed to a chamber and allowed to freely explore before one or more noxious foot shocks are administered. If the animal is then placed back into that same environment later, they will exhibit increased levels of freezing, a defensive behavior that indicates they ascribe some level of threat to this environment now. Decades of work have interrogated the neural circuits that mediate this effect, and using CFC, engram cells were identified in the DG granule cells (Denny et al. 2014; Ramirez et al. 2013; Liu et al. 2012; Ryan et al. 2015).

The hippocampus appears necessary for the initial encoding of memory. As the memory is consolidated, memory storage gradually shifts from the hippocampus to other brain regions. After this, engram cells are thought to be in a "silent" state. That is, the hippocampal engram cells are not reactivated by natural cues 2 weeks after CFC, but exogenous stimulation using optogenetics does induce freezing behavior (Kitamura et al. 2017). It is not yet known how these hippocampal engram cells become silent nor how long they persist in a silent state (Tonegawa et al. 2018).

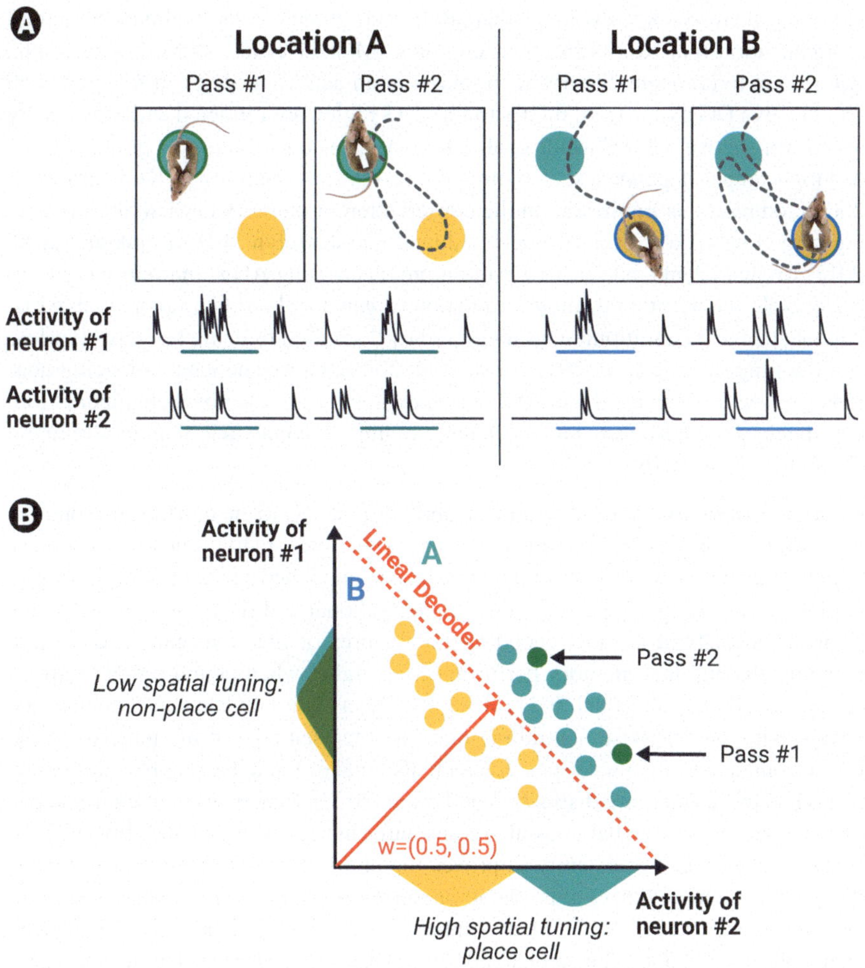


**Fig. 2** Model for how mixed selectivity contributes to encoding environmental features such as position. (**a**) The activity of two simulated neurons as a function of time. Neuron #1 is a simulated place cell with different activity distributions for different locations in the arena. However, the distributions partially overlap, so the activity is only partially predictive of the animal's place in the arena. Neuron #2 has the same activity distribution when the animal is in two different locations of the arena, has low spatial tuning, and is not selective for position. Different responses for the two neurons are elicited by other differences in features, such as direction of motion. (**b**) An example of how place cells and non-place cells can be equally important for encoding the position of the animal. The x-axis represents the average activity of simulated Neuron #1 and the y-axis represents the average activity of simulated Neuron #2. Each dot on the scatterplot represents the average population response for the two neurons during one pass through either Location A (green) or Location B (yellow). The distributions of activities for the two neurons across locations either partially (Neuron #1) or almost completely (Neuron #2) overlap. Additionally, there is variability in responses for each pass. The neural representations at the population level (here, only two neurons for simplicity) are still well separated if correlations between the activities of the two neurons are taken into account. This makes it possible for a linear decoder (red dashed line) to accurately discriminate the two positions based on activity from these two neurons. The resulting decoder's weight vector has two equal components corresponding to the importance of the two neurons in the encoding position. In this example, both neurons are important for encoding position despite their very different tuning properties, and the linear decoder would assign equal weights (w) to the two neurons. (Figure and legend adapted from Stefanini et al. (2020))

However, there is some evidence that adult-born granule cells facilitate the circuit reorganization that results in silent engrams (Meltzer et al. 2005). For example, iGCs seem to compete directly with mGCs in an activity-dependent manner, such that inactive DG axons from mGCs to the CA3 region are retracted and replaced by axons from active iGCs (Yasuda et al. 2011). AHN also facilitates the gradual decay of hippocampal dependency of remote contextual fear memories (Kitamura et al. 2009) and has been implicated in the conservation of memory capacity by promoting forgetting (Akers et al. 2014; Frankland and Josselyn 2016; Frankland et al. 2013; Ko and Frankland 2021). Furthermore, decreasing AHN and removing 4- to 6-week-old iGCs, either through irradiation or genetic ablation, impairs contextual but not cued fear conditioning, as measured by decreased freezing one day after fear conditioning (Denny et al. 2012; Saxe et al. 2006). Optogenetically silencing neurons ≤6 weeks old in the dorsal DG during exploration of the novel chamber before the shock was delivered was sufficient to impair contextual fear conditioning (Danielson et al. 2016).

***Memory Specificity*** AHN supports the ability of the DG to encode the environment for specific episodes by making the neural signature of different contexts more unique (Fig. 3). This has been thought to be accomplished via overall inhibition of the DG, resulting in a "sparser" code. Computational and theoretical models have posited that cells in the DG encoding for features of the environment constitute engram cells and that memory performance is higher when neural representations are sparse (Benna and Fusi 2021). Empirically, as animals learn to associate the reward with an otherwise neutral odorant, the representative firing patterns in the DG become more and more sparse, which also makes the neural representation for two odors more and more distinct (Woods et al. 2020). Reactivation of engram cells via exposure to contextual cues also transiently increases the excitability of these engram cells, which transiently improves the ability to discriminate similar cues, though the relationship of this phenomenon to population sparseness is not yet known (Pignatelli et al. 2019). Many studies have shown that sparseness is enhanced by the presence of iGCs (Drew et al. 2016; Anacker et al. 2018; McHugh et al. 2022; Lacefield et al. 2012; Ikrar et al. 2013). However, ablation or inhibition of adult-born granule cells does not always alter the overall firing rate of the DG (Tuncdemir et al. 2023; Park et al. 2015), and iGCs may excite mGCs in some circumstances (Luna et al. 2019). The impact of iGCs on DG activity is likely more complicated than just increasing sparseness and may depend on the experience and current environment (Luna et al. 2019; Park et al. 2015). We continue to learn how complex neural coding supports complex cognitive tasks (Stefanini et al. 2020; Rigotti et al. 2013; Fusi et al. 2016), and AHN appears to exert multiple influences on DG firing to facilitate the decorrelation of neural activity patterns, a process required for pattern separation—the topic of the next section.

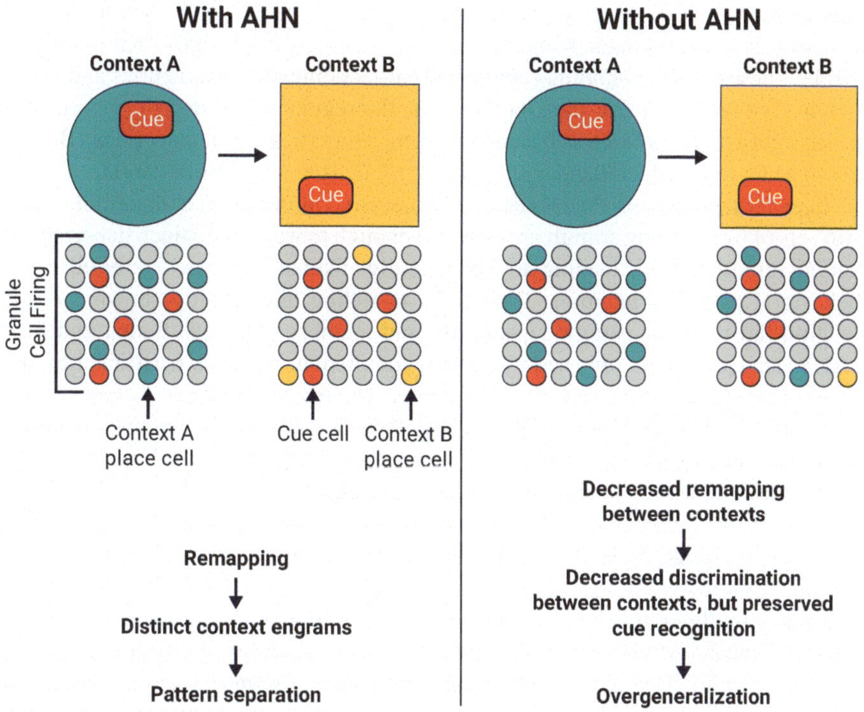


**Fig. 3** A hypothesized model of how adult hippocampal neurogenesis (AHN) supports pattern separation in the DG. On the left, in the presence of active immature granule cells, there is remapping of place cells between contexts A and B. Separate cue cells fire to the same cue, even when presented in a different location of a different context. The combination of green and red circles in the left panel represents the engram of an experience in Context A. The combination of yellow and red circles in the left panel represents the engram of an experience in Context B. Despite an overlap of cue cells between the two contexts, the engrams are distinct enough to support pattern separation. In the right panel, when AHN is absent or immature granule cells are inactive, there is decreased remapping of place cells between the two contexts, represented by green circles in both contexts A and B. Again, cue cells are preserved between contexts, and in fact, exhibit higher firing rates. The engrams for experiences in contexts A and B are less distinct, leading to difficulty differentiating between them and overgeneralization. (Adapted from Tuncdemir et al. (2023))

## AHN and Pattern Separation

The brain has a limited capacity for the storage of information. Moreover, the ability to rapidly store and access information can be improved by organizing memories or other types of information, such as by category or time-period. An important cognitive process that supports memory storage is termed pattern separation. This is the process of separating similar or overlapping information such as contexts, cues, objects, and experiences into distinct neural representations. Pattern separation is thought to support efficient memory storage by decreasing interference (Wiskott et al. 2006). This function is now ascribed to the DG, initially based on rodent work

(Leutgeb et al. 2007; Aimone et al. 2011; Yassa and Stark 2011; Neunuebel and Knierim 2014; Hainmueller and Bartos 2020; Gilbert et al. 2001; Kim and Lee 2011). These empirical findings confirmed earlier computational theories and simulations that predicted that the DG would be the region best suited to perform this process based on its large cell number, firing properties and connections (Becker 2005; Marr 1971; Mcclelland and Goddard 1996; Treves and Rolls 1994).

In pattern separation, the DG uses environmental information to determine when two sets of overlapping stimuli are different enough to warrant distinct, decorrelated neural representation, leading to behaviorally distinct responses referred to as behavioral pattern separation. Theoretical modeling has recently shown a direct relationship between pattern separability and memory capacity (Benna and Fusi 2021). This process is opposed by pattern completion, which is performed by the downstream target of the DG, the CA3 region. The balance between pattern separation and pattern completion might be heavily influenced by the animal's environment and prior experiences, which has interesting implications for human clinical populations, discussed in another section of this chapter.

To measure behavioral pattern separation in rodents, there are both appetitive and aversive tasks. Appetitive tasks include variations on delayed non-match to sample tasks or spatial separation tasks. The link between AHN and pattern separation was initially established by showing that inhibition of AHN impairs the performance of these tasks (Creer et al. 2010; Clelland et al. 2009; Winocur et al. 2006; Wojtowicz et al. 2008). However, a major motivation for understanding the neural circuits underlying behavioral pattern separation is that impaired pattern separation is thought to underly overgeneralization of negative responses seen in many neuropsychiatric disorders (Kheirbek et al. 2013). Therefore, a behavioral pattern separation task involving aversive stimuli may have increased translational potential.

One such task that is frequently used is the fear discrimination task. Following CFC with a foot-shock, rodents are serially presented with that same shock-associated context or a very similar context, where some aspects of the environment have been changed. As with CFC, the strength of the initial encoding is measured by defensive or avoidant behaviors, most commonly freezing. When contexts are similar, rodents will initially generalize and exhibit similar amounts of freezing time in both contexts. However, after repeated presentations of the two contexts together, one paired with and the other without shock, the animals will start to show divergent behaviors, with higher freezing levels in the shock-associated context and lower freezing in the similar, non-shock-associated context. The task's difficulty can be tailored depending on the traits of the study subjects and the size and direction of the expected experimental effects. The task is rendered more difficult, and the slope of diverging freezing behaviors shallower by making the two contexts more similar, randomizing the order that the two contexts are presented, or lowering the intensity of the foot shock. This type of fear discrimination task has been shown to be DG-dependent and sensitive to manipulations of AHN (Sahay et al. 2011; Wu et al. 2015; Wu and Hen 2014; Hill et al. 2015; McHugh et al. 2007; Kheirbek et al. 2012; Miller and Sahay 2019). For example, Danielson et al. (2016) acutely inhibited iGCs using optogenetics in the dorsal DG during a fear discrimination task.

When iGCs were inhibited during exposure to a similar, neutral context, there was impaired discrimination in freezing. However, iGC inhibition while exploring the shock-associated context did not alter pattern separation, and there was no effect when a similar number of mature GCs were silenced in either context (Danielson et al. 2016). Hypotheses for why iGCs are necessary just in the neutral context but not the aversive one have not been fully tested, but it is possible that the salient and daily foot shocks in the aversive context allowed for learning despite decreased AHN activity or that there is a dissociation in dorsal vs. ventral DG-dependent representations for the neutral vs. aversive contexts, which remains to be tested.

Interestingly, iGCs were recently shown to exert differential influence on the remapping of place cells and cue cells in the DG during a contextual discrimination task (Tuncdemir et al. 2023). In normal conditions when iGCs are active, a contextual change results in global remapping of place cells and a modest rate remapping of cue cells; in contrast, when iGCs are silenced, place cell remapping is impaired and cue cells remain unchanged. As a result, in the absence of AHN, contextual representations are dominated by cues at the expense of the spatial location of these cues or other neutral contextual details. This is a situation reminiscent of the overgeneralization of negative cues often observed in anxiety and trauma-related disorders, which are discussed below.

***Related Cognitive Processes and Behaviors*** Context encoding, including pattern separation, is likely part of a cascade of additional cognitive processes that has widespread impacts on other behaviors (Kempermann 2022). AHN correlates with novelty-seeking and attention to novelty (Denny et al. 2012; McHugh et al. 2022; Lemaire et al. 1999; Weeden et al. 2019), which may rely on pattern separation to successfully distinguish new experiences from old ones. Related to novelty-seeking is roaming entropy, an index of exploratory behavior, which is probably also influenced by the ability to decorrelate old versus new experiences. Rodents usually display increased variance in roaming entropy over time, and cyclin D2 knockout mice with reduced AHN lack this increased variance (Lopes et al. 2023), which may lead to more rigid individual adaptation to the environment. Reduced context discrimination might also explain the reduced flexibility in learning strategies (Garthe and Kempermann 2013; Garthe et al. 2016; Dupret et al. 2008; Appleby et al. 2011) and impaired reversal learning (Burghardt et al. 2012; Park et al. 2015; Garthe et al. 2009; Swan et al. 2014) that has been observed after inhibition of AHN. Individual differences in AHN that cause behavioral variation could in turn promote or suppress ongoing AHN in a self-perpetuating cycle. Voluntary behavior such as wheel running has a well-established and dramatic effect on DG cell proliferation (Van Praag et al. 1999; Schoenfeld and Swanson 2021; Gao et al. 2023; Cooper et al. 2018). In genetically identical animals living in a large enclosure, 22% of the variance in AHN was attributed to individual behavioral trajectories, with some effect of locomotor distance but an even larger effect of roaming entropy (Freund et al. 2013).

The extensive research that has been conducted on AHN in rodent models has provided insights into how the DG encodes contextual information needed for episodic memories. This contextual information can then be used to distinguish between situations with overlapping features (Fig. 3). This chapter is not meant to be an exhaustive summary of all the purported roles of AHN, and we have chosen to focus on these functions because of their clinical and translational implications (Anacker and Hen 2017), which we will now discuss.

## Clinical and Translational Implications

There is currently no safe and reliable method for quantifying in vivo AHN in humans, so direct analysis of AHN in humans has relied on studies of postmortem brain tissue, which are sensitive to experimental procedures (Moreno-Jiménez et al. 2021; Flor-García et al. 2020; Tartt et al. 2018). On the other hand, non-invasive functional neuroimaging has greatly expanded our understanding of the neural correlates of cognitive processes. However, a vast majority of functional MRI (fMRI) studies using commonly available 3 Tesla (T) scanners cannot resolve the DG from the CA2 and CA3 regions of the hippocampus. Methods for achieving higher resolution in fMRI to localize blood-oxygen-level-dependent (BOLD) activation to the DG and other hippocampal subfields have been reported (Suthana et al. 2015) but are not yet widely adopted. Nonetheless, fMRI studies have provided evidence for a conserved role of the DG (combined with CA2/3) in humans regarding contextual encoding and behavioral pattern separation (Yassa and Stark 2011; Bakker et al. 2008; Yassa et al. 2011; Stark et al. 2013; Maren et al. 2013; Lang et al. 2009; Alvarez et al. 2008; Marschner et al. 2008; Lacy et al. 2010; Kirwan et al. 2007). Using what we have learned about AHN from animal models, combined with work from human postmortem brain samples and functional neuroimaging studies of cognition, we can generate hypotheses for the pathophysiology of neuropsychiatric illnesses and in turn identify potential biomarkers and therapeutic targets.

***Stress*** Stress is a major risk factor for several neuropsychiatric disorders. Importantly, stress decreases AHN (Gould et al. 1997, 1998; McEwen and Magarinos 2001; Tanapat et al. 1998; Lee et al. 2006; Malberg and Duman 2003; Thomas et al. 2007; Yoshioka et al. 2022), while experimentally increasing AHN may increase resilience to chronic stress (Anacker et al. 2018; Hill et al. 2015; Planchez et al. 2021; Culig et al. 2017). Glucocorticoid stress hormones decrease the sparseness of GC activity and promote fear generalization (Lesuis et al. 2021). Acute stress in healthy individuals, on the other hand, may actually enhance discrimination of similar stimuli, though only with negative, not neutral, stimuli (Segal and Cahill 2009; Cunningham et al. 2018). Stress is of course a nonspecific term referring to a wide range of experiences. Conceptualization of the elements and characteristics of stressors that affect AHN continue to evolve. Undoubtedly though, a mechanistic understanding of neuropsychiatric illness will need to include the complicated and bidirectional effects of stress on emotional processing and cognition.

***Mood, Anxiety, and Trauma-Related Disorders*** Alterations in contextual encoding and deficits in pattern separation may contribute to overgeneralization of negative emotions seen in anxiety, depression, and trauma-related disorders (Kheirbek et al. 2012; Webler et al. 2021; Lissek et al. 2010; Lange et al. 2017; Leal and Yassa 2018; Belzung et al. 2015; Shelton and Kirwan 2013). First, the connections between anxiety and altered contextual encoding or pattern separation seem to be particularly robust. Previous studies in adults have shown a relationship between anxiety symptom severity and impaired memory of negative stimuli (Toffalini et al. 2015; Inaba and Ohira 2009). Study participants with panic and anxiety disorders are more likely to overgeneralize a conditioned-fear response, startling to both neutral and mild shock-associated stimuli (Lissek et al. 2010; Dunsmoor and Paz 2015; Lissek et al. 2014; Fraunfelter et al. 2022; Dymond et al. 2015; Bouton et al. 2001). Individuals with social anxiety disorder show elevated fear conditioning to socially relevant unconditioned stimuli (Lissek et al. 2008). Additionally, impaired pattern separation is observed in patients with anxiety disorders (Balderston et al. 2017) as well as those with sub-threshold anxiety symptoms (Lenaert et al. 2014). In an adolescent sample of subjects 9–14 years old using emotional discrimination task (Leal et al. 2014a, b), there was a significant correlation between overgeneralization of negative stimuli and anxiety severity (McMakin et al. 2022). A recent report indicated that, within a sample of patients with depression, anxiety, and comorbid anxiety and depression, anxiety symptoms were associated with impaired recognition of negative, but not neutral stimuli (Granger et al. 2022). Even in young, healthy subjects, there is a negative correlation between behavioral pattern separation performance and generalization of threat expectancy (Lange et al. 2017).

Aside from performance on memory and pattern separation tasks, there is also evidence that neural activity during the tasks is altered in patients with mood and anxiety symptoms (Yassa and Stark 2011; Leal et al. 2014a; Kirwan and Stark 2007). This suggests that in real-world situations and over extended periods of time, individuals with mood and anxiety symptoms may be exhibiting more subtle impairments in discrimination or less able to keep up when tasks are more difficult. Indeed, the correlation between lower behavioral pattern separation performance and anxiety and depression symptoms seems to be dynamic and more pronounced during acute stress or acute negative affective states (Shelton and Kirwan 2013; Bernstein and McNally 2018; Segal et al. 2012).

Second, there is also an association between behavioral pattern separation and symptoms of depression, which is highly comorbid with anxiety (Shelton and Kirwan 2013; Leal et al. 2014a, b, 2017a; Déry et al. 2013). Using behavioral pattern separation tasks that include aversive and neutral stimuli, subjects with depression demonstrate decreased discrimination of neutral stimuli but greater discrimination (Leal et al. 2014a) and heightened recognition (Bai et al. 2017; Hamilton and Gotlib 2008) of negative stimuli. In subjects with late-life depression, one study has shown that there is increased negative discrimination with more depressive symptoms (Leal et al. 2017a). Depressed patients exhibit negative affective biases and memory disturbances (Leal and Yassa 2018; Chamberlain and

Sakakian 2006; Lam et al. 2014; Pittenger 2013; Murrough et al. 2011; Conradi et al. 2011; Hasselbalch et al. 2011; Dillon and Pizzagalli 2018), and increased discrimination in emotional pattern separation tasks observed in depressed patients may then be due to heightened attention to negative stimuli. Notably, current versions of the rodent fear discrimination task use a neutral and negative context, so it is not yet known whether rodent models relevant to depression increase discrimination between negative stimuli, as is seen in humans with depression. Adapting the rodent tasks to test for this specific effect will be helpful for ongoing translational studies.

Third, regarding trauma-related disorders such as post-traumatic stress disorder (PTSD), extensive research has been conducted on impairments in contextual fear learning during and following trauma, which is suspected to be involved in the etiology of PTSD (Glenn et al. 2014, 2018; Acheson et al. 2012; Liberzon and Abelson 2016; Risbrough et al. 2016). Impaired pattern separation is also implicated in the overgeneralization of fear, which is a hallmark of PTSD (Kheirbek et al. 2012; Besnard and Sahay 2016). The links between AHN and PTSD from human studies are more indirect and speculative than with major depressive disorder. However, the behavioral paradigms that have been used to study fear learning and fear discrimination in laboratory animals (threatening event → vigilance behavior → overgeneralization) more closely resemble circumstances that lead to PTSD symptoms in humans, and the neural circuits that govern this process have therefore been analyzed in detail. Manipulations of AHN and/or behavioral pattern separation have long been posited as a promising target for PTSD treatments (Kheirbek et al. 2012; Besnard and Sahay 2016). Interestingly, the disproportionate conservation of cue encoding with impaired contextual encoding when iGCs are inhibited in mice that has been recently reported (Tuncdemir et al. 2023) has strong implications for the pathophysiology of PTSD, since overemphasis of negatively associated cues in a safe or neutral context is consistent with the type of overgeneralization seen in patients with PTSD.

Many medications used to treat depression, anxiety, and trauma-related affective dysregulation also increase AHN in rodents. This includes serotonin reuptake inhibitors (Santarelli et al. 2003), tricyclic antidepressants (Santarelli et al. 2003; Sairanen et al. 2005) and monoamine oxidase inhibitors (Malberg et al. 2000). This effect has been shown to partially depend on serotonin 5HT1a receptors (Santarelli et al. 2003; Samuels et al. 2015), but treatment with antidepressants with no known serotonergic activity, such as norepinephrine-selective reuptake inhibitors and electroconvulsive stimulation have also demonstrated enhanced AHN (Malberg et al. 2000). Studies of postmortem brain tissue from donors with major depressive disorder find fewer neural progenitor cells and mGCs in unmedicated donors, and treatment with antidepressant medications was correlated with an increased number of neural progenitor cells and mGCs (Boldrini et al. 2009, 2012, 2019; Boldrini and Arango 2010).

Taken together, while there is growing evidence linking AHN-sensitive processing with these psychiatric disorders, it is unlikely that patients with PTSD, unipolar or bipolar depression, anxiety disorders, and panic attacks categorically demonstrate impaired behavioral pattern separation at a group level. Instead, we conceptualize this as more likely to be a cognitive domain that is affected in a subset of patients that interact with and exacerbate other affective symptoms. PTSD, panic disorder, social anxiety and generalized anxiety disorder all share a common theme: overgeneralized fear into a broad range of contexts that impairs functioning. For depression, there appears to be an overgeneralization of negative appraisals of events and situations and a negative bias in memory (Chamberlain and Sakakian 2006; Murrough et al. 2011). Humans are subject to the added feature (and flaw) of their advanced cognition by being able to manufacture negative experiences and hypothetical situations that have never actually occurred, so this overgeneralized fear and negative emotion can be associated with an actual aversive experience or even just the idea of a possible aversive experience. In sum, the potential for treatments that augment AHN continues to be a compelling area of study for the treatment of these psychiatric conditions.

***Aging, Cognitive Decline, and Dementia*** Aged rodents have dramatically decreased levels of hippocampal neurogenesis (Knoth et al. 2010; Kempermann et al. 1998; Amrein et al. 2011; Bondolfi et al. 2004; Kuhn et al. 1996) and also show an age-related deficit in behavioral pattern separation (Wu et al. 2015; Creer et al. 2010; Cès et al. 2018; Nguyen et al. 2020). The rate of AHN in humans also decreases with age (Moreno-Jiménez et al. 2019; Knoth et al. 2010; Boldrini et al. 2018; Tobin et al. 2019; Walgrave et al. 2021; Berger et al. 2020), and the number of neural stem cells identified with canonical markers in human samples after puberty is low (Terreros-Roncal et al. 2021). However, it also seems that humans have an extremely prolonged maturation course for adult-born neurons. The net effect of this extended developmental time course is that, despite fewer stem cells being activated to become adult-born neurons, the total proportion of adult-born cells within the DG during adulthood might still be high (Zhou et al. 2022; Tosoni et al. 2023; Kempermann et al. 2018). If these cells retain the unique firing properties of immature neurons that impact the DG circuit, their impact on the network can still be substantial.

Behavioral pattern separation performance decreases with age in humans (Yassa et al. 2011; Toner et al. 2009; Leal et al. 2017b). People with mild cognitive impairment have even worse behavioral pattern separation (Yassa et al. 2011; Stark et al. 2013; Leal and Yassa 2013), which is further deteriorated in Alzheimer's disease (Leal and Yassa 2013; Wesnes et al. 2014). Pattern separation deficits and impaired cognitive discrimination, due in part to low AHN, have thus emerged as a possible target for prevention and treatment of age-related cognitive decline and/or dementia (Moreno-Jiménez et al. 2019; Knoth et al. 2010; Tobin et al. 2019; Stark et al. 2013; Leal and Yassa 2013, 2018; Wesnes et al. 2014; McAvoy and Sahay 2017; Choi and Tanzi 2019). AHN is significantly decreased in patients with Alzheimer's disease (Moreno-Jiménez et al. 2019; Tobin et al. 2019; Choi and Tanzi 2019, 2023), but Alzheimer's disease is characterized by a massive loss of neurons and neural tissue resulting in global cerebral atrophy (Fox and Schott 2004). Trying to enhance AHN in this condition may then seem futile. Still, the impact of iGCs on hippocampal circuitry may proportionally increase in the setting of widespread cell death (Salta et al. 2023; Choi et al. 2018; Briley et al. 2016). Additionally, the DG is among the first brain regions to be affected in Alzheimer's disease (Braak et al. 1993), so the development of early detection and interventions in this circuit could slow disease progression. In mouse models for Alzheimer's disease, enhancing AHN has yielded improvements in cognitive function (Walgrave et al. 2021; Choi and Tanzi 2023; Choi et al. 2018; Mishra et al. 2022; Richetin et al. 2015). Thus, although there is a plethora of shortcomings to rodent models for Alzheimer's disease (Götz et al. 2018), and challenges to translating these promising findings, manipulating AHN has emerged as a viable candidate to address cognitive impairments seen in aging and dementia.

***Therapeutic Strategies*** Neuropsychiatric disorders, including the ones discussed here, are heterogeneous, with multifactorial risk factors, a broad range of symptoms, and relatively widespread observable neural changes, both functionally and anatomically. One approach to identifying effective therapies is to stratify patients into groups to reduce this heterogeneity, possibly narrowing the number of dysfunctional pathways. This can aid in the identification of pathophysiological mechanisms and allow us to observe whether a single therapeutic strategy is having a significant impact. Behavioral pattern separation has emerged as a compelling marker for stratification in mood, anxiety, and trauma-related disorders because it allows for the quantification of cognitive performance for both neutral and negative stimuli and is also tied to the function of well-delineated brain areas (Leal and Yassa 2018). For neurodegenerative disorders and age-related cognitive decline, behavioral pattern separation may be more sensitive to early impairments than recognition-based memory tasks (Leal and Yassa 2018), and the entorhinal-hippocampal circuit is one of the first sites of observed cytopathology in aging (Stranahan et al. 2011) and Alzheimer's disease (Braak and Braak 1997; Khan et al. 2014).

Reversal of a pathophysiological mechanism with treatment is typically the goal when developing interventions, but there is also room for therapeutic effects through separate pathways. In other chronic illnesses such as cardiovascular disease or Type 2 diabetes, medications do not necessarily reverse disease-causing mechanisms. Likewise, while there is group-level evidence that individuals with mood, anxiety, trauma-related, or cognitive disorders may have decreased levels of AHN, enhancement of AHN with pharmacotherapy may be helpful for patients with seemingly "normal" AHN. Additionally, enhancement of AHN and pattern separation may also help alleviate symptoms even if pattern separation is not significantly outside of a normative range, or in the case of dementia, even if pattern separation is not restored to a normative range. As potential therapeutics are tested, the time course of positive effects cannot be assumed a priori, and a wide range of possibilities should be explored, as is feasible. For instance, it may be that improvements in pattern separation lead to decreases in mood or anxiety symptoms over time and that there is a delay between maximal clinical symptom improvement and detectable pattern separation changes. In the case of dementia, AHN manipulations may be most effective at early stages of illness to prolong the course of neurodegeneration or to flatten the slope of decline and disability. Pharmacological agents or neurostimulation to enhance AHN and pattern separation may also be best used to assist other types of psychotherapy and behavioral interventions. For example, AHN enhancers might increase the efficiency or effectiveness of exposure therapy for PTSD patients, cognitive behavioral therapy for patients with depression or anxiety, or occupational therapy for patients with dementia. In sum, basic science research and preclinical studies will continue to generate hypotheses about DG function to test in human populations. Based on the current knowledge, AHN, memory encoding, and pattern separation continue to be compelling targets for novel therapeutic development.

## References


Abrous DN, Koehl M, Lemoine M (2022) A Baldwin interpretation of adult hippocampal neurogenesis: from functional relevance to physiopathology. Mol Psychiatry 27:383–402

Acheson DT, Gresack JE, Risbrough VB (2012) Hippocampal dysfunction effects on context memory: possible etiology for posttraumatic stress disorder. Neuropharmacology 62:674–685

Aimone JB, Deng W, Gage FH (2011) Resolving new memories: a critical look at the dentate gyrus, adult neurogenesis, and pattern separation. Neuron 70:589–596

Akers KG et al (2014) Hippocampal neurogenesis regulates forgetting during adulthood and infancy. Science 344:598–602

Altman J, Das GD (1965) Autoradiographic and histological evidence of postnatal hippocampal neurogenesis in rats. J Comp Neurol 124:319–335

Alvarez RP, Biggs A, Chen G, Pine DS, Grillon C (2008) Contextual fear conditioning in humans: cortical-hippocampal and amygdala contributions. J Neurosci 28:6211–6219

Ammothumkandy A et al (2022) Altered adult neurogenesis and gliogenesis in patients with mesial temporal lobe epilepsy. Nat Neurosci 25:493–503

Amrein I, Isler K, Lipp HP (2011) Comparing adult hippocampal neurogenesis in mammalian species and orders: influence of chronological age and life history stage. Eur J Neurosci 34:978–987

Anacker C, Hen R (2017) Adult hippocampal neurogenesis and cognitive flexibility – linking memory and mood. Nat Rev Neurosci 18:335–346

Anacker C et al (2018) Hippocampal neurogenesis confers stress resilience by inhibiting the ventral dentate gyrus. Nature 559:98–102

Appleby PA, Kempermann G, Wiskott L (2011) The role of additive neurogenesis and synaptic plasticity in a hippocampal memory model with grid-cell like input. PLoS Comput Biol 7:e1001063

Babcock KR, Page JS, Fallon JR, Webb AE (2021) Adult hippocampal neurogenesis in aging and Alzheimer's disease. Stem Cell Rep 16:681–693. https://doi.org/10.1016/j.stemcr.2021.01.019

Bai T et al (2017) Electroconvulsive therapy regulates emotional memory bias of depressed patients. Psychiatry Res 257:296–302

Bakker A, Kirwan CB, Miller M, Stark CEL (2008) Pattern separation in the human hippocampal CA3 and dentate gyrus. Science 319:1640–1642

Balderston NL et al (2017) Effect of anxiety on behavioural pattern separation in humans. Cogn Emot 31:238–248

Bannerman DM et al (2002) Double dissociation of function within the hippocampus: spatial memory and hyponeophagia. Behav Neurosci 116(5):884–901. https://doi.org/10.1037/0735-7044.116.5.884

Bannerman DM et al (2004) Regional dissociations within the hippocampus – memory and anxiety. Neurosci Biobehav Rev 28:273–283

Bao AM, Swaab DF (2018) The art of matching brain tissue from patients and controls for postmortem research. Handb Clin Neurol 150:197–217

Becker S (2005) A computational principle for hippocampal learning and neurogenesis. Hippocampus 15:722–738

Belzung C, Willner P, Philippot P (2015) Depression: from psychopathology to pathophysiology. Curr Opin Neurobiol 30:24–30

Benna MK, Fusi S (2021) Place cells may simply be memory cells: memory compression leads to spatial tuning and history dependence. Proc Natl Acad Sci USA 118:e2018422118

Bergami M et al (2015) A critical period for experience-dependent remodeling of adult-born neuron connectivity. Neuron 85:710–717

Berger T, Lee H, Young AH, Aarsland D, Thuret S (2020) Adult hippocampal neurogenesis in major depressive disorder and Alzheimer's disease. Trends Mol Med 26:803–818

Bernstein EE, McNally RJ (2018) Exploring behavioral pattern separation and risk for emotional disorders. J Anxiety Disord 59:27–33

Besnard A, Sahay A (2016) Adult hippocampal neurogenesis, fear generalization, and stress. Neuropsychopharmacology 41:24–44

Boldrini M, Arango V (2010) Antidepressants, age, and neuroprogenitors. Neuropsychopharmacology 35:351–352

Boldrini M et al (2009) Antidepressants increase neural progenitor cells in the human hippocampus. Neuropsychopharmacology 34:2376–2389

Boldrini M et al (2012) Hippocampal angiogenesis and progenitor cell proliferation are increased with antidepressant use in major depression. Biol Psychiatry 72:562–571

Boldrini M et al (2018) Human hippocampal neurogenesis persists throughout aging. Cell Stem Cell 22:589–599.e5

Boldrini M et al (2019) Resilience is associated with larger dentate gyrus, while suicide decedents with major depressive disorder have fewer granule neurons. Biol Psychiatry 85:850–862

Bondolfi L, Ermini F, Long JM, Ingram DK, Jucker M (2004) Impact of age and caloric restriction on neurogenesis in the dentate gyrus of C57BL/6 mice. Neurobiol Aging 25:333–340

Bouton ME, Mineka S, Barlow DH (2001) A modern learning theory perspective on the etiology of panic disorder. Psychol Rev 108:4–32

Braak H, Braak E (1997) Frequency of stages of Alzheimer-related lesions in different age categories. Neurobiol Aging 18:351–357

Braak H, Braak E, Bohl J (1993) Staging of Alzheimer-related cortical destruction. Eur Neurol 33:403–408

Briley D et al (2016) Preserved neurogenesis in non-demented individuals with AD neuropathology. Sci Rep 6:27812

Burgess N, O'Keefe J (1996) Neuronal computations underlying the firing of place cells and their role in navigation. Hippocampus 6:749–762

Burghardt NS, Park EH, Hen R, Fenton AA (2012) Adult-born hippocampal neurons promote cognitive flexibility in mice. Hippocampus 22:1795–1808

Cahill SP, Martinovic A, Cole JD, Seib DR, Snyder JS (2019) A combination of running and memantine increases neurogenesis and reduces activation of developmentally-born dentate granule neurons in rats. Behav Brain Res 372:112005

Campos AC et al (2013) The anxiolytic effect of cannabidiol on chronically stressed mice depends on hippocampal neurogenesis: involvement of the endocannabinoid system. Int J Neuropsychopharmacol 16:1407–1419

Castilla-Ortega E, Santín LJ (2020) Adult hippocampal neurogenesis as a target for cocaine addiction: a review of recent developments. Curr Opin Pharmacol 50:109–116

Ceccarelli M, D'Andrea G, Micheli L, Tirone F (2020) Interaction between neurogenic stimuli and the gene network controlling the activation of stem cells of the adult neurogenic niches, in physiological and pathological conditions. Front Cell Dev Biol 8:211

Cès A et al (2018) Age-related vulnerability of pattern separation in C57BL/6J mice. Neurobiol Aging 62:120–129
Chamberlain SR, Sakakian BJ (2006) The neuropsychology of mood disorders. Curr Psychiatry Rep 8:458–463
Choi SH, Tanzi RE (2019) Is Alzheimer's disease a neurogenesis disorder? Cell Stem Cell 25:7–8
Choi SH, Tanzi RE (2023) Adult neurogenesis in Alzheimer's disease. Hippocampus 33:307–321
Choi SH et al (2018) Combined adult neurogenesis and BDNF mimic exercise effects on cognition in an Alzheimer's mouse model. Science 361:eaan8821
Christian KM, Song H, Ming GL (2014) Functions and dysfunctions of adult hippocampal neurogenesis. Annu Rev Neurosci 37:243–262
Cipriani A et al (2018) Comparative efficacy and acceptability of 21 antidepressant drugs for the acute treatment of adults with major depressive disorder: a systematic review and network meta-analysis. Lancet 391:1357–1366
Clark PJ, Brzezinska WJ, Puchalski EK, Krone DA, Rhodes JS (2009) Functional analysis of neurovascular adaptations to exercise in the dentate gyrus of young adult mice associated with cognitive gain. Hippocampus 19:937–950
Clelland CD et al (2009) A functional role for adult hippocampal neurogenesis in spatial pattern separation. Science 325:210–213
Clemenson GD et al (2015) Enrichment rescues contextual discrimination deficit associated with immediate shock. Hippocampus 25:385–392
Cole JD et al (2020) Adult-born hippocampal neurons undergo extended development and are morphologically distinct from neonatally-born neurons. J Neurosci 40:5740–5756
Colgin LL, Moser EI, Moser MB (2008) Understanding memory through hippocampal remapping. Trends Neurosci 31:469–477
Conradi HJ, Ormel J, De Jonge P (2011) Presence of individual (residual) symptoms during depressive episodes and periods of remission: a 3-year prospective study. Psychol Med 41:1165–1174
Cooper C, Moon HY, Van Praag H (2018) On the run for hippocampal plasticity. Cold Spring Harb Perspect Med 8:a029736
Creer DJ, Romberg C, Saksida LM, Van Praag H, Bussey TJ (2010) Running enhances spatial pattern separation in mice. Proc Natl Acad Sci USA 107:2367–2372
Culig L et al (2017) Increasing adult hippocampal neurogenesis in mice after exposure to unpredictable chronic mild stress may counteract some of the effects of stress. Neuropharmacology 126:179–189
Cunningham TJ, Leal SL, Yassa MA, Payne JD (2018) Post-encoding stress enhances mnemonic discrimination of negative stimuli. Learn Mem 25:611–619
Danielson NBB et al (2016) Distinct contribution of adult-born hippocampal granule cells to context encoding. Neuron 90:101–112
Denny CA, Burghardt NS, Schachter DM, Hen R, Drew MR (2012) 4- to 6-week-old adult-born hippocampal neurons influence novelty-evoked exploration and contextual fear conditioning. Hippocampus 22:1188–1201
Denny CA et al (2014) Hippocampal memory traces are differentially modulated by experience, time, and adult neurogenesis. Neuron 83:189–201
Déry N et al (2013) Adult hippocampal neurogenesis reduces memory interference in humans: opposing effects of aerobic exercise and depression. Front Neurosci 7:66
Deshmukh SS, Knierim JJ (2013) Influence of local objects on hippocampal representations: landmark vectors and memory. Hippocampus 23:253–267
Dieni CV et al (2016) Low excitatory innervation balances high intrinsic excitability of immature dentate neurons. Nat Commun 7:11313
Dillon DG, Pizzagalli DA (2018) Mechanisms of memory disruption in depression. Trends Neurosci 41:137–149
Dong HW, Swanson LW, Chen L, Fanselow MS, Toga AW (2009) Genomic-anatomic evidence for distinct functional domains in hippocampal field CA1. Proc Natl Acad Sci USA 106:11794–11799

Dranovsky A, Hen R (2006) Hippocampal neurogenesis: regulation by stress and antidepressants. Biol Psychiatry 59:1136–1143
Dranovsky A et al (2011) Experience dictates stem cell fate in the adult hippocampus. Neuron 70:908
Drew M, Hen R (2007) Adult hippocampal neurogenesis as target for the treatment of depression. CNS Neurol Disord Drug Targets 6:205–218
Drew LJ, Fusi S, Hen R (2013) Adult neurogenesis in the mammalian hippocampus: why the dentate gyrus? Learn Mem 20:710–729
Drew LJ et al (2016) Activation of local inhibitory circuits in the dentate gyrus by adult-born neurons. Hippocampus 26:763–778
Dunsmoor JE, Paz R (2015) Fear generalization and anxiety: behavioral and neural mechanisms. Biol Psychiatry 78:336–343
Dupret D et al (2008) Spatial relational memory requires hippocampal adult neurogenesis. PLoS One 3:e1959
Dymond S, Dunsmoor JE, Vervliet B, Roche B, Hermans D (2015) Fear generalization in humans: systematic review and implications for anxiety disorder research. Behav Ther 46:561–582
Eichenbaum H (2018) Barlow versus Hebb: when is it time to abandon the notion of feature detectors and adopt the cell assembly as the unit of cognition? Neurosci Lett 680:88–93
Eriksson PS et al (1998) Neurogenesis in the adult human hippocampus. Nat Med 4:1313–1317
Ernst A et al (2014) Neurogenesis in the striatum of the adult human brain. Cell 156:1072–1083
Espósito MS et al (2005) Neuronal differentiation in the adult hippocampus recapitulates embryonic development. J Neurosci 25:10074–10086
Fanselow MS, Dong H (2010) Are the dorsal and ventral hippocampus functionally distinct structures? Neuron 65:7–19
Ferbinteanu J, McDonald RJ (2001) Dorsal/ventral hippocampus, fornix, and conditioned place preference. Hippocampus 11:187–200
Filipkowski RK, Kaczmarek L (2018) Severely impaired adult brain neurogenesis in cyclin D2 knockout mice produces very limited phenotypic changes. Prog Neuropsychopharmacol Biol Psychiatry 80:63–67
Flor-García M et al (2020) Unraveling human adult hippocampal neurogenesis. Nat Protoc 15:668–693
Foster TC (2006) Biological markers of age-related memory deficits: treatment of senescent physiology. CNS Drugs 20:153–166
Fox NC, Schott JM (2004) Imaging cerebral atrophy: normal ageing to Alzheimer's disease. Lancet 363:392–394
Frankland PW, Josselyn SA (2016) Hippocampal neurogenesis and memory clearance. Neuropsychopharmacology 41:382–383
Frankland PW, Köhler S, Josselyn SA (2013) Hippocampal neurogenesis and forgetting. Trends Neurosci 36:497–503
Fraunfelter L, Gerdes ABM, Alpers GW (2022) Fear one, fear them all: a systematic review and meta-analysis of fear generalization in pathological anxiety. Neurosci Biobehav Rev 139:104707
Freund J et al (2013) Emergence of individuality in genetically identical mice. Science (80-. ) 340:756–759
Fusi S, Miller EK, Rigotti M (2016) Why neurons mix: high dimensionality for higher cognition. Curr Opin Neurobiol 37:66–74
Galea LAM et al (2013) Sex, hormones and neurogenesis in the hippocampus: hormonal modulation of neurogenesis and potential functional implications. J Neuroendocrinol 25:1039–1061
Gallardo-Caballero M et al (2023) Prolonged fixation and postmortem delay impede the study of adult neurogenesis in mice. Commun Biol 61(6):1–12
Gao Y, Syed M, Zhao X (2023) Mechanisms underlying the effect of voluntary running on adult hippocampal neurogenesis. Hippocampus 33:373–390

Garrett L, Lie DC, Hrabé de Angelis M, Wurst W, Hölter SM (2012) Voluntary wheel running in mice increases the rate of neurogenesis without affecting anxiety-related behaviour in single tests. BMC Neurosci 13:61
Garthe A, Kempermann G (2013) An old test for new neurons: refining the Morris water maze to study the functional relevance of adult hippocampal neurogenesis. Front Neurosci 7:63. https://doi.org/10.3389/FNINS.2013.00063
Garthe A, Behr J, Kempermann G (2009) Adult-generated hippocampal neurons allow the flexible use of spatially precise learning strategies. PLoS One 4:e5464
Garthe A, Huang Z, Kaczmarek L, Filipkowski RK, Kempermann G (2014) Not all water mazes are created equal: cyclin D2 knockout mice with constitutively suppressed adult hippocampal neurogenesis do show specific spatial learning deficits. Genes Brain Behav 13:357–364
Garthe A, Roeder I, Kempermann G (2016) Mice in an enriched environment learn more flexibly because of adult hippocampal neurogenesis. Hippocampus 26:261–271
Ge S et al (2006) GABA regulates synaptic integration of newly generated neurons in the adult brain. Nature 439:589–593
Ge S, Yang C, Hsu K, Ming G, Song H (2007) A critical period for enhanced synaptic plasticity in newly generated neurons of the adult brain. Neuron 54:559–566
Geiller T, Fattahi M, Choi JS, Royer S (2017) Place cells are more strongly tied to landmarks in deep than in superficial CA1. Nat Commun 81(8):1–11
Gilbert PE, Kesner RP, Lee I (2001) Dissociating hippocampal subregions: double dissociation between dentate gyrus and CA1. Hippocampus 11:626–636
Glenn DE, Minor TR, Vervliet B, Craske MG (2014) The effect of glucose on hippocampal-dependent contextual fear conditioning. Biol Psychiatry 75:847–854
Glenn DE, Risbrough VB, Simmons AN, Acheson DT, Stout DM (2018) The future of contextual fear learning for PTSD research: a methodological review of neuroimaging studies. Curr Top Behav Neurosci 38:207–228
GoodSmith D et al (2022) Flexible encoding of objects and space in single cells of the dentate gyrus. Curr Biol 32:1088–1101.e5
Götz J, Bodea LG, Goedert M (2018) Rodent models for Alzheimer disease. Nat Rev Neurosci 1910(19):583–598
Gould E, Tanapat P (1999) Stress and hippocampal neurogenesis. Biol Psychiatry 46:1472–1479
Gould E, McEwen BS, Tanapat P, Galea LAM, Fuchs E (1997) Neurogenesis in the dentate gyrus of the adult tree shrew is regulated by psychosocial stress and NMDA receptor activation. J Neurosci 17:2492–2498
Gould E, Tanapat P, Mcewen BS, Flügge G, Fuchs E (1998) Proliferation of granule cell precursors in the dentate gyrus of adult monkeys is diminished by stress. Proc Natl Acad Sci USA 95:3168–3171
Gould E, Beylin A, Tanapat P, Reeves A, Shors TJ (1999) Learning enhances adult neurogenesis in the hippocampal formation. Nat Neurosci 2:260–265
Granger SJ et al (2022) Latent anxiety in clinical depression is associated with worse recognition of emotional stimuli. J Affect Disord 301:368–377
Gu Y et al (2012) Optical controlling reveals time-dependent roles for adult-born dentate granule cells. Nat Neurosci 15:1700
Hainmueller T, Bartos M (2018) Parallel emergence of stable and dynamic memory engrams in the hippocampus. Nature 558:292–296
Hainmueller T, Bartos M (2020) Dentate gyrus circuits for encoding, retrieval and discrimination of episodic memories. Nat Rev Neurosci 21:153–168
Hamilton JP, Gotlib IH (2008) Neural substrates of increased memory sensitivity for negative stimuli in major depression. Biol Psychiatry 63:1155–1162
Hargreaves EL, Rao G, Lee I, Knierim JJ (2005) Major dissociation between medial and lateral entorhinal input to dorsal hippocampus. Science 308:1792–1794
Hasselbalch BJ, Knorr U, Kessing LV (2011) Cognitive impairment in the remitted state of unipolar depressive disorder: a systematic review. J Affect Disord 134:20–31

Heine VM, Zareno J, Maslam S, Joëls M, Lucassen PJ (2005) Chronic stress in the adult dentate gyrus reduces cell proliferation near the vasculature and VEGF and Flk-1 protein expression. Eur J Neurosci 21:1304–1314

Hill AS, Sahay A, Hen R (2015) Increasing adult hippocampal neurogenesis is sufficient to reduce anxiety and depression-like behaviors. Neuropsychopharmacology 40(10):2368–2378

Høydal ØA, Skytøen ER, Andersson SO, Moser MB, Moser EI (2019) Object-vector coding in the medial entorhinal cortex. Nature 568(7752):400–404

Huckleberry KA, Shansky RM (2021) The unique plasticity of hippocampal adult-born neurons: contributing to a heterogeneous dentate. Hippocampus 31:543–556

Huckleberry KA et al (2018) Dorsal and ventral hippocampal adult-born neurons contribute to context fear memory. Neuropsychopharmacology 43(12):2487–2496

Iglesias JE et al (2015) A computational atlas of the hippocampal formation using ex vivo, ultra-high resolution MRI: application to adaptive segmentation of in vivo MRI. Neuroimage 115:117–137

Ikrar T et al (2013) Adult neurogenesis modifies excitability of the dentate gyrus. Front Neural Circuits 7:204

Inaba M, Ohira H (2009) Reduced recollective memory about negative items in high trait anxiety individuals: an ERP study. Int J Psychophysiol 74:106–113

Jessberger S, Kempermann G (2003) Adult-born hippocampal neurons mature into activity-dependent responsiveness. Eur J Neurosci 18:2707–2712

Jimenez JC et al (2020) Contextual fear memory retrieval by correlated ensembles of ventral CA1 neurons. Nat Commun 111(11):1–11

Johansson CB, Svensson M, Wallstedt L, Janson AM, Frisén J (1999a) Neural stem cells in the adult human brain. Exp Cell Res 253:733–736

Johansson CB et al (1999b) Identification of a neural stem cell in the adult mammalian central nervous system. Cell 96:25–34

Jung D et al (2019) Dentate granule and mossy cells exhibit distinct spatiotemporal responses to local change in a one-dimensional landscape of visual-tactile cues. Sci Rep 9:20225

Jungenitz T, Radic T, Jedlicka P, Schwarzacher SW (2014) High-frequency stimulation induces gradual immediate early gene expression in maturing adult-generated hippocampal granule cells. Cereb Cortex 24:1845–1857

Kempermann G (2022) What is adult hippocampal neurogenesis good for? Front Neurosci 16:852680

Kempermann G, Kuhn HG, Gage FH (1997) More hippocampal neurons in adult mice living in an enriched environment. Nature 386:493–495

Kempermann G, Kuhn HG, Gage FH (1998) Experience-induced neurogenesis in the senescent dentate gyrus. J Neurosci 18:3206–3212

Kempermann G, Gast D, Gage FH (2002) Neuroplasticity in old age: sustained fivefold induction of hippocampal neurogenesis by long-term environmental enrichment. Ann Neurol 52:135–143

Kempermann G et al (2018) Human adult neurogenesis: evidence and remaining questions. Cell Stem Cell 23:25–30

Kerloch T, Clavreul S, Goron A, Abrous DN, Pacary E (2019) Dentate granule neurons generated during perinatal life display distinct morphological features compared with later-born neurons in the mouse hippocampus. Cereb Cortex 29:3527–3539

Kesner RP (2013) An analysis of the dentate gyrus function. Behav Brain Res 254:1–7

Khan UA et al (2014) Molecular drivers and cortical spread of lateral entorhinal cortex dysfunction in preclinical Alzheimer's disease. Nat Neurosci 17:304–311

Kheirbek MA, Klemenhagen KC, Sahay A, Hen R (2012) Neurogenesis and generalization: a new approach to stratify and treat anxiety disorders. Nat Neurosci 15:1613

Kheirbek MA et al (2013) Differential control of learning and anxiety along the dorsoventral axis of the dentate gyrus. Neuron 77:955–968

Kim J, Lee I (2011) Hippocampus is necessary for spatial discrimination using distal cue-configuration. Hippocampus 21:609–621

Kim TA, Syty MD, Wu K, Ge S (2022) Adult hippocampal neurogenesis and its impairment in Alzheimer's disease. Zool Res 43:481–496

Kirshenbaum GS, Lieberman SR, Briner TJ, David Leonardo E, Dranovsky A (2014) Adolescent but not adult-born neurons are critical for susceptibility to chronic social defeat. Front Behav Neurosci 8:289

Kirwan CB, Stark CEL (2007) Overcoming interference: an fMRI investigation of pattern separation in the medial temporal lobe. Learn Mem 14:625–633

Kirwan CB, Jones CK, Miller MI, Stark CEL (2007) High-resolution fMRI investigation of the medial temporal lobe. Hum Brain Mapp 28:959–966

Kitamura T et al (2009) Adult neurogenesis modulates the hippocampus-dependent period of associative fear memory. Cell 139:814–827

Kitamura T et al (2017) Engrams and circuits crucial for systems consolidation of a memory. Science (80-. ) 356:73–78

Kjelstrup KG et al (2002) Reduced fear expression after lesions of the ventral hippocampus. Proc Natl Acad Sci USA 99:10825–10830

Knierim JJ, Neunuebel JP, Deshmukh SS (2013) Functional correlates of the lateral and medial entorhinal cortex: objects, path integration and local-global reference frames. Philos Trans R Soc Lond Ser B Biol Sci 369:20130369

Knoth R et al (2010) Murine features of neurogenesis in the human hippocampus across the lifespan from 0 to 100 years. PLoS One 5:e8809

Ko SY, Frankland PW (2021) Neurogenesis-dependent transformation of hippocampal engrams. Neurosci Lett 762:136176

Kobilo T et al (2011) Running is the neurogenic and neurotrophic stimulus in environmental enrichment. Learn Mem 18:605–609

Kowalczyk A et al (2004) The critical role of cyclin D2 in adult neurogenesis. J Cell Biol 167:209–213

Kubie JL, Levy ERJ, Fenton AA (2020) Is hippocampal remapping the physiological basis for context? Hippocampus 30:851–864

Kuhn HG, Dickinson-Anson H, Gage FH (1996) Neurogenesis in the dentate gyrus of the adult rat: age-related decrease of neuronal progenitor proliferation. J Neurosci 16:2027–2033

Lacefield CO, Itskov V, Reardon T, Hen R, Gordon JA (2012) Effects of adult-generated granule cells on coordinated network activity in the dentate gyrus. Hippocampus 22:106–116

Lacy JW, Yassa MA, Stark SM, Muftuler LT, Stark CEL (2010) Distinct pattern separation related transfer functions in human CA3/dentate and CA1 revealed using high-resolution fMRI and variable mnemonic similarity. Learn Mem 18:15–18

Lam RW, Kennedy SH, McIntyre RS, Khullar A (2014) Cognitive dysfunction in major depressive disorder: effects on psychosocial functioning and implications for treatment. Can J Psychiatr 59:649–654

Lang S et al (2009) Context conditioning and extinction in humans: differential contribution of the hippocampus, amygdala and prefrontal cortex. Eur J Neurosci 29:823–832

Lange I et al (2017) Behavioral pattern separation and its link to the neural mechanisms of fear generalization. Soc Cogn Affect Neurosci 12:1720–1729

Latuske P, Kornienko O, Kohler L, Allen K (2018) Hippocampal remapping and its entorhinal origin. Front Behav Neurosci 11:253

Leal SL, Yassa MA (2013) Perturbations of neural circuitry in aging, mild cognitive impairment, and Alzheimer's disease. Ageing Res Rev 12:823–831

Leal SL, Yassa MA (2018) Integrating new findings and examining clinical applications of pattern separation. Nat Neurosci 21:163–173

Leal SL, Tighe SK, Jones CK, Yassa MA (2014a) Pattern separation of emotional information in hippocampal dentate and CA3. Hippocampus 24:1146–1155

Leal SL, Tighe SK, Yassa MA (2014b) Asymmetric effects of emotion on mnemonic interference. Neurobiol Learn Mem 111:41–48

Leal SL, Noche JA, Murray EA, Yassa MA (2017a) Disruption of amygdala–entorhinal–hippocampal network in late-life depression. Hippocampus 27:464–476
Leal SL, Noche JA, Murray EA, Yassa MA (2017b) Age-related individual variability in memory performance is associated with amygdala-hippocampal circuit function and emotional pattern separation. Neurobiol Aging 49:9–19
Lee JW, Jung MW (2017) Separation or binding? Role of the dentate gyrus in hippocampal mnemonic processing. Neurosci Biobehav Rev 75:183–194
Lee KJ et al (2006) Chronic mild stress decreases survival, but not proliferation, of new-born cells in adult rat hippocampus. Exp Mol Med 38:44–54
Lee JW, Kim WR, Sun W, Jung MW (2009) Role of dentate gyrus in aligning internal spatial map to external landmark. Learn Mem 16:530–536
Lehmann ML, Brachman RA, Martinowich K, Schloesser RJ, Herkenham M (2013) Glucocorticoids orchestrate divergent effects on mood through adult neurogenesis. J Neurosci 33:2961–2972
Leiter O, Kempermann G, Walker TL (2016) A common language: how neuroimmunological cross talk regulates adult hippocampal neurogenesis. Stem Cells Int 2016:1–13
Lemaire V, Aurousseau C, Le Moal M, Abrous DN (1999) Behavioural trait of reactivity to novelty is related to hippocampal neurogenesis. Eur J Neurosci 11:4006–4014
Lemaire V et al (2012) Long-lasting plasticity of hippocampal adult-born neurons. J Neurosci 32:3101–3108
Lenaert B et al (2014) Aversive learning and generalization predict subclinical levels of anxiety: a six-month longitudinal study. J Anxiety Disord 28:747–753
Leonardo ED, Richardson-Jones JW, Sibille E, Kottman A, Hen R (2006) Molecular heterogeneity along the dorsal-ventral axis of the murine hippocampal CA1 field: a microarray analysis of gene expression. Neuroscience 137:177–186
Lesuis SL et al (2021) Glucocorticoids promote fear generalization by increasing the size of a dentate gyrus engram cell population. Biol Psychiatry 90:494–504
Leutgeb JK, Leutgeb S, Moser MB, Moser EI (2007) Pattern separation in the dentate gyrus and CA3 of the hippocampus. Science (80-. ) 315:961–966
Li Y, Zhao X (2014) Concise review: fragile X proteins in stem cell maintenance and differentiation. Stem Cells 32:1724–1733
Li Y et al (2017) Satb2 ablation impairs hippocampus-based long-term spatial memory and short-term working memory and immediate early genes (IEGs)-mediated hippocampal synaptic plasticity. Mol Neurobiol 2017. https://doi.org/10.1007/s12035-017-0531-5
Liberzon I, Abelson JL (2016) Context processing and the neurobiology of post-traumatic stress disorder. Neuron 92:14–30
Lissek S et al (2008) Elevated fear conditioning to socially relevant unconditioned stimuli in social anxiety disorder. Am J Psychiatry 165:124–132
Lissek S et al (2010) Overgeneralization of conditioned fear as a pathogenic marker of panic disorder. Am J Psychiatry 167:47–55
Lissek S et al (2014) Generalized anxiety disorder is associated with overgeneralization of classically conditioned fear. Biol Psychiatry 75:909–915
Liu X et al (2012) Optogenetic stimulation of a hippocampal engram activates fear memory recall. Nature 484:381–385
Lopes JB, Małz M, Senko AN, Zocher S, Kempermann G (2023) Loss of individualized behavioral trajectories in adult neurogenesis-deficient cyclin D2 knockout mice. Hippocampus 33:360–372
Luna VM et al (2019) Adult-born hippocampal neurons bidirectionally modulate entorhinal inputs into the dentate gyrus. Science (80-. ) 364:578–583
Mahmoud R, Wainwright SR, Galea LAM (2016) Sex hormones and adult hippocampal neurogenesis: regulation, implications, and potential mechanisms. Front Neuroendocrinol 41:129–152
Malberg JE, Duman RS (2003) Cell proliferation in adult hippocampus is decreased by inescapable stress: reversal by fluoxetine treatment. Neuropsychopharmacology 28:1562–1571

Malberg JE, Eisch AJ, Nestler EJ, Duman RS (2000) Chronic antidepressant treatment increases neurogenesis in adult rat hippocampus. J Neurosci 20:9104–9110
Maren S, Holt WG (2004) Hippocampus and Pavlovian fear conditioning in rats: muscimol infusions into the ventral, but not dorsal, hippocampus impair the acquisition of conditional freezing to an auditory conditional stimulus. Behav Neurosci 118:97–110
Maren S, Phan KL, Liberzon I (2013) The contextual brain: implications for fear conditioning, extinction and psychopathology. Nat Rev Neurosci 14:417–428
Marín-Burgin A, Mongiat LA, Pardi MB, Schinder AF (2012) Unique processing during a period of high excitation/inhibition balance in adult-born neurons. Science (80-. ) 335:1238–1242
Marr D (1971) Simple memory: a theory for archicortex. Philos Trans R Soc Lond Ser B Biol Sci 262:23–81
Marschner A, Kalisch R, Vervliet B, Vansteenwegen D, Büchel C (2008) Dissociable roles for the hippocampus and the amygdala in human cued versus context fear conditioning. J Neurosci 28:9030–9036
Masachs N et al (2021) The temporal origin of dentate granule neurons dictates their role in spatial memory. Mol Psychiatry 2612(26):7130–7140
McAvoy KM, Sahay A (2017) Targeting adult neurogenesis to optimize hippocampal circuits in aging. Neurotherapeutics 14:630–645
Mcclelland JL, Goddard NH (1996) Considerations arising from a complementary learning systems perspective on hippocampus and neocortex. Hippocampus 6:654–665
McEwen BS, Magarinos AM (2001) Stress and hippocampal plasticity: implications for the pathophysiology of affective disorders. Hum Psychopharmacol 16:S7–S19
McHugh TJ et al (2007) Dentate gyrus NMDA receptors mediate rapid pattern separation in the hippocampal network. Science 317:94–99
McHugh SB et al (2022) Adult-born dentate granule cells promote hippocampal population sparsity. Nat Neurosci 2511(25):1481–1491
McMakin DL, Kimbler A, Tustison NJ, Pettit JW, Mattfeld AT (2022) Negative overgeneralization is associated with anxiety and mechanisms of pattern completion in peripubertal youth. Soc Cogn Affect Neurosci 17:231–240
Meltzer LA, Yabaluri R, Deisseroth K (2005) A role for circuit homeostasis in adult neurogenesis. Trends Neurosci 28:653–660
Miller SM, Sahay A (2019) Functions of adult-born neurons in hippocampal memory interference and indexing. Nat Neurosci 22:1565–1575
Mirescu C, Gould E (2006) Stress and adult neurogenesis. Hippocampus 16:233–238
Mishra R et al (2022) Augmenting neurogenesis rescues memory impairments in Alzheimer's disease by restoring the memory-storing neurons. J Exp Med 219:e20220391
Mongiat LA, Espósito MS, Lombardi G, Schinder AF (2009) Reliable activation of immature neurons in the adult hippocampus. PLoS One 4:e5320
Moreno-Jiménez EP et al (2019) Adult hippocampal neurogenesis is abundant in neurologically healthy subjects and drops sharply in patients with Alzheimer's disease. Nat Med 25:554–560
Moreno-Jiménez EP, Terreros-Roncal J, Flor-García M, Rábano A, Llorens-Martín M (2021) Evidences for adult hippocampal neurogenesis in humans. J Neurosci 41:2541–2553
Moser MB, Moser EI (1998) Functional differentiation in the hippocampus. Hippocampus 8:608–619
Moser MB, Moser EI, Forrest E, Andersen P, Morris RGM (1995) Spatial learning with a minislab in the dorsal hippocampus. Proc Natl Acad Sci USA 92:9697–9701
Murrough JW, Iacoviello B, Neumeister A, Charney DS, Iosifescu DV (2011) Cognitive dysfunction in depression: neurocircuitry and new therapeutic strategies. Neurobiol Learn Mem 96:553–563
Mustroph ML et al (2012) Aerobic exercise is the critical variable in an enriched environment that increases hippocampal neurogenesis and water maze learning in male C57BL/6J mice. Neuroscience 219:62–71

Nakashiba T et al (2012) Young dentate granule cells mediate pattern separation, whereas old granule cells facilitate pattern completion. Cell 149:188–201
Neunuebel JP, Knierim JJ (2014) CA3 retrieves coherent representations from degraded input: direct evidence for CA3 pattern completion and dentate gyrus pattern separation. Neuron 81:416–427
Nguyen PT et al (2020) Microglial remodeling of the extracellular matrix promotes synapse plasticity. Cell 182:388–403.e15
O'Keefe J, Dostrovsky J (1971) The hippocampus as a spatial map. Preliminary evidence from unit activity in the freely-moving rat. Brain Res 34:171–175
Olariu A, Cleaver KM, Cameron HA (2007) Decreased neurogenesis in aged rats results from loss of granule cell precursors without lengthening of the cell cycle. J Comp Neurol 501:659–667
Overstreet LS et al (2004) A transgenic marker for newly born granule cells in dentate gyrus. J Neurosci 24:3251–3259
Parent JM, Tada E, Fike JR, Lowenstein DH (1999) Inhibition of dentate granule cell neurogenesis with brain irradiation does not prevent seizure-induced mossy fiber synaptic reorganization in the rat. J Neurosci 19:4508–4519
Park EH, Burghardt NS, Dvorak D, Hen R, Fenton AA (2015) Experience-dependent regulation of dentate gyrus excitability by adult-born granule cells. J Neurosci 35:11656–11666
Parylak SL et al (2023) Neuronal activity-related transcription is blunted in immature compared to mature dentate granule cells. Hippocampus 33:412–423
Patzlaff NE, Shen M, Zhao X (2018) Regulation of adult neurogenesis by the fragile X family of RNA binding proteins. Brain Plast 3:205–223
Pignatelli M et al (2019) Engram cell excitability state determines the efficacy of memory retrieval. Neuron 101:274–284.e5
Pittenger C (2013) Disorders of memory and plasticity in psychiatric disease. Dialogues Clin Neurosci 15:455–463
Planchez B et al (2021) Increasing adult hippocampal neurogenesis promotes resilience in a mouse model of depression. Cells 10:972
Rai KS, Hattiangady B, Shetty AK (2007) Enhanced production and dendritic growth of new dentate granule cells in the middle-aged hippocampus following intracerebroventricular FGF-2 infusions. Eur J Neurosci 26:1765–1779
Ramirez S et al (2013) Creating a false memory in the hippocampus. Science 341:387–391
Regensburger M, Prots I, Winner B (2014) Adult hippocampal neurogenesis in Parkinson's disease: impact on neuronal survival and plasticity. Neural Plast 2014:454696
Richetin K et al (2015) Genetic manipulation of adult-born hippocampal neurons rescues memory in a mouse model of Alzheimer's disease. Brain 138:440–455
Rigotti M et al (2013) The importance of mixed selectivity in complex cognitive tasks. Nature 497:585–590
Risbrough VB, Glenn DE, Baker DG (2016) On the road to translation for PTSD treatment: theoretical and practical considerations of the use of human models of conditioned fear for drug development. Curr Top Behav Neurosci 28:173–196
Ryan TJ, Roy DS, Pignatelli M, Arons A, Tonegawa S (2015) Memory. Engram cells retain memory under retrograde amnesia. Science 348:1007–1013
Sahay A, Hen R (2007) Adult hippocampal neurogenesis in depression. Nat Neurosci 10:1110–1115
Sahay A et al (2011) Increasing adult hippocampal neurogenesis is sufficient to improve pattern separation HHS Public Access. Nature 472:466–470
Sairanen M, Lucas G, Ernfors P, Castrén M, Castrén E (2005) Brain-derived neurotrophic factor and antidepressant drugs have different but coordinated effects on neuronal turnover, proliferation, and survival in the adult dentate gyrus. J Neurosci 25:1089–1094
Salta E et al (2023) Adult hippocampal neurogenesis in Alzheimer's disease: a roadmap to clinical relevance. Cell Stem Cell 30:120–136
Samuels BA et al (2015) 5-HT1A receptors on mature dentate gyrus granule cells are critical for the antidepressant response. Nat Neurosci 18:1606–1616

Sanai N et al (2011) Corridors of migrating neurons in the human brain and their decline during infancy. Nature 478:382–386
Santarelli L et al (2003) Requirement of hippocampal neurogenesis for the behavioral effects of antidepressants. Science (80-. ) 301:805–809
Saxe MD et al (2006) Ablation of hippocampal neurogenesis impairs contextual fear conditioning and synaptic plasticity in the dentate gyrus. Proc Natl Acad Sci USA 103:17501–17506
Schmidt-Hieber C, Jones P, Bischofberger J (2004) Enhanced synaptic plasticity in newly generated granule cells of the adult hippocampus. Nature 429:184–187
Schoenfeld TJ, Swanson C (2021) A runner's high for new neurons? Potential role for endorphins in exercise effects on adult neurogenesis. Biomol Ther 11:1077
Segal SK, Cahill L (2009) Endogenous noradrenergic activation and memory for emotional material in men and women. Psychoneuroendocrinology 34:1263–1271
Segal SK, Stark SM, Kattan D, Stark CE, Yassa MA (2012) Norepinephrine-mediated emotional arousal facilitates subsequent pattern separation. Neurobiol Learn Mem 97:465–469
Shelton DJ, Kirwan CB (2013) A possible negative influence of depression on the ability to overcome memory interference. Behav Brain Res 256:20–26
Snyder JS, Kee N, Wojtowicz JM (2001) Effects of adult neurogenesis on synaptic plasticity in the rat dentate gyrus. J Neurophysiol 85:2423–2431
Sorrells SF et al (2018) Human hippocampal neurogenesis drops sharply in children to undetectable levels in adults. Nature 555:377–381
Sorrells SF et al (2021) Positive controls in adults and children support that very few, if any, new neurons are born in the adult human hippocampus. J Neurosci 41:2554–2565
Spalding KL et al (2013) Dynamics of hippocampal neurogenesis in adult humans. Cell 153:1219
Speisman RB, Kumar A, Rani A, Foster TC, Ormerod BK (2013) Daily exercise improves memory, stimulates hippocampal neurogenesis and modulates immune and neuroimmune cytokines in aging rats. Brain Behav Immun 28:25–43
Stark SM, Yassa MA, Lacy JW, Stark CEL (2013) A task to assess behavioral pattern separation (BPS) in humans: data from healthy aging and mild cognitive impairment. Neuropsychologia 51:2442–2449
Stefanini F et al (2020) A distributed neural code in the dentate gyrus and in CA1. Neuron 107:703–716.e4
Stranahan AM, Haberman RP, Gallagher M (2011) Cognitive decline is associated with reduced reelin expression in the entorhinal cortex of aged rats. Cereb Cortex 21:392–400
Strange BA, Witter MP, Lein ES, Moser EI (2014) Functional organization of the hippocampal longitudinal axis. Nat Rev Neurosci 15:655–669
Surget A, Belzung C (2022) Adult hippocampal neurogenesis shapes adaptation and improves stress response: a mechanistic and integrative perspective. Mol Psychiatry 27:403–421
Suthana NA et al (2015) High-resolution 7T fMRI of human hippocampal subfields during associative learning. J Cogn Neurosci 27:1194
Swan AA et al (2014) Characterization of the role of adult neurogenesis in touch-screen discrimination learning. Hippocampus 24:1581–1591
Synder JS et al (2009) Adult-born hippocampal neurons are more numerous, faster maturing, and more involved in behavior in rats than in mice. J Neurosci 29:14484–14495
Tanapat P, Galea LA, Gould E (1998) Stress inhibits the proliferation of granule cell precursors in the developing dentate gyrus. Int J Dev Neurosci 16:235–239
Tannenholz L, Jimenez JC, Kheirbek MA (2014) Local and regional heterogeneity underlying hippocampal modulation of cognition and mood. Front Behav Neurosci 8:147
Tanti A, Belzung C (2013) Neurogenesis along the septo-temporal axis of the hippocampus: are depression and the action of antidepressants region-specific? Neuroscience 252:234–252
Tanti A, Rainer Q, Minier F, Surget A, Belzung C (2012) Differential environmental regulation of neurogenesis along the septo-temporal axis of the hippocampus. Neuropharmacology 63:374–384

Tanti A et al (2013) Region-dependent and stage-specific effects of stress, environmental enrichment, and antidepressant treatment on hippocampal neurogenesis. Hippocampus 23:797–811
Tartt AN et al (2018) Considerations for assessing the extent of hippocampal neurogenesis in the adult and aging human brain. Cell Stem Cell 23:782–783
Tartt AN, Mariani MB, Hen R, Mann JJ, Boldrini M (2022) Dysregulation of adult hippocampal neuroplasticity in major depression: pathogenesis and therapeutic implications. Mol Psychiatry 27:2689–2699
Terreros-Roncal J et al (2021) Impact of neurodegenerative diseases on human adult hippocampal neurogenesis. Science 374:1106–1113
Thomas RM, Hotsenpiller G, Peterson DA (2007) Acute psychosocial stress reduces cell survival in adult hippocampal neurogenesis without altering proliferation. J Neurosci 27:2734–2743
Tobin MK et al (2019) Human hippocampal neurogenesis persists in aged adults and Alzheimer's disease patients. Cell Stem Cell 24:974–982.e3
Toffalini E, Mirandola C, Coli T, Cornoldi C (2015) High trait anxiety increases inferential false memories for negative (but not positive) emotional events. Personal Individ Differ 75:201–204
Tonegawa S, Morrissey MD, Kitamura T (2018) The role of engram cells in the systems consolidation of memory. Nat Rev Neurosci 198(19):485–498
Toner CK, Pirogovsky E, Kirwan CB, Gilbert PE (2009) Visual object pattern separation deficits in nondemented older adults. Learn Mem 16:338–342
Toni N et al (2007) Synapse formation on neurons born in the adult hippocampus. Nat Neurosci 10:727–734
Toni N et al (2008) Neurons born in the adult dentate gyrus form functional synapses with target cells. Nat Neurosci 11:901–907
Tosoni G et al (2023) Mapping human adult hippocampal neurogenesis with single-cell transcriptomics: reconciling controversy or fueling the debate? Neuron 111:1714–1731.e3
Treves A, Rolls ET (1994) Computational analysis of the role of the hippocampus in memory. Hippocampus 4:374–391
Tronel S et al (2010) Spatial learning sculpts the dendritic arbor of adult-born hippocampal neurons. Proc Natl Acad Sci USA 107:7963–7968
Tronel S, Lemaire V, Charrier V, Montaron MF, Abrous DN (2015) Influence of ontogenetic age on the role of dentate granule neurons. Brain Struct Funct 220:645–661
Tuncdemir SN et al (2022) Parallel processing of sensory cue and spatial information in the dentate gyrus. Cell Rep 38:110257
Tuncdemir SN et al (2023) Adult-born granule cells facilitate remapping of spatial and non-spatial representations in the dentate gyrus. Neuron 111:4024. https://doi.org/10.1016/J.NEURON.2023.09.016
Urbach A et al (2013) Cyclin D2 knockout mice with depleted adult neurogenesis learn Barnes maze task. Behav Neurosci 127:1–8
Van Praag H, Kempermann G, Gage FH (1999) Running increases cell proliferation and neurogenesis in the adult mouse dentate gyrus. Nat Neurosci 2:266–270
Van Praag H et al (2002) Functional neurogenesis in the adult hippocampus. Nature 415:1030–1034
Van Praag H, Shubert T, Zhao C, Gage FH (2005) Exercise enhances learning and hippocampal neurogenesis in aged mice. J Neurosci 25:8680–8685
Veena J, Rao BSS, Srikumar BN (2011) Regulation of adult neurogenesis in the hippocampus by stress, acetylcholine and dopamine. J Nat Sci Biol Med 2:26–37
Walgrave H et al (2021) Restoring miR-132 expression rescues adult hippocampal neurogenesis and memory deficits in Alzheimer's disease. Cell Stem Cell 28:1805–1821.e8
Webler RD et al (2021) The neurobiology of human fear generalization: meta-analysis and working neural model. Neurosci Biobehav Rev 128:421–436
Weeden CSS, Mercurio JC, Cameron HA (2019) A role for hippocampal adult neurogenesis in shifting attention toward novel stimuli. Behav Brain Res 376:112152
Wesnes KA, Annas P, Basun H, Edgar C, Blennow K (2014) Performance on a pattern separation task by Alzheimer's patients shows possible links between disrupted dentate gyrus activity

and apolipoprotein E ∈4 status and cerebrospinal fluid amyloid-β42 levels. Alzheimers Res Ther 6:20
Winocur G, Wojtowicz JM, Sekeres M, Snyder JS, Wang S (2006) Inhibition of neurogenesis interferes with hippocampus-dependent memory function. Hippocampus 16:296–304
Wiskott L, Rasch MJ, Kempermann G (2006) A functional hypothesis for adult hippocampal neurogenesis: avoidance of catastrophic interference in the dentate gyrus. Hippocampus 16:329–343
Wojtowicz JM, Askew ML, Winocur G (2008) The effects of running and of inhibiting adult neurogenesis on learning and memory in rats. Eur J Neurosci 27:1494–1502
Woods NI et al (2020) The dentate gyrus classifies cortical representations of learned stimuli. Neuron 107:173
Wu MV, Hen R (2014) Functional dissociation of adult-born neurons along the dorsoventral axis of the dentate gyrus. Hippocampus 24:751–761
Wu MV, Luna VM, Hen R (2015) Running rescues a fear-based contextual discrimination deficit in aged mice. Front Syst Neurosci 9:114
Yang SM, Alvarez DD, Schinder AF (2015) Reliable genetic labeling of adult-born dentate granule cells using Ascl1 CreERT2 and Glast CreERT2 murine lines. J Neurosci 35:15379–15390
Yassa MA, Stark CEL (2011) Pattern separation in the hippocampus. Trends Neurosci 34:515–525
Yassa MA et al (2011) Pattern separation deficits associated with increased hippocampal CA3 and dentate gyrus activity in nondemented older adults. Hippocampus 21:968–979
Yasuda M et al (2011) Multiple forms of activity-dependent competition refine hippocampal circuits in vivo. Neuron 70:1128–1142
Yoshioka T, Yamada D, Kobayashi R, Segi-Nishida E, Saitoh A (2022) Chronic vicarious social defeat stress attenuates new-born neuronal cell survival in mouse hippocampus. Behav Brain Res 416:113536
Zhang Y, Loh HH, Law PY (2016) Effect of opioid on adult hippocampal neurogenesis. ScientificWorldJournal 2016:2601264
Zhou Y et al (2022) Molecular landscapes of human hippocampal immature neurons across lifespan. Nature 607(7919):527–533

# What Can Hippocampal Engrams Tell Us About Encoding Spatial Navigation?

**Lorena Andreoli and Kazumasa Z. Tanaka**

**Abstract** The hippocampus is indispensable for episodic memories, but its particular role in the process is still unclear. This chapter briefly overviews past studies focusing on place cells and memory engrams, highlighting their potential roles in spatial navigation. Future work reconciling these two lines of studies would provide a comprehensive view of the specific contribution of the hippocampus and a better understanding of how memory engrams support memory.

**Keywords** Place cells · Spatial navigation · Navigational strategy · Engram stability

From the most mundane tasks of going to the grocery store to the more complex behaviors of playing piano, reading a book, or conducting research, they all use memory. Right now, as you read this book, you are constantly retrieving knowledge acquired through your studies of previous books and scientific papers or even remembering the specific class in which you first heard about a certain topic. Similarly, when going to the grocery, you must remember the path to the supermarket, which is often determined with the aid of landmarks along with constant updates of self-positioning and movement.

Whether it is to read a book or to navigate to a destination, the ability to acquire, store, and retrieve information is fundamental for humans and other animals. Memory is a complex process that allows us to perform this myriad of tasks, but a more accurate word to describe it would be memories, as it is now known that different tasks use different types of memories. We can, for example, actively recall a specific experience, which is called episodic memory. But alternatively, without using the memory of a specific event, one can, for example, get better at a motor task, such as drawing a cartoon. Retrieving information to perform a skill involves

L. Andreoli · K. Z. Tanaka (✉)
Okinawa Institute of Science and Technology Graduate University (OIST),
Memory Research Unit, Okinawa, Japan
e-mail: lorena.andreoli@oist.jp; KAZUMASA.TANAKA@OIST.JP



J. Gräff, S. Ramirez (eds.), *Engrams*, Advances in Neurobiology 38,
https://doi.org/10.1007/978-3-031-62983-9_11

procedural memory. Interestingly enough, different types of memory are supported by different brain regions as well. As you will see in this chapter, several clinical and animal studies led to the current view that, in one way or another, the hippocampus supports episodic memory (Squire 2004; Henke 2010; Yonelinas et al. 2019).

The hippocampus is also widely accepted to be implicated in navigation. Initial lesion studies trying to link hippocampal damage to learning or memory deficits in animals surprisingly found that such lesions affected the ability to traverse different types of mazes in rodents. The link between the hippocampus and navigation became more evident with the discovery of a specific and important type of neuron in the rodent hippocampus: place cells (O'Keefe and Dostrovsky 1971; O'Keefe 1976). These cells were found to be selectively responsive to the animal's location in the environment, firing more frequently when the animal was in a particular location. Since then, other cells related to navigation have been discovered in the hippocampus, leading to the view that this area is also essential for navigation.

Modern neuroscience posits memory and navigation as the main roles of the hippocampus, but what is the relationship between the two? Past studies provided several views to link memory and navigation and proposed a specific role of the hippocampus for both cognitive processes. Nevertheless, recent studies revealed physiological properties of memory engram cells that are not necessarily congruent with predictions from these conventional ideas. The goal of this chapter is to elucidate the physiological basis of hippocampal memory and navigation, as well as try to bridge the gap between these two fields.

## Hippocampus and Memory

The hippocampus is one of the most studied brain areas. A quick search on PubMed using the term "hippocampus" will return more than 180,000 results, which is more than the results of other well-researched brain areas like the "prefrontal cortex" (66,000) and "cerebellum" (97,000) combined. In the middle to late twentieth century, this brain area was linked to one of the most fundamental cognitive processes that make us who we are: memory. And it is due to its role in memory that the hippocampus' anatomy, physiology, and behavioral correlates are so well-studied. In this section, we will explore the history of hippocampal physiology, along with important physiological processes that occur in the hippocampus and are thought to support memory.

## *Hippocampus Function and History*

### History

This section aims to comprehend where we are today with memory, navigation, and hippocampus studies by revisiting the history of this area (for a comprehensive historical perspective on hippocampal studies, please refer to Andersen et al. 2007). In the early 1900s, the first case linking the hippocampus with memory was published, describing two patients with memory deficits in which their brain autopsy revealed softening of the hippocampus and neighboring structures. However, before being associated with memory, and more specifically declarative memory formation, the hippocampus had many hypothesized functions. Due to an apparent observation that the hippocampus would be larger in animals with highly developed olfactory mucosa and olfactory bulbs, in the early 1930s, the hippocampus was considered to be part of the olfactory system. Later on, the hippocampus was thought to be part of the Papez circuit, which was thought to underpin emotions and take part in attention control. Finally, clinical observations of patients who underwent either hippocampal ablation or stimulation were important pieces to connect the hippocampus to what we currently believe to be its role in memory.

### Human Studies

In 1957, neurosurgeon William Scoville and psychologist Brenda Milner explored this connection between the hippocampus and memory in their report on ten patients who underwent bilateral medial temporal-lobe resection (Scoville and Milner 1957). They observed that resection of the medial temporal lobe led to anterograde (and some level of retrograde) amnesia. In 1963, Wilder Penfield and Phanon Perot also contributed to ascribing the role of the hippocampus to memory. In their report of a series of clinical cases, they describe what they called experiential hallucinations or experiential responses when patients experienced a seizure or when an electrical stimulation was applied to the temporal lobe. These phenomena were described as hallucinations of things previously experienced by the patient before the seizure or surgery (Penfield and Perot 1963). These early findings were the basis for connecting the hippocampus with declarative memory.

While lesions of the hippocampus seemed to affect declarative memory, other types of nondeclarative memory were unaffected in patients that underwent medial temporal lobe resection. Procedural memory (motor skills and habits) appeared to be intact in such patients. This was demonstrated by Milner with the famous patient H. M. performing a mirror-drawing test. In this task, H. M. had to draw a five-pointed star by looking at its reflection in a mirror. H. M. was not aware he had performed the task before, but his accuracy in drawing the stars improved over sessions (Milner 1970).

This early work with patients undergoing resection or stimulation of brain areas was extremely important in showing the role of the hippocampus on memory but also supported the notion that there is not only one area of the brain responsible for memory. This led to the concept of memory as composed of multiple systems with different neuroanatomy and neurophysiology. From the memory systems perspective, declarative memory comprises learning relationships between things or events, and it is supported by the medial temporal lobe (Gasbarri et al. 2014), while other types of memories, such as procedural memories and habits, for example, are supported by the basal ganglia. In H. M.'s case, episodic memory, a type of declarative memory, was affected by the resection of part of the hippocampus, but his ability to improve motor skills (procedural memory) was unaffected, as there was no damage to his basal ganglia.

## Long-Term Potentiation

Once the relationship between the hippocampus and memory was established, one outstanding question that arose was how memories are stored in the brain at a cellular level. In his famous book "The Organization of Behavior," Donald O. Hebb, a Canadian psychologist, suggested a possible mechanism through which learning could occur at a cellular level:

> When an axon of a cell A is near enough to excite a cell B and repeatedly or persistently takes part in firing it, some growth process or metabolic change takes place in one or both cells such that A's efficiency, as one of the cells firing B, is increased. (Hebb 1949)

This assumption implies that if a neuron repeatedly excites another neuron, some cellular change will take place in both cells in a way that will augment the efficiency of the first cell to excite the second one. But this was yet to be proved in an experimental setting. It was in 1973 that Bliss and Lømo (1973) demonstrated this principle in the hippocampus in vivo. In their work, rabbits were anesthetized, and a stimulation electrode was placed in the perforant path fibers that have axons that run in the molecular layer of the dentate gyrus (DG), where they had the recording electrode. They observed an increase in the excitatory postsynaptic potential (EPSP) of the granule cells in the DG compared to the baseline EPSP after repetitive stimulation. This long-lasting potentiation, as they called it at the time, now known as long-term potentiation (LTP), was shown to last up to 6 h (see Lømo (2018) for a critical review of LTP findings).

Due to its cytoarchitecture, the basic organization of the main connections of the hippocampus is well preserved in brain slices. After the publication by Bliss and Lømo, the long-lasting potentiation of synapses was demonstrated in vitro by stimulating the Schaffer collaterals and recording from the CA1 in hippocampal slices. They observed that the repetitive stimulation led to an increase in the amplitude of the population spike of the recorded area, which was maintained up to several hours after stimulation (Schwartzkroin and Wester 1975).

Since then, LTP has been well studied, and some of its properties, aside from its conceptual idea, are consistent with the idea of its involvement in memory processes. The first property is input specificity: LTP is restricted to the synapses that were stimulated, and it does not occur in all synapses of a given cell (which could be a useful process to store specific information at the cellular level) (Purves et al. 2004). The second one is its associativity: if a pathway is weakly activated (which is not enough to induce LTP) at the same time as another pathway is strongly activated, both synapses will be strengthened. The latter principle is thought to be behind associative or classical conditioning, as well as involved in other cognitive processes where the association between sets of information is necessary.

Key molecular changes happen in pre- and postsynaptic neurons undergoing LTP. The induction of LTP seems to depend on intracellular calcium concentration increase (Byrne 2008). Additionally, it involves glutamate as the pre-synaptic transmitters and the postsynaptic glutamate receptors N-methyl-D-aspartate (NMDA) and α-amino-3-hydroxy-5-methyl-4-isoxazolepropionic (AMPA) receptors (Lømo 2018). Regarding NMDA receptors, several pharmacological studies have now consistently shown that blocking or using competitive antagonists of NMDA can prevent the induction of LTP (Byrne 2008). Meanwhile, the LTP expression in the CA3-CA1 synapses has been associated with both an increase in the number of functional AMPA receptors and with the phosphorylation of AMPA receptors in the postsynaptic neuron (Byrne 2008).

The discovery of long-term potentiation provided an elegant cellular and molecular mechanism through which synapses could encode information. The building knowledge in this area fortified the belief that memory could be encoded over these modified synapses. But how synaptic change effectively leads to memory formation remained an open question.

## Engrams and Memory

Immediate early genes (IEGs) contributed to testing the idea that a memory is encoded within the neuronal ensemble containing synapses modified by the acquisition of the memory. The increase in the expression of particular IEGs, such as c-Fos, Arc, Homer1, and Zif268, is induced by the increase in neuronal activity (Sagar et al. 1988; Greenberg and Ziff 1984; Müller et al. 1984; Kruijer et al. 1984; Cochran et al. 1984), patterned stimulations known to induce long-term potentiation (LTP) (e.g., Cole et al. 1989), and behavioral training (Guzowski et al. 2005). Early studies tagged IEG-positive cells during the behavioral task and examined their reactivation (i.e., re-expression of IEG) later in time (Guzowski et al. 1999). In multiple brain areas, significant (yet not 100%) reactivation of neuronal ensembles was observed during a memory test. Genetic approaches to control the window of neuronal tagging using IEGs further advanced our understanding of engram cells. In the pioneering studies, the neuronal ensemble that is IEG positive during training was tagged with either an optogenetic or a chemogenetic probe, and their activities were later activated or inactivated to determine their causal contribution to memory-related behaviors (Liu et al. 2012; Garner et al. 2012; Denny et al. 2014; Tanaka et al. 2014). These lines of studies strengthen the view that learning is mediated by synaptic modulation introduced during the encoding of memory (see also Ryan et al. (2015) for an alternative explanation).

A particular role of the hippocampal engram is suggested to represent context. Lesion studies elegantly revealed that the constellation of multimodal sensory cues is encoded within the hippocampus as a unitary representation (Rudy and O'Reilly 2001). The encoding of the contextual information requires functional NMDA receptors (NMDARs) in the hippocampus because pharmacological blockade of the receptor impairs the formation of the contextual fear memory when NMDARs antagonists are injected during the pre-exposure phase but not when injected during the immediate shock phase (Matus-Amat et al. 2007). A more recent study demonstrated the functional contribution of engram cells in this process (Ohkawa et al. 2015). Ohkawa and colleagues demonstrated that simultaneous optogenetic activation of engram cells in the hippocampus formed during the context pre-exposure and engram in the basolateral amygdala formed during the immediate shock produces an association of the two, which have never been presented together (Ohkawa et al. 2015). In line with the view of contextual representation, the increase of c-Fos expression is found when animals find a novel spatial arrangement of familiar cues (Jenkins et al. 2004). Therefore, the literature in the engram and IEG mapping studies is in support of the notion that the hippocampus stores contextual information for episodic experiences.

## Hippocampus and Space

In addition to its role in memory, another main assumed function of the hippocampus is navigation. The idea of the brain containing a map of an environment has a rich history. In his famous paper describing how rats resolve and navigate through mazes, Tolman (1948) hypothesized that animals use a cognitive map of the environment when navigating. This map would have routes, paths, and other aspects of the environment, which will determine the behavior of the animal in that maze.

With the discovery of hippocampal cells that fire preferably when the animal is in a specific location, known as "place cells" (O'Keefe and Dostrovsky 1971; O'Keefe 1976), the view of the hippocampus as an ideal candidate to be the neurobiological structure for this cognitive map became attractive. In the following section, we expand the concept of "place cells" as well as provide a historical perspective, describe the main characteristics, and discuss the current state of the place cell field of study.

## *Place Cells*

### History and Characteristics of Place Cells

Because of the effects of temporal lobe resection found in human patients (Scoville and Milner 1957; Penfield and Perot 1963), several lesion studies in animals took place in the late 50s and 60s trying to connect hippocampal function with memory and learning (for a comprehensive summary of such studies, see contemporary review by Douglas (1967)). In humans, the main effect of hippocampal damage was related to memory and learning, and most animal studies at the time tried to link hippocampal damage to learning deficits. Surprisingly, it was found that such lesions did not interfere with learning in some types of problems. Rather, hippocampal lesions seemed to lead to decreased behavior flexibility, increased difficulty in maze learning, and increased activity when exploring an environment (Douglas 1967).

After the findings that related hippocampal lesions to deficits in spatial-related tasks, O'Keefe and Dostrovsky (1971) hypothesized that the hippocampus could be the brain area that provides the animal with a cognitive spatial map of the environment. In this classic study, they electrophysiologically recorded spike activity from cells in the *cornu ammonis* 1 (CA1), CA4, and dentate gyrus of rats' hippocampus (O'Keefe and Dostrovsky 1971). They observed that a subset of cells recorded in those sites fired either exclusively or maximally when the animal was in a specific part of a rectangular-shaped platform and facing a specific direction. Modifications to the environment, such as rotating the platform, removing objects from the room, and turning off the lights, altered the activity of some of the recorded cells, leading them to believe that the cells' activity was related to spatial orientation in a more abstract manner. This study provided the first evidence of hippocampal cells having a preferred spatial orientation, but animals were mainly restrained, and the cells seemed to respond to other modalities of stimuli, such as the tactile stimulation generated during restraining.

In a later work, O'Keefe described what was then called a "place unit," and now it is known as a place cell (Fig. 1a, b): a neuron that would have its maximal firing when the animal was in a specific position of the environment (O'Keefe 1976). Approximately half of the cells recorded in this experiment showed spatial tunning and were classified as place units. He recorded from CA1 cells of rats exploring a T-shaped maze where rewards were randomly placed at the end of one of the arms. O'Keefe hypothesized that the hippocampus was not simply responding to sensory stimuli but rather was involved in a more abstract representation of the physical space. Because of that, in this study, he manipulated environmental factors such as illumination, odors, and sounds and found out that the place units did not seem to respond to a specific sensory stimulus. Such observations further strengthened the idea that the hippocampus, now through this place cells' activity, gives rise to the cognitive map.

More than 40 years have passed since place cells were first described, providing elegant experimental evidence for hippocampal involvement in the cognitive map

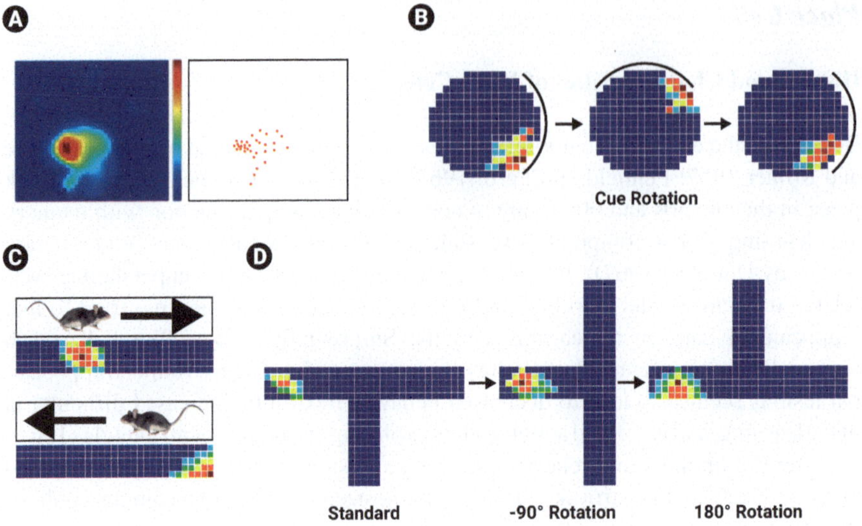


**Fig. 1** Properties of place cells. (**a**) A heat map showing the increased firing of place cells at a specific location within an open field (left) and actual spike positions (right). (**b**) Place fields can be direction-specific. When the animal takes a linear trajectory, the same place cell exhibits distinct patterns of place fields depending on the direction. (**c**, **d**) Place fields are anchored to distal cues in the environment. (B adapted from Muller and Kubie (1987) and D adapted from Cressant et al. (2002), respectively)

hypothesized years before by Tolman. Since then, studies have not only shown that place cells can also be found in humans (Ekstrom et al. 2003; Miller et al. 2013) but also that the hippocampus and parahippocampal areas contain other navigational-related cells such as head direction cells (Taube et al. 1990a, b), border cells (Lever et al. 2009), and grid cells (Hafting et al. 2005).

Studies in the neuroscience of navigation field have worked toward the characterization of place cells and the broader understanding of their role in navigation. If place cells are related to the animal's position in the environment, it would be expected that the place field's (position in space where those cells fire) location would be influenced by environmental factors. Indeed, place fields were shown to be influenced by distal cues (O'Keefe and Conway 1978; Muller and Kubie 1987; Cressant et al. 2002). For example, Muller and Kubie (1987) showed that the place fields follow a salient cue (white square) in the wall of a maze if you rotate the wall and the cue or just the cue (Fig. 1c). And Cressant et al. (2002) demonstrated that a place field remains somewhat fixed to the distal cues in a rotating T-maze (Fig. 1d). The discovery of these cells that responded to the environment was a beautiful demonstration of the cognitive map hypothesized by Tolman years before. As more studies started characterizing these cells, and what type of alterations in the environment they responded to, there was still the need to understand the underlying physiology behind the cognitive map.

## Physiology of Place Cells

In his famous paper "Cognitive Maps in Rats and Men," from 1948, Edward Tolman wrote:

> Before the learning of a specific maze, the connecting switches (synapses according to the physiologist) are closed in one set of ways and produce the primarily exploratory responses which appear in the early trials. Learning, according to this view, consists in the respective strengthening and weakening of various of these connections; those connections which result in the animal's going down the true path become relatively more open to the passage of nervous impulses, whereas those which lead him into the blinds become relatively less open.

These words were written more than 20 years before the first publication on long-term potentiation, but Tolman already briefly explored theoretically the notion that synapses had to be strengthened for the animal to learn a path in a maze.

As discussed in the previous section of this chapter, place cell research has shown that animals can rapidly form stable representations of new environments. A natural follow-up question would be, what are the cellular and molecular mechanisms through which these place fields are formed and maintained? Since LTP was shown to happen at the Schaffer collaterals, which connect CA3 to CA1, and both these areas have been shown to have place cells, one candidate mechanism is the long-term potentiation.

An important work toward understanding the role of synaptic plasticity applied to physiological studies of navigation was done by Tsien et al. (1996) characterizing spatial navigational deficits in animals with knockout NMDA receptors (Tsien et al. 1996). In this paper, they selectively knocked out the NMDAR1 gene in pyramidal cells of the CA1 of mice hippocampus. This affected synaptic plasticity by ablating the slow component of post-synaptic current (EPSC), which is mediated by NMDARs. In these animals, they also failed to induce LTP, short-term potentiation, and produce LTD. The changes in synaptic transmission and plasticity are thought to be behind these animals' deficits in navigational tasks. Mice lacking the NMDAR1 in pyramidal cells in the CA1 performed poorly in a Water Maze task compared to controls by having higher escape latency when the platform was not visible. In a transfer test where the platform is removed from the water maze, they also had a shorter search time in the quadrant where the platform was previously at. Higher search time in the same quadrant is used as an indicative that the animal is using a spatial strategy, which does not seem to be the case for these NMDAR1 knockout mutants.

Because of this deficit in navigation, McHugh et al. (1996) investigated the influence of synaptic plasticity on the physiology of place cells of animals doing a hippocampal-dependent spatial learning task. By recording extracellular electrophysiological activity of animals freely exploring different environments, McHugh et al. (1996) found that several properties of place cells, such as firing rate and complex spike index, were the same in NMDAR1 knockout and control cells. They also showed that a similar proportion of putative pyramidal cells showed significant spatial tuning by forming place fields in the environment in both groups of animals, and even having no NMDA-dependent plasticity, place cells of these animals showed stability across re-exposure to the same environment as control mice's place cells. Other properties of place cells, such as directionality, were also preserved in the NMDAR1 knockout mice, but the place cells of these animals seem to be less specific, having place fields more likely to be broad, diffuse, and with multiple peaks (while place fields from control animals usually had one salient peak).

Building on these works relating LTP to place cell formation and activity, Kentros et al. (1998) blocked NMDA receptors pharmacologically and assessed the effects on place cell activity of animals exposed to a novel and familiar environment, both short (hours) and long term (1 day after). They mainly evaluated the stability of the place fields by comparing the firing patterns between days in a specific space of a cylindrical maze. They found that hippocampal representation of a familiar environment was not affected by NMDA receptors blockage. Blocking NMDA receptors did not affect hippocampal representation in a familiar environment nor did it affect remapping to a novel environment and maintaining the new representation in the short-term (approximately 1.5 h). What blocking NMDA receptors affected was the long-term maintenance of these new firing fields in a novel environment when the animal was re-exposed to the "novel" environment after 24 h. This result suggests that the long-term stabilization of firing fields and long-term maintenance of LTP mediated by NMDA receptors might be related.

## Stability of Place Cells

For learning to occur, through the formation of a memory of a specific space, plastic processes such as LTP are important. But not only do animals have to learn about new environments, they must also be able to consistently navigate to the same location over days, months, or even years. In order to have stable navigational behavior, it is expected that the neuronal representation of that space is also stable.

The initial place cell studies already demonstrated the stability of firing in the same place field within a session, but how stable are these hippocampal space maps over several days? In 1990, Thompson and Best published a study in which they claimed that place fields were stable for up to 153 days (Thompson and Best 1990). They assessed the stability of the place field of ten putative pyramidal cells in the CA1 of rats navigating through a 6-arm radial maze. From the same cell, place fields observed in different sessions were significantly closer to one another than would be expected by chance, demonstrating the stability of place cells across days. This was a pioneering effort to demonstrate the stable properties of place cells, but it is worth noting that this study was performed using extracellular electrophysiological recordings. Even using rigorous methods to infer whether the activity recorded is from the same neuron, a definitive demonstration that the same cell continues to fire in the same place over time can only be done via imaging studies or with denser electrophysiology recording channels that allow correction of electrode drifting (Steinmetz et al. 2021).

With the advent of newer electrophysiological recording methods, more precise identification of neurons was possible using tetrode recordings. Utilizing this advantage, Mankin et al. (2012) could sort different neurons based on their peak size in each of the channels of a single tetrode. Although not as precise as imaging the activity of those neurons would be, this was a more reliable approach to identifying the same unit over different recordings. They exposed rats to two different arenas and compared the place cell activity in CA1 and CA3 between those sessions and sessions that happened from 15 min to 31 h after the first recording (Mankin et al. 2012). They also demonstrated that the hippocampus represents the environment in a stable manner even after 24 h, but they found that this effect was different between CA1 and CA3. While CA3 place cell population showed a reliable firing pattern over longer periods, CA1 population had more variability of firing across different recording sessions in the same arena.

Taking advantage of newer imaging methods, Ziv and colleagues imaged place cell activity of freely behaving mice to check the stability of place cells over prolonged periods of time (Ziv et al. 2013). They expressed calcium indicators in the hippocampus pyramidal cells and recorded cell activity with a miniature microscope for 45 days while animals visited a familiar linear track. Most of the cells imaged were active in one or two sessions, and only 2.8% of cells were recorded in all ten sessions (interestingly, in every session, roughly 30% of the total recorded cells were active). Additionally, 15–25% of the cells showed overlap of place field between any two sessions (even when 30 days apart). And when cells showed place fields in more than one session, the locations of these place fields were identical. To show that this overlap of fields between days was enough to retain a stable representation of the space, they used a decoding algorithm and showed that it could decode the trajectory of the mouse in a session using neuronal data from another session that happened 30 days prior. From these results, they concluded that place-coding ensembles have a "fluctuating membership," i.e., they change over time, but the small (15–25%) overlap of place fields is enough to retain spatial information in the CA1. And even though there is the stability of place fields across sessions, there also seems to exist a unique "signature" from each experience in the arena, given that 75–85% of cells do not overlap between two sessions.

In 2020, another pivotal work utilizing miniature microscopes was published and helped us understand the dynamics of the long-term stability of place cells. In this work, Sheintuch and colleagues exposed mice to a linear track where they could run back and forth to get a reward in the extremities while recording calcium activity from their CA1 pyramidal cells (Sheintuch et al. 2020). Mice underwent several different sessions a day, spaced by an intertrial interval in an opaque bucket near the linear track. They studied the activity of the same place cell across different sessions on the same day, as well as sessions that occurred even 29 days after the first recording. In each session, animals could freely run between extremities, and the spatial representation of each cell was computed, meaning the location where the cell would preferentially fire. Since animals were consistently exposed to the same experimental environment, they thought the cell would have only one spatial representation or map and consistently fire in the same place. But interestingly, although some animals had only one map of the space, other animals presented two maps, so in some sessions of the same day, that cell's place field would be in some part of the track, and in other sessions, it would be in another part of the track. They found that most place cells maintained their place fields event after more than 10 days. The number of maps did not change across days, but the probability that the same map would occur in two different trials decreased as a function of the elapsed time. Interestingly, the switch between maps would occur between sessions and not within the same session, so they made a slight change that abolished the existence of multiple maps: they changed the opaque intertrial bucket to a transparent one. When animals could see where they were in the environment during the intertrial interval, no animal showed multiple maps over different sessions in a day.

Since a considerable number of mice presented multiple maps for the same environment, this raises the question of what is the functionality of each map. One possibility is for redundancy of information to be sure that the map for that environment is stored more reliably. Another possibility is that this reflects that animals were paying attention to different parts of the environment since the room in which they recorded had several distal cues. And an important possibility is that the different maps reflect different navigational strategies that the animal used to solve the task. In the following section, we will explore the different strategies animals can use when navigating toward a goal.

## Navigational Strategies

Now let us go back to our example of grocery shopping. When going to the supermarket, we can pay attention to landmarks and buildings in order to find your way. But once we've have gone enough times from the house to the supermarket, there is no need to pay attention to the landmarks. Instead, we can almost automatically go to the store by turning right and left here and there while ignoring the landmarks. The different strategies we can use to navigate to a goal location can be divided between allocentric (when related to the environment) and egocentric (when related to self-motion).

The relationship between memory and navigation has recently gained another interesting connection. Similar to the memory systems perspective (where different types of memories are supported by different parts of the brain), these two navigational strategies were also shown to be supported by different brain structures. The hippocampus, which supports episodic memory, has also been shown to mediate allocentric navigation. Alternatively, egocentric navigation has been shown to be dependent on the basal ganglia, and more specifically, the dorsolateral striatum. The influence of hippocampus and dorsolateral striatum in allocentric and egocentric navigation, respectively, has been shown through lesion (Packard and McGaugh 1996; Yin and Knowlton 2004) and stimulation (Packard 1999) studies in both areas of animals navigating through mazes. This relationship has also been shown with the study of the expression of IEGs in each area after animals perform navigational tasks (which we will further explore in section "Immediate Early Genes and Navigation").

## Engrams and Navigation

In section "Engrams and Memory", we made the point that the hippocampal engram is thought to encode contextual representation, which in animal studies is an analog to episodic memory. The fact that the hippocampus supports both episodic memory engrams and allocentric navigation leads to interesting discussions on what are the relationships between both processes: memory and navigation. We believe that combining both approaches in navigational studies, along with properly eliciting allocentric navigation in place cell studies, will help bridge the gap between both fields and allow for a better understanding of the role of the hippocampus in behavior. In this chapter, we will introduce the research that has been pioneering the combination of engram technology with navigational studies for a better understanding of hippocampal physiology.

### *Immediate Early Genes and Navigation*

One of the first works trying to link the expression of IEGs to place cells was done in 1999. The group found that the percentage of arc-expressing cells in the CA1 following exploration of an environment was similar to the percentage of activated neurons in single-unit recordings in similar conditions (Guzowski et al. 1999). The similarity of the size of these two populations led them to hypothesize that, in the CA1, IEG expression could be induced in neurons after a cell expresses a "place field."

Since then, there have been a plethora of papers linking IEG expression in the hippocampus and navigation, even if not explicitly linking to place cell activity. Passino et al. (2002) showed that C57 mice undergoing a probe trial in a dual-solution plus maze had a higher number of Fos-positive cells in the dentate gyrus (DG), CA1, and CA3 when compared to control animals that did not perform a probe trial. The increase of c-Fos in those areas occurred regardless of the animals being trained for a short or long period. They also observed more c-Fos expressing cells in the CA1 of animals doing a probe trial after short training when compared to long-training animals. Two interesting points can be raised from their results. First, the difference in c-Fos expressing cells in the CA1 between short and long-training animals can be due to a difference in navigational strategy used in both groups since it has been shown that in a dual solution plus maze animals trained for a shorter period are more likely to use allocentric hippocampal strategy (Packard and McGaugh 1996). Secondly, when the animals were being released from the novel arm in the probe trial, the increase in c-Fos in hippocampal areas could be due to the introduction to a novel context, the opposite arm to which they did not have access before.

Both points were addressed by another study with c-Fos expression in the dual-solution plus maze by Colombo, Brightwell, and Countryman. In this study,

Colombo et al. (2003) also demonstrated that animals engaged in a navigation task in the dual-solution plus maze had more c-Fos expressing neurons in the hippocampus than control rats matched for motor activity (Colombo et al. 2003). But they identified which strategy (allocentric or egocentric) the animals were taking during a probe trial. Animals that used an allocentric approach had more c-Fos-positive cells in DG and CA1 when compared to animals using an egocentric approach (Colombo et al. 2003). By showing this, they also ruled out the possibility of the new environment driving c-Fos expression since both allo- and egocentric animals were released from the new arm.

## *Engrams and Navigation*

These first few works laid the foundation for the study of the relationship between hippocampal engram and place cells, but it was not until 2018 that an explicit link between those two populations was first demonstrated. Tanaka et al. (2018) used opto-tagging of engram cells in the hippocampus to demonstrate the link between contextual memory representation by engram cells and spatial representation by place cells (Fig. 2a). The authors first exposed mice to context A while recording spike activity and induced the expression of channelrhodopsin in engram cells (c-Fos positive). They identified engram (c-Fos positive) and non-engram (c-Fos negative) cells by opto-stimulating c-Fos-positive neurons that expressed channelrhodopsin and later re-exposed the animals to either context A or a new context B. Even though it was predicted that IEG expression would be induced in neurons after a cell expressed a place field (Guzowski et al. 1999), the picture was at the time incomplete. From the recorded putative pyramidal cells in Tanaka's study, only approximately 19% were c-Fos positive. But importantly, approximately 80% of c-Fos-positive cells did have a place field during the first exposure to context A. If engram cells are used to represent stable information, c-Fos-positive cells should be more stable and remap less than c-Fos negative when animals are re-exposed to context A. This was also not the case, and c-Fos-positive cells shifted their firing location more than c-Fos-negative cells when re-exposed to context A. Engram cells also seemed to be more important for discriminating between contexts, and they were mostly silent when animals were exposed to context B, while the spatial information carried by c-Fos-negative cells was unchanged in context B.

In a slightly different approach, Pettit et al. (2022) also elucidated the link between hippocampal engram cells and place cells (Fig. 2b). They recorded place cell activity through two-photon imaging of a red-shifted calcium indicator and could later assess c-Fos expression in the same neurons imaging short half-life eGFP under the control of Fos promoter. Animals would run in a virtual reality (VR) linear track of two meters, with different cues along the track and a reward zone of 20 cm where animals received a liquid reward. Place cells were imaged while the animals traversed the track, and Fos reporter was imaged 3 h after the beginning of each experiment. Contrary to Tanaka et al. (2018), they found that cells with higher

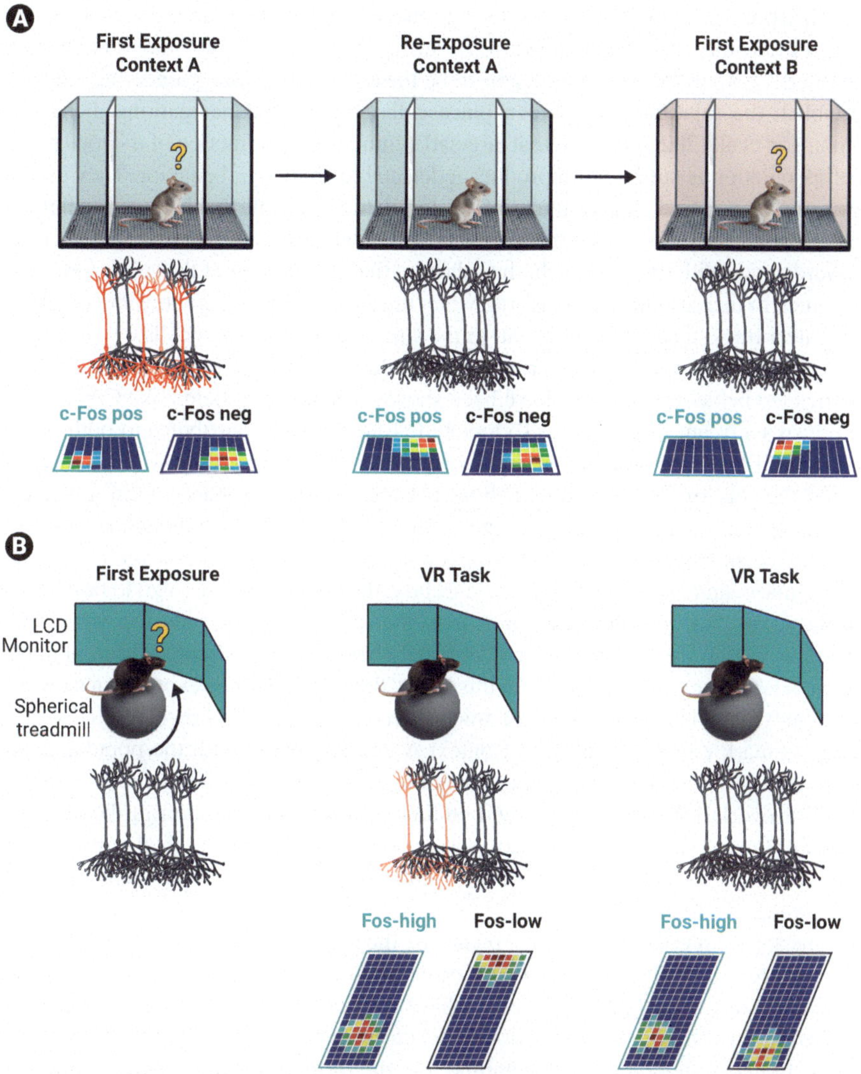


**Fig. 2** A double dissociation between c-Fos expression and task demand in the stability of place cells. (**a**) Engram cells formed during exploration in the novel context show preferential remapping when the animal re-explored the same context, while the other place cells exhibit higher stability in their firing locations. These two types of place cells respond differently to another novel context. (Adapted from Tanaka et al. (2018)) (**b**) After training in a VR linear track to find the reward location, cells expressing a high degree of c-Fos during an ongoing session show higher stability across two VR sessions than other place cells. (Adapted from Pettit et al. (2022))

Fos expression had more spatial information, that most cells with place fields were highly expressing Fos cells, and that Fos-high cells' place fields were more stable across days. In both studies, expression of c-Fos in the cells was associated with a bigger place field size.

The stimulation of a fraction of the neuronal ensemble of engram cells that encode a specific context has been shown to be sufficient to produce a behavioral change. Similarly, with place cells, this seems to be the case as well. In a pioneering study to establish the relationship between place cells activity and navigational behavior, Robinson et al. (2020) showed that targeted stimulation of a fraction of the total place cell population is sufficient to produce a detectable change in behavior. They used a two-photon targeted photostimulation to stimulate place cells that had a place field in the reward zone of a linear track (reward-PCs) when the animals were not in the reward zone. With this approach, they showed that stimulation of the reward-PCs led the mice to increase their licking rate when the animal was in a nonrewarded part of the linear track, a result very similar to the stimulation of engram cells.

In addition to studies of engram cell stimulation in the DG leading to memory-associated behavior, DG cells have been shown to have place fields like CA1, albeit more sparsely but more stable. To reconcile how the DG contributes to both novel contextual learning and spatial representations, Wilmerding et al. (2023) investigated the role of DG in distinguishing between multiple routes within a T-maze during spatial navigation. They trained animals in a delayed-nonmatch to position (DNMP) task where during a sampling, animals first were forced to run to one side of a T-maze, and after 15 s delay in a start box, they could choose to go to both sides of the maze, but only the side opposite to the sampling phase side was rewarded. Animals were trained in one maze (Context T) but tested in a slightly different maze (Context A) doing the same task. In this study, they tagged c-Fos expressing neurons by removing doxycycline food of mice injected with AAV-cFos-tTA-TRE-eYFP virus to mark engram cells from Context A and compared with the population of endogenous c-Fos of animals either doing the same task or not. They found a higher level of overlap between tagged and endogenous c-Fos population in the animals that were re-exposed to the context than expected by chance and then animals that were either exposed to only half of the T-maze (Context B) or to an open field with no rewards (Context C).

The apparent conflicting results regarding the role of engram cells in spatial navigation offer an opportunity for memory research to advance in the coming years. Two important aspects that we believe have not been fully addressed in the papers discussed thus far are the distinction between context and space and the role of the navigational strategy used by the animal. For the former, it is important to highlight the differences in the timing of c-Fos tagging and imaging. While Tanaka et al. (2018) tagged c-Fos expressing cells after animals were exposed to a new context, Pettit et al. (2022) and Wilmerding et al. (2023) investigated the activity of c-Fos expressing cells after animals were trained in a navigation task. Another important aspect is the role of the strategy used by the animal in the moment of c-Fos tagging or imaging. Since Pettit et al. (2022) imaged c-Fos expressing neurons of highly trained animals, it is possible that the animals no longer relied on allocentric strategy since they have been extensively trained. Studies investigating the role of hippocampal engram cells in navigation should be aware and ideally probe what type of navigational strategy the animals are using, as it has been shown that some types of navigation do not require hippocampal activity.

## Conclusions

In this chapter, we have explored the role of the hippocampus in contextual or episodic memory and navigation. Both these pieces of literature provide robust demonstrations of the link between the hippocampus and these processes. One possible discussion based on the difference between these two fields is what the hippocampus' actual job really is during navigation. It has been argued that the role of the hippocampus related to navigation is purely memory itself (for a better construction of this argument, refer to Eichenbaum and Cohen (2014) and Eichenbaum (2017)). Although this is an important discussion, we do not believe that there is enough evidence to pinpoint the role of the hippocampus in navigation to simply memory or to navigation as a separate process. Rather, we propose that what is needed to unravel the role of the hippocampus in both processes is to combine these fields and the technology/techniques used by them.

Future studies should focus on the relationship between engram and place cells, and behavioral designs should cover the difference between the concepts of space and context, which is not clear from current studies. When engram cells are tagged in a specific context, is that information different than the spatial information provided by place cells? There are several ways in which engram cells can interact with place cells. Firstly, place cells and engram cells can be composed of the same population, and the experimental evidence thus far cannot completely rule out this possibility, as discussed in section "Engrams and Navigation". Another possibility is that place cells provide spatial information for engram cells to form a more abstract concept of context. And lastly, engram cells can provide top-down information about the context for place cells (and other navigation-related cells in parahippocampal regions) to build the spatial representation of an environment, though this hypothesis is difficult to reconcile with current experimental data given how fast spatial maps are formed by place cells.

Another important step toward a better understanding of the role of the hippocampus is to properly verify and probe which type of navigation the animal is using. As suggested by Sheintuch et al., the hippocampus might utilize distinct maps for different spatial strategies. This proposition is in line with the view that the instability of place fields in engram cells observed during the first exposure to a novel context reflects a higher degree of freedom to incorporate possible strategies or behaviors that the animal might employ later in the same environment (Micou and O'Leary 2023). After the strategy is fixed, the engram cells no longer need to have room for other strategies hence losing their stability and fixing the map.

Much has been done over the last few years toward a better understanding of the hippocampus in both memory and navigation fields. But the complete understanding of the role of this well-studied brain area can only be achieved by integrating the knowledge between these two fields. Learning what is the relationship between engram cells and place cells is, to us, one of the final puzzle pieces to better understand the neuronal underpinnings of memory and navigation.

## References

Andersen P, Morris R, Amaral D, Bliss T, O'Keefe J (2007) Historical perspective: proposed functions, biological characteristics, and neurobiological models of the hippocampus. In: Andersen P et al (eds) The hippocampus book. Oxford Academic, New York, pp 9–36

Bliss TV, Lømo T (1973) Long-lasting potentiation of synaptic transmission in the dentate area of the anaesthetized rabbit following stimulation of the perforant path. J Physiol 232(2):331–356

Byrne JH (2008) Learning and memory: basic mechanisms. In: Squire L, Berg D, Bloom F, du Lac S, Ghosh A, Spitzer N (eds) Fundamental neuroscience, 3rd edn. Academic Press

Cochran BH, Zullo J, Verma IM, Stiles CD (1984) Expression of the c-fos gene and of an fos-related gene is stimulated by platelet-derived growth factor. Science 226(4678):1080–1082

Cole AJ, Saffen DW, Baraban JM, Worley PF (1989) Rapid increase of an immediate early gene messenger RNA in hippocampal neurons by synaptic NMDA receptor activation. Nature 340(6233):474–476

Colombo PJ, Brightwell JJ, Countryman RA (2003) Cognitive strategy-specific increases in phosphorylated cAMP response element-binding protein and c-fos in the hippocampus and dorsal striatum. J Neurosci 23(8):3547–3554

Cressant A, Muller RU, Poucet B (2002) Remapping of place cell firing patterns after maze rotations. Exp Brain Res 143:470–479

Denny CA, Kheirbek MA, Alba EL, Tanaka KF, Brachman RA, Laughman KB, Tomm NK, Turi GF, Losonczy A, Hen R (2014) Hippocampal memory traces are differentially modulated by experience, time, and adult neurogenesis. Neuron 83(1):189–201

Douglas RJ (1967) The hippocampus and behavior. Psychol Bull 67(6):416

Eichenbaum H (2017) The role of the hippocampus in navigation is memory. J Neurophysiol 117(4):1785–1796

Eichenbaum H, Cohen NJ (2014) Can we reconcile the declarative memory and spatial navigation views on hippocampal function? Neuron 83(4):764–770

Ekstrom AD, Kahana MJ, Caplan JB, Fields TA, Isham EA, Newman EL, Fried I (2003) Cellular networks underlying human spatial navigation. Nature 425(6954):184–188

Garner AR, Rowland DC, Hwang SY, Baumgaertel K, Roth BL, Kentros C, Mayford M (2012) Generation of a synthetic memory trace. Science 335(6075):1513–1516

Gasbarri A, Pompili A, Packard MG, Tomaz C (2014) Habit learning and memory in mammals: behavioral and neural characteristics. Neurobiol Learn Mem 114:198–208

Greenberg ME, Ziff EB (1984) Stimulation of 3T3 cells induces transcription of the c-fos proto-oncogene. Nature 311(5985):433–438

Guzowski JF, McNaughton BL, Barnes CA, Worley PF (1999) Environment-specific expression of the immediate-early gene arc in hippocampal neuronal ensembles. Nat Neurosci 2(12):1120–1124

Guzowski JF, Timlin JA, Roysam B, McNaughton BL, Worley PF, Barnes CA (2005) Mapping behaviorally relevant neural circuits with immediate-early gene expression. Curr Opin Neurobiol 15(5):599–606

Hafting T, Fyhn M, Molden S, Moser MB, Moser EI (2005) Microstructure of a spatial map in the entorhinal cortex. Nature 436(7052):801–806

Hebb DO (1949) The organization of behavior. Wiley

Henke K (2010) A model for memory systems based on processing modes rather than consciousness. Nat Rev Neurosci 11(7):523–532

Jenkins TA, Amin E, Pearce JM, Brown MW, Aggleton JP (2004) Novel spatial arrangements of familiar visual stimuli promote activity in the rat hippocampal formation but not the Parahippocampal cortices: a c-fos expression study. Neuroscience 124(1):43–52

Kentros C, Hargreaves E, Hawkins RD, Kandel ER, Shapiro M, Muller RV (1998) Abolition of long-term stability of new hippocampal place cell maps by NMDA receptor blockade. Science 280(5372):2121–2126

Kruijer W, Cooper JA, Hunter T, Verma IM (1984) Platelet-derived growth factor induces rapid but transient expression of the c-fos gene and protein. Nature 312(5996):711–716

Lever C, Burton S, Jeewajee A, O'Keefe J, Burgess N (2009) Boundary vector cells in the subiculum of the hippocampal formation. J Neurosci 29(31):9771–9777

Liu X, Ramirez S, Pang PT, Puryear CB, Govindarajan A, Deisseroth K, Tonegawa S (2012) Optogenetic stimulation of a hippocampal engram activates fear memory recall. Nature 484(7394):381–385

Lømo T (2018) Discovering long-term potentiation (LTP)–recollections and reflections on what came after. Acta Physiol 222(2):e12921

Mankin EA, Sparks FT, Slayyeh B, Sutherland RJ, Leutgeb S, Leutgeb JK (2012) Neuronal code for extended time in the hippocampus. Proc Natl Acad Sci 109(47):19462–19467

Matus-Amat P, Higgins EA, Sprunger D, Wright-Hardesty K, Rudy JW (2007) The role of dorsal hippocampus and basolateral amygdala NMDA receptors in the acquisition and retrieval of context and contextual fear memories. Behav Neurosci 121(4):721–731

McHugh TJ, Blum KI, Tsien JZ, Tonegawa S, Wilson MA (1996) Impaired hippocampal representation of space in CA1-specific NMDAR1 knockout mice. Cell 87(7):1339–1349

Micou C, O'Leary T (2023) Representational drift as a window into neural and behavioural plasticity. Curr Opin Neurobiol 81:102746

Miller JF, Neufang M, Solway A, Brandt A, Trippel M, Mader I et al (2013) Neural activity in human hippocampal formation reveals the spatial context of retrieved memories. Science 342(6162):1111–1114

Milner B (1970) Memory and the medial temporal regions of the brain. Biol Mem 23:31–59

Muller RU, Kubie JL (1987) The effects of changes in the environment on the spatial firing of hippocampal complex-spike cells. J Neurosci 7(7):1951–1968

Müller R, Bravo R, Burckhardt J, Curran T (1984) Induction of c-fos gene and protein by growth factors precedes activation of c-myc. Nature 312(5996):716–720

O'Keefe J (1976) Place units in the hippocampus of the freely moving rat. Exp Neurol 51(1):78–109

O'Keefe J, Conway DH (1978) Hippocampal place units in the freely moving rat: why they fire where they fire. Exp Brain Res 31:573–590

O'Keefe J, Dostrovsky J (1971) The hippocampus as a spatial map: preliminary evidence from unit activity in the freely-moving rat. Brain Res 34:171

Ohkawa N, Saitoh Y, Suzuki A, Tsujimura S, Murayama E, Kosugi S, Nishizono H et al (2015) Artificial association of pre-stored information to generate a qualitatively new memory. Cell Rep 11:261. https://doi.org/10.1016/j.celrep.2015.03.017

Packard MG (1999) Glutamate infused posttraining into the hippocampus or caudate-putamen differentially strengthens place and response learning. Proc Natl Acad Sci 96(22):12881–12886

Packard MG, McGaugh JL (1996) Inactivation of hippocampus or caudate nucleus with lidocaine differentially affects expression of place and response learning. Neurobiol Learn Mem 65(1):65–72

Passino E, Middei S, Restivo L, Bertaina-Anglade V, Ammassari-Teule M (2002) Genetic approach to variability of memory systems: analysis of place vs. response learning and fos-related expression in hippocampal and striatal areas of C57BL/6 and DBA/2 mice. Hippocampus 12(1):63–75

Penfield W, Perot P (1963) The brain's record of auditory and visual experience: a final summary and discussion. Brain 86(4):595–696

Pettit NL, Yap EL, Greenberg ME, Harvey CD (2022) Fos ensembles encode and shape stable spatial maps in the hippocampus. Nature 609(7926):327–334

Purves D, Augustine GJ, Fitzpatrick D, Hall WC, LaMantia AS, McNamara JO, Williams SM (2004) Plasticity of mature synapses and circuits. In: Neuroscience, 3rd edn. Sinauer Associates, Inc, Sunderland

Robinson NT, Descamps LA, Russell LE, Buchholz MO, Bicknell BA, Antonov GK et al (2020) Targeted activation of hippocampal place cells drives memory-guided spatial behavior. Cell 183(6):1586–1599

Rudy JW, O'Reilly RC (2001) Conjunctive representations, the hippocampus, and contextual fear conditioning. Cogn Affect Behav Neurosci 1(1):66–82
Ryan TJ, Roy DS, Pignatelli M, Arons A, Tonegawa S (2015) Engram cells retain memory under retrograde amnesia. Science 348(6238):1007–1013
Sagar SM, Sharp FR, Curran T (1988) Expression of c-fos protein in brain: metabolic mapping at the cellular level. Science 240(4857):1328–1331
Schwartzkroin PA, Wester K (1975) Long-lasting facilitation of a synaptic potential following tetanization in thein vitro hippocampal slice. Brain Res 89(1):107–119
Scoville WB, Milner B (1957) Loss of recent memory after bilateral hippocampal lesions. J Neurol Neurosurg Psychiatry 20(1):11
Sheintuch L, Geva N, Baumer H, Rechavi Y, Rubin A, Ziv Y (2020) Multiple maps of the same spatial context can stably coexist in the mouse hippocampus. Curr Biol 30(8):1467–1476
Squire LR (2004) Memory systems of the brain: a brief history and current perspective. Neurobiol Learn Mem 82(3):171–177
Steinmetz NA, Aydin C, Lebedeva A, Okun M, Pachitariu M, Bauza M et al (2021) Neuropixels 2.0: a miniaturized high-density probe for stable, long-term brain recordings. Science 372(6539):eabf4588
Tanaka KZ, Pevzner A, Hamidi AB, Nakazawa Y, Graham J, Wiltgen BJ (2014) Cortical representations are reinstated by the hippocampus during memory retrieval. Neuron 84(2):347–354
Tanaka KZ, He H, Tomar A, Niisato K, Huang AJ, McHugh TJ (2018) The hippocampal engram maps experience but not place. Science 361(6400):392–397
Taube JS, Muller RU, Ranck JB (1990a) Head-direction cells recorded from the postsubiculum in freely moving rats. I. Description and quantitative analysis. J Neurosci 10(2):420–435
Taube JS, Muller RU, Ranck JB (1990b) Head-direction cells recorded from the postsubiculum in freely moving rats. II. Effects of environmental manipulations. J Neurosci 10(2):436–447
Thompson LT, Best PJ (1990) Long-term stability of the place-field activity of single units recorded from the dorsal hippocampus of freely behaving rats. Brain Res 509(2):299–308
Tolman EC (1948) Cognitive maps in rats and men. Psychol Rev 55(4):189
Tsien JZ, Huerta PT, Tonegawa S (1996) The essential role of hippocampal CA1 NMDA receptor–dependent synaptic plasticity in spatial memory. Cell 87(7):1327–1338
Wilmerding LK, Kondratyev I, Ramirez S, Hasselmo ME (2023) Route-dependent spatial engram tagging in mouse dentate gyrus. Neurobiol Learn Mem 200:107738
Yin HH, Knowlton BJ (2004) Contributions of striatal subregions to place and response learning. Learn Mem 11(4):459
Yonelinas AP, Ranganath C, Ekstrom AD, Wiltgen BJ (2019) A contextual binding theory of episodic memory: systems consolidation reconsidered. Nat Rev Neurosci 20(6):364–375
Ziv Y, Burns LD, Cocker ED, Hamel EO, Ghosh KK, Kitch LJ et al (2013) Long-term dynamics of CA1 hippocampal place codes. Nat Neurosci 16(3):264–266

# Understanding Others' Distress Through Past Experiences: The Role of Memory Engram Cells in Observational Fear

**Takashi Kitamura, Kritika Ramesh, and Joseph I. Terranova**

**Abstract** For individuals to survive and function in society, it is essential that they recognize, interact with, and learn from other conspecifics. Observational fear (OF) is the well-conserved empathic ability of individuals to understand the other's aversive situation. While it is widely known that factors such as prior similar aversive experience and social familiarity with the demonstrator facilitate OF, the neural circuit mechanisms that explicitly regulate experience-dependent OF (Exp OF) were unclear. In this review, we examine the neural circuit mechanisms that regulate OF, with an emphasis on rodent models, and then discuss emerging evidence for the role of fear memory engram cells in the regulation of Exp OF. First, we examine the neural circuit mechanisms that underlie Naive OF, which is when an observer lacks prior experiences relevant to OF. In particular, the anterior cingulate cortex to basolateral amygdala (BLA) neural circuit is essential for Naive OF. Next, we discuss a recent study that developed a behavioral paradigm in mice to examine the neural circuit mechanisms that underlie Exp OF. This study found that fear memory engram cells in the BLA of observers, which form during a prior similar aversive experience with shock, are reactivated by ventral hippocampal neurons in response to shock delivery to the familiar demonstrator to elicit Exp OF. Finally, we discuss

---

T. Kitamura (✉)
Department of Psychiatry, University of Texas Southwestern Medical Center, Dallas, TX, USA

Department of Neuroscience, University of Texas Southwestern Medical Center, Dallas, TX, USA

Peter O'Donnell Jr. Brain Institute, University of Texas Southwestern Medical Center, Dallas, TX, USA
e-mail: Takashi.Kitamura@UTSouthwestern.edu

K. Ramesh
Department of Psychiatry, University of Texas Southwestern Medical Center, Dallas, TX, USA

J. I. Terranova (✉)
Department of Anatomy, Midwestern University, IL, Downers Grove, USA
e-mail: jterra@midwestern.edu

 
J. Gräff, S. Ramirez (eds.), *Engrams*, Advances in Neurobiology 38,
https://doi.org/10.1007/978-3-031-62983-9_12

the implications of fear memory engram cells in Exp OF and directions of future research that are of both translational and basic interest.

**Keywords** Observational fear · Naive OF · Experience-dependent OF · Exp OF · Engram · Neural ensemble · Empathy · Perception-emotion coupling · Prosocial behavior · Simulation theory

## Introduction

Fear, a powerful and evolutionarily ancient emotion that is elicited through perceived danger or threat, is essential for survival (Adolphs 2013; LeDoux 2012). While there is much discussion about how to precisely define fear in humans and other animal species (Mobbs et al. 2019), fear can be operationally defined as an aversive functional state that arises in individuals in response to threatening stimuli (Adolphs 2013), thus enabling animals to avoid potentially life-threating situations. Fear can be experienced directly, such as by exposure to a dangerous environment (Marks et al. 2022), social defeat by another conspecific (Diaz and Lin 2020), or an encounter with a predator (Clinchy et al. 2013). Alternatively, fear can be acquired indirectly via social observation, such as witnessing another conspecific respond with fear to aversive environments or situations, underscoring the social nature of fear (Olsson and Phelps 2007; Barrett et al. 2007). The empathic process by which an individual can vicariously experience the other's aversive events is called observational fear (OF) (Debiec and Olsson 2017; Keum and Shin 2016, 2019a; Panksepp and Lahvis 2011; Kim et al. 2019, 2021; Morozov and Ito 2019), which serves as a foundation for observational learning. Rodents can learn to fear a context where they witnessed a conspecific in an aversive situation (Jeon et al. 2010; Jeon and Shin 2011; Kim et al. 2012; Keum et al. 2016; Lidhar et al. 2017; Smith et al. 2021; Fendt et al. 2021; Twining et al. 2017; Yusufishaq and Rosenkranz 2013; Shi et al. 2022a; Terranova et al. 2023) or to a previously neutral cue after witnessing another conspecific respond with fear to that cue (Fendt et al. 2021; Bruchey et al. 2010; Monfils and Agee 2019; Jones et al. 2018). Other animal species like cats (John et al. 1968), cows (Munksgaard et al. 2001), nonhuman primates (Mineka et al. 1984), and humans (Olsson et al. 2007, 2016a; Hooker et al. 2008; Skversky-Blocq et al. 2021, 2022; Bilodeau-Houle et al. 2020, 2023; Dou et al. 2023; Kaźmierowska et al. 2022; García-Moreno et al. 2021; Silvers et al. 2021; Marin et al. 2020; Szczepanik et al. 2020; Selbing and Olsson 2019) exhibit OF and thus engage in observational learning, indicating that OF is well conserved. Empathic responses like OF can facilitate prosocial or helping behaviors in rodents such as consolatory behaviors, reduction in harm to others, and freeing a trapped conspecific (Selbing and Olsson 2019; Phillips et al. 2023; Ben-Ami Bartal et al. 2011, 2014; Burkett et al. 2016; Church 1959; Hernandez-Lallement et al. 2020; Sato et al. 2015). Moreover, in humans, empathic responses like OF drive prosocial behavior toward others in distress (Gallo et al. 2018; Vaish et al. 2009). Thus, OF is a foundational empathic response through which an individual can acquire fear via social observation.

## Neural Circuit Mechanisms of Naive OF

Several studies have established rodent models that enable the examination of the neural circuit mechanisms that regulate OF. In these behavioral models, an observer witnesses aversive stimuli delivered to a demonstrator and responds with fear behavior to the demonstrator's distress (Debiec and Olsson 2017; Panksepp and Lahvis 2011; Jeon et al. 2010; Jeon and Shin 2011; Keum and Shin 2019b; Terranova et al. 2022; Jung et al. 2013). Several studies use a modified two chamber enclosure in which there is an insulated observer chamber and a demonstrator chamber with an exposed shock grid floor that are separated by a transparent partition (Debiec and Olsson 2017; Panksepp and Lahvis 2011; Jeon et al. 2010; Jeon and Shin 2011; Keum and Shin 2019b; Terranova et al. 2022; Jung et al. 2013). After a habituation period in which the observer and demonstrator explore their respective chambers, aversive stimuli (e.g., shocks) are delivered to the demonstrator, causing the demonstrator to respond with fear, which subsequently induces the OF response in the observer. One group of studies uses an OF model where the observer is completely naive to the testing procedure and is unfamiliar with the demonstrator, which we refer as Naive OF (Fig. 1a) (Jeon et al. 2010; Keum et al. 2016; Terranova et al. 2022). This OF model delivers strong aversive stimuli to the demonstrator, like strong shocks, to elicit fear behavior in the observer and requires a visual perception stimulus, such that blocking the observer's view of the demonstrator chamber with an opaque partition diminishes the observer's OF response (Jeon et al. 2010; Keum et al. 2016; Terranova et al. 2022; Hong and Choi 2018). Moreover, replacing the demonstrator with an inanimate object and replicating the demonstrator's fear behavior, e.g., jumping or running does not elicit the Naive OF response in the observer (Terranova et al. 2022), indicating that Naive OF is a socially induced empathic response. Finally, relative dominance status of the observer or demonstrator in their respective home cages does not affect observer freezing response in Naive OF (Park et al. 2022).

The anterior cingulate cortex (ACC) and basolateral amygdala (BLA) have emerged as essential brain regions that regulate Naive OF (Fig. 1b). Neural activity is elevated in both ACC and BLA during Naive OF, and both brain regions are necessary for Naive OF (Jeon et al. 2010; Twining et al. 2017; Terranova et al. 2022). Inactivation of the midline and intralaminar thalamic nuclei, such as the parafascicular and mediodorsal thalamus, which project to ACC, reduces OF response in observers (Jeon et al. 2010). Within ACC, reduced dopamine signaling via D2 receptor antagonism decreases OF response in observers, whereas elevated serotonergic signaling decreases OF response (Kim et al. 2014). Inhibition of somatostatin interneurons within ACC increases OF response in observers, whereas stimulation

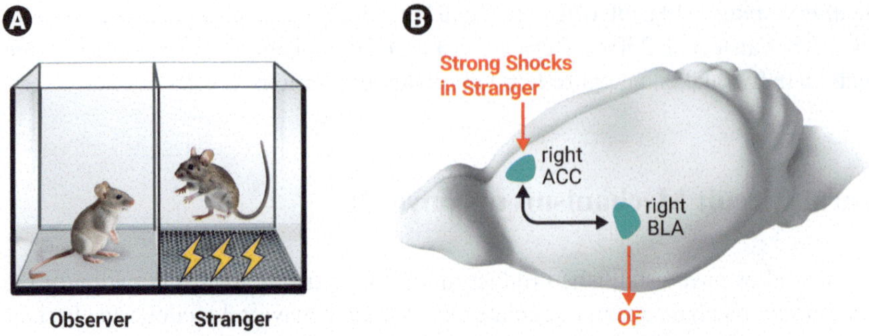


**Fig. 1** Neural circuit mechanisms of Naive observational fear (OF). (**a**) Behavioral paradigm for Naive OF. A fear conditioning apparatus is modified to create the OF apparatus, which contains an insulated observer chamber and a demonstrator chamber with an exposed shock grid floor, both of which are separated by a transparent plastic partition. The observer and demonstrator chambers are identical, except the observer chamber has an opaque plastic floor. The observer is completely naive to the testing procedure, and the demonstrator is an unfamiliar conspecific. After a habituation period, strong aversive shock stimuli are delivered to the demonstrator, which causes a robust behavioral response from the demonstrator. The observer expresses OF by responding to the demonstrator's aversive situation with fear behaviors like freezing. (**b**) Naive OF neural circuit mechanisms. Neural connections between the ACC and BLA, specifically reciprocal projections between the ACC and BLA in the right hemisphere, are essential for observers to express Naive OF. ACC, anterior cingulate cortex; BLA, basolateral amygdala

of somatostatin interneurons within ACC decreases OF response, suggesting that interneurons within ACC inhibit the OF response of observers in Naive OF (Keum et al. 2018). The role of the ACC to BLA pathway in Naive OF has also been more clearly examined. Functional connectivity and neural oscillatory activity at the theta rhythm frequency (5–7 Hz) is increased in ACC and BLA in Naive OF, and the ACC to BLA pathway is essential for Naive OF (Jeon et al. 2010; Terranova et al. 2022). Further studies have considered the role laterality of brain regions that regulate Naive OF, including the ACC, BLA, and thalamus. Inactivation of the right ACC, but not left ACC, reduced OF response in observers, whereas inactivation of the thalamus in either hemisphere reduced OF response in observers, indicating a role of laterality in ACC and not the thalamus in the regulation of Naive OF (Kim et al. 2012). A recent study by Kim et al. found that reciprocal projections between the right ACC and right BLA are essential for Naive OF (Kim et al. 2023). This study demonstrated a causal link between right ACC and right BLA neural oscillatory activity at the theta rhythm frequency and expression of OF by observers by selectively disrupting theta oscillations during the demonstrator shock period using a closed-loop manipulation (Kim et al. 2023). Finally, this study showed that hippocampal theta oscillations are synchronized with right ACC and right BLA theta oscillations and that wide range modulation of these hippocampal theta oscillations, in contrast to local chemogenetic inhibition of the dorsal or ventral hippocampus (Terranova et al. 2022), modulates the fear response of observers in Naive OF (Kim et al. 2023). Thus, synchronization of the ACC to BLA pathway at the theta rhythm frequency is an essential neural circuit mechanism that regulates Naive OF.

## Neural Circuit Mechanisms of Experience-Dependent OF

Empathic response, including OF, are facilitated by both prior similar experience in the observer and social familiarity with the demonstrator, which can help individuals to better understand the demonstrator's aversive situation if the demonstrator's reaction is ambiguous. In humans, empathy toward others is positively associated with prior similar aversive experience, such that individuals are more likely to self-report higher levels of empathy for a negative situation with another person if they have had previous experience with a similar negative situation (Eklund et al. 2009). Humans are also more likely to exhibit empathic responses like OF toward socially familiar individuals (Cao et al. 2015; Han 2018). Empathic responses like OF are facilitated by prior experiences in other animal species, including elephants (Plotnik and de Waal 2014; Raviv et al. 2023; Pérez-Manrique and Gomila 2018), dogs (Pérez-Manrique and Gomila 2018; Quervel-Chaumette et al. 2016; Van Bourg et al. 2020), and nonhuman primates (Mineka et al. 1984; Campbell and de Waal 2014; Sato et al. 2019). Empathic responses in rodents are also modulated by prior experiences, including OF. For example, observers with prior experience, including prior similar aversive experience with shock and/or social familiarity with the demonstrator, exhibit enhanced OF (Jeon et al. 2010; Lidhar et al. 2017; Sanders et al. 2013; Chen et al. 2009; Pereira et al. 2012; Atsak et al. 2011; Gonzalez-Liencres et al. 2014; Ueno et al. 2020; Nomura et al. 2019).

Recent studies in rodents have examined the neural circuit mechanisms of how prior experiences can facilitate OF. Acute systemic injection of an oxytocin (OT) antagonist decreases OF response in observers tested with a familiar demonstrator, whereas acute intranasal application of OT increases OF response in observers tested with an unfamiliar demonstrator and is associated with higher neural activity in the ACC (Pisansky et al. 2017a). Chronic intranasal application of OT also increases OF response in observers OF response in observers tested with an unfamiliar demonstrator (Pisansky et al. 2017a). Chemogenetic activation of OT neurons in the paraventricular nucleus of the hypothalamus (PVN) facilitates observer freezing response to an unfamiliar demonstrator, further indicating an important role of OT (Pisansky et al. 2017a). Moreover, elevated neural activity of OT neurons in the PVN is associated with increased OF response in observers that were prenatally treated with valproic acid (Takayama et al. 2022). OF with a familiar cagemate demonstrator induces neuroplastic changes within the ACC and potentiates the ACC to BLA pathway, thus enabling synaptic plastic changes that may enhance consolidation of future fear memories (Ito and Morozov 2019; Liu et al. 2017). Stimulation of parvalbumin interneurons within ACC decreases OF response in observers tested with a familiar cagemate demonstrator that received strong shock stimuli (Zhou et al. 2018). A priming shock administered to observers 30 min before OF testing activates a subpopulation of BLA-projecting ACC neurons that is reactivated by shock delivery to the demonstrator during OF, suggesting that the subpopulation of ACC to BLA neurons that regulates the observer's fear response to their own aversive shock experience may also regulate the observer's fear response to the demonstrator's aversive shock experience (Sakaguchi et al. 2018). Another recent study examined neuroplastic changes within the insular cortex and found that OF with a familiar demonstrator induced long-lasting synaptic changes in the insular cortex of observers (Shi et al. 2022b). Finally, a recent study by Huang, Chung et al. examined the neural activity of the ventromedial prefrontal cortex (vmPFC) during shock delivery to a familiar demonstrator of observers that received prior shock experience in another context (Huang et al. 2023). Optogenetic inhibition of vmPFC neurons did not affect observer freezing response but decreased escape behavior away from the demonstrator chamber and increased gazing behavior toward the demonstrator chamber (Huang et al. 2023). Moreover, optogenetic inhibition of vmPFC-projecting ACC or BLA neurons facilitated escape behavior away from the demonstrator chamber, suggesting that escape behavior exhibited by the observer during OF is regulated by an inhibitory interneuron circuit (Huang et al. 2023).

Additional studies have examined single-unit neural activity in cortical regions like ACC in OF and found the presence of emotional mirror neurons, or neurons that respond to both when the observer witnesses a demonstrator experience pain and when the observer itself experiences pain (Carrillo et al. 2019; Wu et al. 2023). Both the ACC and insular cortex have mirror neurons and anti-mirror neurons, which are neurons that exhibit opposite firing patterns to other's pain and own pain. Interestingly, there is overlap between the subpopulation of ACC and insular cortex neurons that fire during a prosocial helping task and the subpopulation of ACC and insular cortex neuron that are mirror neurons, suggesting that the neural mechanisms that regulate empathic responses like OF and prosocial behaviors are linked (Wu et al. 2023). This finding is consistent with a recent report that found that optogenetic inhibition of the rostral ACC attenuated expression of OF and consolatory behavior in the observer (Phillips et al. 2023). The stability of cortical mirror (and anti-mirror) neurons during repeated OF testing sessions and inputs to these cortical neurons are interesting questions that will further elucidate the role of these neurons in the regulation of empathic and prosocial behaviors.

Some studies have sought to directly manipulate the ACC to BLA pathway in observers with prior experience in OF (Smith et al. 2021; Allsop et al. 2018). Surprisingly, in contrast to Naive OF, the ACC to BLA pathway is dispensable for observers with prior experience, although inhibition of the ACC to BLA pathway in observers with prior experience disrupts the fear of observers to a cue that is associated with shock delivery to the demonstrator and recall of observational contextual fear memory (Smith et al. 2021; Allsop et al. 2018). A major limitation of these and other previous studies that have examined the effect of prior experiences on the neural circuit mechanisms that regulate OF, however, is that these studies use strong shock protocols that would be sufficient to induce OF in observers that lack prior experience (Jeon and Shin 2011). Therefore, it is impossible to determine from these studies the neural circuit mechanisms that regulate OF when strong aversive stimuli are delivered to the demonstrator, the neural circuit mechanisms that regulate OF when observers have prior experience, and the neural circuit mechanisms that regulate both.

Terranova et al. addressed this limitation by designing a behavioral paradigm that delivers a weak shock protocol to the demonstrator, thus requiring observers to have both prior similar aversive experience with shock and social familiarity with the demonstrator to express OF, which they refer to as experience-dependent OF (Exp OF) (Fig. 2a). Like Naive OF, Exp OF requires a social stimulus (Terranova et al. 2022; Ueno et al. 2020). In contrast to Naive OF, however, Exp OF requires a

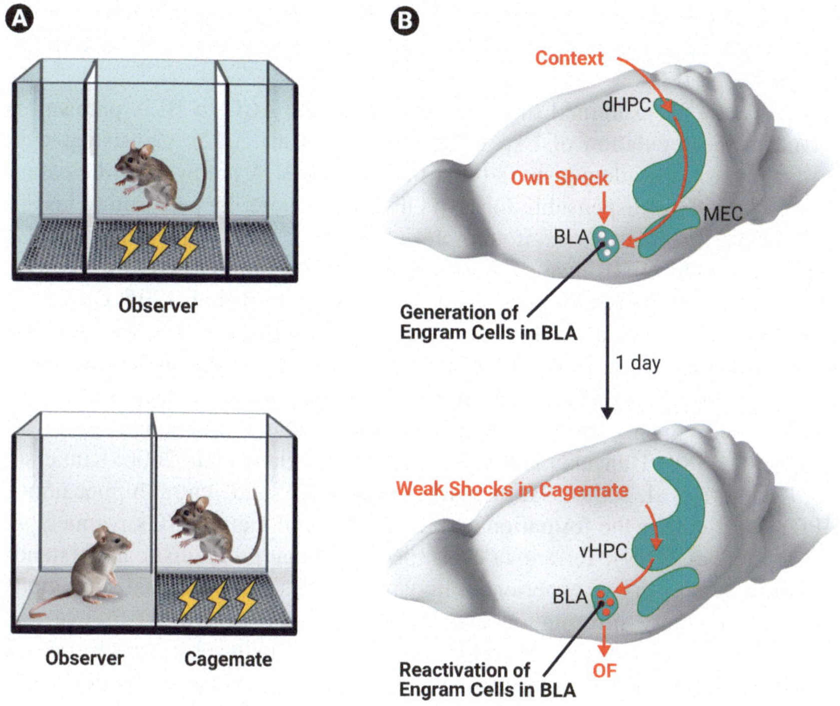


**Fig. 2** Neural circuit mechanisms of experience-dependent observational fear (Exp OF). (**a**) Behavioral paradigm for Exp OF. An observer is first subjected to a fear conditioning paradigm in a context that is distinct from the OF apparatus. The observer receives a single shock and associates that shock with the context. One day later, the observer with shock experience and a familiar cagemate demonstrator are placed into the OF apparatus. After a habituation period, weak aversive shock stimuli are delivered to the demonstrator, which produces a more subtle fear response in the demonstrator, in contrast to the strong shock protocol used in Naive OF that produces a robust fear response in the demonstrator. Observers express Exp OF only if they (1) have prior experience with shock in another context and (2) they witness a familiar cagemate demonstrator receive shocks. (**b**) Exp OF neural circuit mechanisms. During prior similar aversive experiences with shock, fear memory engram cells are generated in the BLA by dHPC, which projects to BLA through regions such as the medial entorhinal cortex (Kitamura et al. 2014, 2017). The next day, when observers witness the familiar cagemate demonstrator receive shocks, vHPC neurons access the BLA fear memory engram cells to elicit Exp OF in the observer. dHPC, dorsal hippocampus; MEC, medial entorhinal cortex; BLA, basolateral amygdala; vHPC, ventral hippocampus

multisensory perception stimulus, including visual cues, auditory cues, and olfactory cues, which we speculate are transmitted by volatile pheromones linked to major urinary proteins (Terranova et al. 2022; Ueno et al. 2020; Cheetham et al. 2007; Roberts et al. 2018; Hurst et al. 2001; Zhao et al. 2023). Moreover, dominance status affects the expression of OF in observers that have prior experience, with subordinate observers exhibiting a stronger OF response when a familiar dominant mouse is used as the demonstrator (Shi et al. 2022a; Park et al. 2022). Critically, observers with aversive experience that is unrelated to the demonstrator's aversive experience fail to show an enhanced OF response, suggesting that prior similar aversive experience and not general stressful experience is essential for Exp OF (Sanders et al. 2013).

Terranova et al. examined the extent to which the ACC to BLA pathway is involved in the regulation of Exp OF (Terranova et al. 2022). Consistent with previous reports, they demonstrated that the ACC to BLA pathway, while essential for Naive OF, is dispensable for Exp OF. Furthermore, by performing loss of function experiments using excitotoxic lesion methods, and by measuring neural activity induced by OF by quantifying expression of the immediate early gene Arc (activity-regulated cytoskeleton-associated protein (Lyford et al. 1995; Guzowski et al. 1999; Bramham et al. 2008)), the authors showed that the ACC was dispensable for Exp OF, whereas the BLA was essential. Since dorsal hippocampus (dHPC) is critical for the acquisition and recall of contextual fear memory to remember one's own contextual fear experience (Maren and Fanselow 1997; Wiltgen et al. 2006; Hunsaker and Kesner 2008; Frankland et al. 2006; Kim et al. 1993; Kitamura et al. 2009, 2015; Kheirbek et al. 2013) and ventral hippocampus (vHPC) is critical for the formation and recall of social memory (Okuyama et al. 2016; Deng et al. 2019), Terranova et al. speculated that prerequisite understanding of the others' aversive experiences may recruit these hippocampal subregions to integrate memory information about prior similar aversive experience and social familiarity with the demonstrator to elicit Exp OF in the observer (Terranova et al. 2022). To test this hypothesis, the authors performed chemogenetic inhibition of dHPC or vHPC and found that dHPC was necessary during prior similar aversive experience with shock for observers to express Exp OF, whereas vHPC was necessary during shock delivery to the familiar demonstrator for observers to express Exp OF. Based on the contributions of dorsoventral HPC and BLA to Exp OF, Terranova et al. examined the roles of dHPC or vHPC on BLA neural activity in observers during prior similar aversive experience with shock or during shock delivery to the familiar demonstrator, respectively. They found differential contributions of dorsoventral HPC on BLA neural activity, such that dHPC was essential for elevated BLA neural activity during prior similar aversive experience with shock, whereas vHPC was essential for elevated BLA neural activity during shock delivery to the familiar demonstrator. Thus, in contrast to Naive OF, which utilizes the ACC to BLA pathway, Terranova et al. demonstrated that Exp OF utilizes a dorsoventral HPC to BLA pathway.

## A BLA Fear Memory Engram Mechanism for Exp OF

How does the BLA integrate information streams from dHPC, which communicates information about prior similar aversive experience with shock, and vHPC, which communicates information about social familiarity of the demonstrator, to elicit Exp OF in the observer? Simulation Theory, a philosophical concept that asserts that an observer envisions themselves in the demonstrator's situation through an unconscious internal simulation in which they recall the memory of their own prior similar experience (Heal 1986; Gordon 1986; Gallese and Goldman 1998), provides a potential framework for Exp OF. While previous reports suggest that a shared neural population, like mirror neurons, may integrate information streams that are necessary for Exp OF (Carrillo et al. 2019; Wu et al. 2023; Gallese and Goldman 1998; Gibson 1958, 1979; Gangopadhyay and Schilbach 2012; Singer et al. 2004), no studies have directly examined whether these shared neural populations are required. Recent technological advances have leveraged activity-dependent cell labeling methods to label and manipulate neural subpopulations that are activated in response to a specific behavioral test (Liu et al. 2012; Sorensen et al. 2016). These technologies have led researchers to discover memory engrams or physical memory traces within the brain, which is an idea that was first proposed in the early twentieth century by the German Biologist Richard Semon (Semon 1911). Memory engrams are subpopulations of neurons that are activated by a salient experience and subsequently undergo biological changes to encode a specific memory episode (Liu et al. 2012; Josselyn and Tonegawa 2020; Kandel et al. 2014; Silva et al. 2009). Since several brain regions, including the BLA and dHPC, contain fear memory engrams encoding contextual information (Kitamura et al. 2017; Tonegawa et al. 2015, 2018; Choi et al. 2018; Marks et al. 2021; Terranova et al. 2019), Terranova et al. hypothesized that fear memory engram cells encoding prior similar fear experience integrate a perception-emotion coupling of own shock experience with the familiar demonstrator's shock experience to elicit Exp OF in the observer (Fig. 2b). To test this hypothesis, the authors performed an activity-dependent cell labeling experiment using the robust activity marking (RAM) system, which links the human c-fos minimal promoter with four repetitions of an enhancer module (PRAM) to the destabilized tetracycline transactivator (Sorensen et al. 2016; Sun et al. 2020). With this approach, the authors could use the inducible doxycycline system to selectively label the subpopulation of neurons that were activated in the BLA or dHPC during prior similar aversive experience with shock and examine the reactivation of these brain regions when observers were subjected to Exp OF or were subjected to recall of their own shock experience by exposure to the conditioned context. While fear memory engram cells in BLA and dHPC were significantly reactivated when observers were exposed to the conditioned context, only fear memory engram cells in BLA were significantly reactivated when observers were exposed to Exp OF. Fear memory engram cells in BLA were essential for observers to recall their shock experience in the conditioned context and for observers to express Exp OF. Furthermore, inhibition of an ensemble of BLA neurons labeled during an unrelated aversive experience (bright light exposure on an elevated platform) failed to affect observer freezing response in Exp OF, indicating that fear memory engram cells in BLA encoding a prior similar aversive experience with shock are specifically necessary for Exp OF.

To examine the temporal dynamics of BLA fear memory engram cell activity in Exp OF, Terranova et al. injected a genetically encoded calcium indicator, GCaMP6f (Chen et al. 2013), and performed longitudinal in vivo calcium imaging through a head-mounted miniaturized microscope during prior similar aversive experience with shock and during Exp OF. The authors identified a subpopulation of shock-responding cells in BLA during prior similar aversive experience with shock, which had previously been shown to be fear memory engram cells (Zhang et al. 2020), and then examined the activity of these BLA fear memory engram cells in response to shock delivery to the familiar demonstrator in Exp OF. Consistent with the activity-dependent cell labeling experiments, the authors found that BLA neurons that fired in response to own shock experience were significantly more likely to fire during shock delivery to the familiar demonstrator. Thus, in concordance with Simulation Theory, BLA fear memory engram cells integrate a perception-emotion coupling of prior similar aversive experience with shock in the observer and shock delivery to the familiar demonstrator to elicit Exp OF in the observer.

## vHPC Neurons Access BLA Fear Memory Engram Cells to Trigger Exp OF

Since vHPC neural activity during shock delivery to the familiar demonstrator, and not during own shock experience, is necessary for observers to express Exp OF, and given the monosynaptic connections between vHPC pyramidal neurons and excitatory neurons in BLA (Bazelot et al. 2015; Hübner et al. 2014; Kim and Cho 2017, 2020), Terranova et al. examined how the vHPC communicates with BLA fear memory engram cells to elicit Exp OF in the observer. They found that vHPC is necessary for the reactivation of BLA fear memory engram cells in Exp OF and that the vHPC also contains distinct subpopulation of neurons that respond to shock delivery to the familiar or unfamiliar demonstrator. This is consistent with the social memory literature, which found that vHPC contains distinct subpopulations of neurons that respond to the presentation of familiar or unfamiliar mice (Okuyama et al. 2016; Deng et al. 2019). An experiment that combined retrograde tracing with Arc immunohistochemistry revealed that BLA-projecting vHPC neurons from the ventral CA1 (vCA1) subregion, but not other region-projecting vCA1 neurons, are preferentially activated when observers are tested for OF with a familiar cagemate demonstrator compared with an unfamiliar stranger demonstrator. The authors further confirmed these findings by performing optogenetic inhibition of BLA-projecting vHPC neurons during shock delivery to the familiar demonstrator, which showed that the vHPC to BLA pathway is necessary during the familiar demonstrator's shock moments for observers to express Exp OF. Moreover, Terranova et al. examined the role of other brain regions the vHPC projects to besides the BLA in Exp OF, including the medial prefrontal cortex, nucleus accumbens, and other amygdalar subregions (Bazelot et al. 2015; Hübner et al. 2014; Kim and Cho 2017; Ciocchi et al. 2015). Consistent with the retrograde tracing and optogenetic data, the authors found that neural regions other than the BLA that received projections from the vHPC were dispensable for Exp OF, indicating the specific importance of the vHPC to BLA pathway in the regulation of Exp OF.

vHPC pyramidal neurons both excite BLA pyramidal neurons by directly projecting to these neurons and inhibit these neurons by projecting to BLA inhibitory neurons, which then project to the BLA pyramidal neurons (Bazelot et al. 2015; Hübner et al. 2014; Kim and Cho 2017, 2020; Graham et al. 2021). Since the feedforward inhibition pathway is engaged when vHPC neurons are activated at the theta frequency (Bazelot et al. 2015; Graham et al. 2021), Terranova et al. examined the population activity of vHPC during Exp OF by measuring local field potentials (LFPs) within vHPC to determine whether the vHPC pyramidal neurons were accessing pyramidal neurons within the BLA via feedforward excitation or inhibition. They found that LFP power at the theta frequency was reduced in vHPC of observers during Exp OF compared with the habituation period, suggesting that vHPC pyramidal neurons directly access BLA pyramidal neurons to elicit Exp OF in the observer. To directly examine whether vHPC neurons access the BLA fear memory engram cells to elicit Exp OF in the observer, Terranova et al. examined the reactivation of BLA fear memory engram cells in observers during Exp OF while also performing optogenetic inhibition during the familiar demonstrator's shock moments. Critically, the authors demonstrated that BLA-projecting vHPC neurons directly access BLA fear memory engram cells during the familiar demonstrator's shock moments to elicit Exp OF. Finally, by testing observers in the Exp OF paradigm and delivering strong shocks to the demonstrator, the authors showed that the dorsoventral HPC to BLA memory engram pathway and ACC to BLA pathway parallelly facilitate OF, which could potentially enhance the empathic ability of individuals in different social scenarios (Morelli et al. 2014).

## Future Directions

While great progress has been made in understanding the neural circuit mechanisms that regulate Naive OF and Exp OF, several additional questions remain that are of great value for future study. First, are there sex differences in the neural mechanisms that regulate Naive OF and Exp OF? Previous reports have suggested that females may have different freezing patterns in OF compared with males (Lidhar et al. 2017; Pisansky et al. 2017a; Han et al. 2020), suggesting that the neural circuit mechanisms in females may be different. Since one study found that there was no difference in freezing levels of female observers to a familiar or unfamiliar demonstrator (Pisansky et al. 2017a), it is possible that BLA fear memory engram cells in female observers are activated by vHPC neurons regardless of the demonstrator's familiarity. Alternatively, female observers may not use BLA fear memory engram cells for Exp OF, instead relying on only the ACC to BLA pathway. Furthermore, sex differences in the signaling of social neuropeptides like OT, or the closely related neuropeptide arginine vasopressin (Song and Albers 2018), may contribute to potential sex differences in OF, especially given the sexually dimorphic effects of these neuropeptides on social behaviors (Borland et al. 2019a, b; Caldwell 2018; Terranova et al. 2016, 2017). In addition, differential effects of estrogen and testosterone may contribute to sex differences in OF (Olsson et al. 2016b). The examination of these neural circuit mechanisms may give insight into sex differences that have been reported for empathy and other social cognitive behaviors and the potential contribution of memory engrams to these sex differences (Proverbio 2021).

Second, social cognitive deficits, including deficits in empathic responses, have been reported in individuals with neuropsychiatric disorders such as autism spectrum disorder (ASD) and schizophrenia (Rum and Perry 2020; Wang et al. 2022; Weinreb et al. 2022). Although behavioral interventions can help improve empathic responses and other social cognitive behaviors in individuals with neuropsychiatric disorders like ASD and schizophrenia (Kurtz et al. 2016; Dandil et al. 2020), pharmacotherapy treatment options can be limited (Vaskinn and Horan 2020; Goh et al. 2021; Sikich et al. 2021; Ford and Young 2022). Therefore, examination of the basic neural circuit mechanisms that regulate empathic responses like Naive OF and Exp OF may reveal novel therapeutic targets that can be leveraged in the treatment of neuropsychiatric disorders. Some studies have begun to examine how OF is impacted in mouse models of ASD (Twining et al. 2017; Keum et al. 2018; Takayama et al. 2022; Pisansky et al. 2017b) and neurodegenerative disorders, including Alzheimer's disease (Choi and Jeong 2017) and frontotemporal dementia (Phillips et al. 2023). One study in particular found that deletion of the Nrxn3 gene, which has been implicated as a risk factor for ASD (Tromp et al. 2021), altered the freezing response of observers in Naive OF (Keum et al. 2018). Thus, OF behavioral models may be useful for examining the genetic and neural circuit mechanisms that underlie alterations in empathic responses that are associated with ASD. Moreover, since the behavioral paradigms for Naive OF and Exp OF are robust, and given the potential role of memory engram cells in neuropsychiatric disorders (Josselyn and Tonegawa 2020; Terranova et al. 2019; Roy et al. 2016), investigating how these responses breakdown in rodent models of neuropsychiatric disorders and the corresponding neural circuit mechanisms, especially mechanisms related to memory engram cells, may produce findings of great translational interest.

Finally, a limitation of OF models is that the observer only passively watches the demonstrator receive shocks and is therefore unable to alleviate the demonstrator's aversive situation. Are the neural circuit mechanisms that regulate Naive OF and Exp OF necessary for prosocial behavior to alleviate the demonstrator's aversive situation? The findings of one study suggest that the BLA fear memory engram cells that are essential for Exp OF may also be necessary for prosocial motivation to alleviate the demonstrator's aversive situation (Wu et al. 2023), but additional mechanistic studies are needed to examine this question.

## Conclusion

The application of activity-dependent cell labeling technologies to behavioral neuroscience research has enabled the interrogation of the role of memory engrams in the recall of past events and the linking of related experiences that are temporally close (Josselyn and Tonegawa 2020). By examining fear memory engrams, Terranova et al. demonstrate that the HPC-BLA network acts as an integrator of a perception-emotion coupling of prior similar aversive experience with shock in the observer and social familiarity with the demonstrator to elicit Exp OF in the observer. Moreover, the authors demonstrate a novel purpose for fear memory engrams in the BLA. Instead of simply being used to recall one's own fear experience, BLA fear memory engrams can be used to understand the familiar other's similar aversive experience. Thus, memory engrams are not only useful for recalling one's past experiences but for understanding the other's aversive situation.

## References

Adolphs R (2013) The biology of fear. Curr Biol 23:R79–R93. https://doi.org/10.1016/j.cub.2012.11.055

Allsop SA et al (2018) Corticoamygdala transfer of socially derived information gates observational learning. Cell 173:1329–1342.e1318. https://doi.org/10.1016/j.cell.2018.04.004

Atsak P et al (2011) Experience modulates vicarious freezing in rats: a model for empathy. PLoS One 6:e21855. https://doi.org/10.1371/journal.pone.0021855

Barrett LF, Mesquita B, Ochsner KN, Gross JJ (2007) The experience of emotion. Annu Rev Psychol 58:373–403. https://doi.org/10.1146/annurev.psych.58.110405.085709

Bazelot M et al (2015) Hippocampal theta input to the amygdala shapes feedforward inhibition to gate heterosynaptic plasticity. Neuron 87:1290–1303. https://doi.org/10.1016/j.neuron.2015.08.024

Ben-Ami Bartal I, Decety J, Mason P (2011) Empathy and pro-social behavior in rats. Science 334:1427–1430. https://doi.org/10.1126/science.1210789

Ben-Ami Bartal I, Rodgers DA, Bernardez Sarria MS, Decety J, Mason P (2014) Pro-social behavior in rats is modulated by social experience. Elife 3:e01385. https://doi.org/10.7554/eLife.01385

Bilodeau-Houle A et al (2020) Anxiety sensitivity moderates the association between father-child relationship security and fear transmission. Front Psychol 11:579514. https://doi.org/10.3389/fpsyg.2020.579514

Bilodeau-Houle A, Morand-Beaulieu S, Bouchard V, Marin MF (2023) Parent-child physiological concordance predicts stronger observational fear learning in children with a less secure relationship with their parent. J Exp Child Psychol 226:105553. https://doi.org/10.1016/j.jecp.2022.105553

Borland JM, Rilling JK, Frantz KJ, Albers HE (2019a) Sex-dependent regulation of social reward by oxytocin: an inverted U hypothesis. Neuropsychopharmacology 44:97–110. https://doi.org/10.1038/s41386-018-0129-2

Borland JM et al (2019b) Sex-dependent regulation of social reward by oxytocin receptors in the ventral tegmental area. Neuropsychopharmacology 44:785–792. https://doi.org/10.1038/s41386-018-0262-y

Bramham CR, Worley PF, Moore MJ, Guzowski JF (2008) The immediate early gene arc/arg3.1: regulation, mechanisms, and function. J Neurosci 28:11760–11767. https://doi.org/10.1523/JNEUROSCI.3864-08.2008
Bruchey AK, Jones CE, Monfils MH (2010) Fear conditioning by-proxy: social transmission of fear during memory retrieval. Behav Brain Res 214:80–84. https://doi.org/10.1016/j.bbr.2010.04.047
Burkett JP et al (2016) Oxytocin-dependent consolation behavior in rodents. Science 351:375–378. https://doi.org/10.1126/science.aac4785
Caldwell HK (2018) Oxytocin and sex differences in behavior. Curr Opin Behav Sci 23:13–20. https://doi.org/10.1016/j.cobeha.2018.02.002
Campbell MW, de Waal FB (2014) Chimpanzees empathize with group mates and humans, but not with baboons or unfamiliar chimpanzees. Proc Biol Sci 281:20140013. https://doi.org/10.1098/rspb.2014.0013
Cao Y, Contreras-Huerta LS, McFadyen J, Cunnington R (2015) Racial bias in neural response to others' pain is reduced with other-race contact. Cortex 70:68–78. https://doi.org/10.1016/j.cortex.2015.02.010
Carrillo M et al (2019) Emotional mirror neurons in the rat's anterior cingulate cortex. Curr Biol 29:1301–1312.e1306. https://doi.org/10.1016/j.cub.2019.03.024
Cheetham SA et al (2007) The genetic basis of individual-recognition signals in the mouse. Curr Biol 17:1771–1777. https://doi.org/10.1016/j.cub.2007.10.007
Chen Q, Panksepp JB, Lahvis GP (2009) Empathy is moderated by genetic background in mice. PLoS One 4:e4387. https://doi.org/10.1371/journal.pone.0004387
Chen TW et al (2013) Ultrasensitive fluorescent proteins for imaging neuronal activity. Nature 499:295–300. https://doi.org/10.1038/nature12354
Choi J, Jeong Y (2017) Elevated emotional contagion in a mouse model of Alzheimer's disease is associated with increased synchronization in the insula and amygdala. Sci Rep 7:46262. https://doi.org/10.1038/srep46262
Choi JH et al (2018) Interregional synaptic maps among engram cells underlie memory formation. Science 360:430–435. https://doi.org/10.1126/science.aas9204
Church RM (1959) Emotional reactions of rats to the pain of others. J Comp Physiol Psychol 52:132–134. https://doi.org/10.1037/h0043531
Ciocchi S, Passecker J, Malagon-Vina H, Mikus N, Klausberger T (2015) Selective information routing by ventral hippocampal CA1 projection neurons. Science 348:560–563. https://doi.org/10.1126/science.aaa3245
Clinchy M, Sheriff MJ, Zanette LY (2013) Predator-induced stress and the ecology of fear. Funct Ecol 27:56–65. https://doi.org/10.1111/1365-2435.12007
Dandil Y, Smith K, Kinnaird E, Toloza C, Tchanturia K (2020) Cognitive remediation interventions in autism spectrum condition: a systematic review. Front Psych 11:722. https://doi.org/10.3389/fpsyt.2020.00722
Debiec J, Olsson A (2017) Social fear learning: from animal models to human function. Trends Cogn Sci 21:546–555. https://doi.org/10.1016/j.tics.2017.04.010
Deng X, Gu L, Sui N, Guo J, Liang J (2019) Parvalbumin interneuron in the ventral hippocampus functions as a discriminator in social memory. Proc Natl Acad Sci USA 116:16583–16592. https://doi.org/10.1073/pnas.1819133116
Diaz V, Lin D (2020) Neural circuits for coping with social defeat. Curr Opin Neurobiol 60:99–107. https://doi.org/10.1016/j.conb.2019.11.016
Dou H, Lei Y, Pan Y, Li H, Astikainen P (2023) Impact of observational and direct learning on fear conditioning generalization in humans. Prog Neuro-Psychopharmacol Biol Psychiatry 121:110650. https://doi.org/10.1016/j.pnpbp.2022.110650
Eklund J, Andersson-Straberg T, Hansen EM (2009) "I've also experienced loss and fear": effects of prior similar experience on empathy. Scand J Psychol 50:65–69. https://doi.org/10.1111/j.1467-9450.2008.00673.x

Fendt M, Gonzalez-Guerrero CP, Kahl E (2021) Observational fear learning in rats: role of trait anxiety and ultrasonic vocalization. Brain Sci 11:423. https://doi.org/10.3390/brainsci11040423
Ford CL, Young LJ (2022) Refining oxytocin therapy for autism: context is key. Nat Rev Neurol 18:67–68. https://doi.org/10.1038/s41582-021-00602-9
Frankland PW et al (2006) Stability of recent and remote contextual fear memory. Learn Mem 13:451–457. https://doi.org/10.1101/lm.183406
Gallese V, Goldman A (1998) Mirror neurons and the simulation theory of mind-reading. Trends Cogn Sci 2:493–501. https://doi.org/10.1016/s1364-6613(98)01262-5
Gallo S et al (2018) The causal role of the somatosensory cortex in prosocial behaviour. Elife 7:e32740. https://doi.org/10.7554/eLife.32740
Gangopadhyay N, Schilbach L (2012) Seeing minds: a neurophilosophical investigation of the role of perception-action coupling in social perception. Soc Neurosci 7:410–423. https://doi.org/10.1080/17470919.2011.633754
García-Moreno JA, Cañadas-Pérez F, García-García J, Roldan-Tapia MD (2021) Cognitive reserve and anxiety interactions play a fundamental role in the response to the stress. Front Psychol 12:673596. https://doi.org/10.3389/fpsyg.2021.673596
Gibson JJ (1958) Visually controlled locomotion and visual orientation in animals. Br J Psychol 49:182–194. https://doi.org/10.1111/j.2044-8295.1958.tb00656.x
Gibson JJ (1979) The ecological approach to visual perception. Houghton, Mifflin and Company
Goh KK, Chen C-H, Lane H-Y (2021) Oxytocin in schizophrenia: pathophysiology and implications for future treatment. Int J Mol Sci 22:2146
Gonzalez-Liencres C, Juckel G, Tas C, Friebe A, Brune M (2014) Emotional contagion in mice: the role of familiarity. Behav Brain Res 263:16–21. https://doi.org/10.1016/j.bbr.2014.01.020
Gordon RM (1986) Folk psychology as simulation. Mind Lang 1:158–171
Graham J et al (2021) High-frequency stimulation of ventral CA1 neurons reduces amygdala activity and inhibits fear. Front Behav Neurosci 15:595049. https://doi.org/10.3389/fnbeh.2021.595049
Guzowski JF, McNaughton BL, Barnes CA, Worley PF (1999) Environment-specific expression of the immediate-early gene Arc in hippocampal neuronal ensembles. Nat Neurosci 2:1120–1124. https://doi.org/10.1038/16046
Han S (2018) Neurocognitive basis of racial ingroup bias in empathy. Trends Cogn Sci 22:400–421. https://doi.org/10.1016/j.tics.2018.02.013
Han Y, Sichterman B, Carrillo M, Gazzola V, Keysers C (2020) Similar levels of emotional contagion in male and female rats. Sci Rep 10:2763. https://doi.org/10.1038/s41598-020-59680-2
Heal J (1986) Replication and functionalism. In: Butterfield J (ed) Language, mind, and logic, vol 14. Cambridge University Press, pp 135–150
Hernandez-Lallement J et al (2020) Harm to others acts as a negative reinforcer in rats. Curr Biol 30:949–961.e947. https://doi.org/10.1016/j.cub.2020.01.017
Hong EH, Choi JS (2018) Observational threat conditioning is induced by circa-strike activity burst but not freezing and requires visual attention. Behav Brain Res 353:161–167. https://doi.org/10.1016/j.bbr.2018.06.034
Hooker CI, Verosky SC, Miyakawa A, Knight RT, D'Esposito M (2008) The influence of personality on neural mechanisms of observational fear and reward learning. Neuropsychologia 46:2709–2724. https://doi.org/10.1016/j.neuropsychologia.2008.05.005
Huang Z et al (2023) Ventromedial prefrontal neurons represent self-states shaped by vicarious fear in male mice. Nat Commun 14:3458. https://doi.org/10.1038/s41467-023-39081-5
Hübner C, Bosch D, Gall A, Lüthi A, Ehrlich I (2014) Ex vivo dissection of optogenetically activated mPFC and hippocampal inputs to neurons in the basolateral amygdala: implications for fear and emotional memory. Front Behav Neurosci 8:64. https://doi.org/10.3389/fnbeh.2014.00064
Hunsaker MR, Kesner RP (2008) Dissociations across the dorsal-ventral axis of CA3 and CA1 for encoding and retrieval of contextual and auditory-cued fear. Neurobiol Learn Mem 89:61–69. https://doi.org/10.1016/j.nlm.2007.08.016

Hurst JL et al (2001) Individual recognition in mice mediated by major urinary proteins. Nature 414:631–634. https://doi.org/10.1038/414631a
Ito W, Morozov A (2019) Prefrontal-amygdala plasticity enabled by observational fear. Neuropsychopharmacology 44:1778–1787. https://doi.org/10.1038/s41386-019-0342-7
Jeon D, Shin HS (2011) A mouse model for observational fear learning and the empathetic response. Curr Protoc Neurosci Chapter 8:Unit 8.27. https://doi.org/10.1002/0471142301.ns0827s57
Jeon D et al (2010) Observational fear learning involves affective pain system and Cav1.2 Ca2+ channels in ACC. Nat Neurosci 13:482–488. https://doi.org/10.1038/nn.2504
John ER, Chesler P, Bartlett F, Victor I (1968) Observation learning in cats. Science 159:1489–1491. https://doi.org/10.1126/science.159.3822.1489
Jones CE, Agee L, Monfils MH (2018) Fear conditioning by proxy: social transmission of fear between interacting conspecifics. Curr Protoc Neurosci 83:e43. https://doi.org/10.1002/cpns.43
Josselyn SA, Tonegawa S (2020) Memory engrams: recalling the past and imagining the future. Science 367:eaaw4325. https://doi.org/10.1126/science.aaw4325
Jung S et al (2013) Social deficits in the AY-9944 mouse model of atypical absence epilepsy. Behav Brain Res 236:23–29. https://doi.org/10.1016/j.bbr.2012.08.029
Kandel ER, Dudai Y, Mayford MR (2014) The molecular and systems biology of memory. Cell 157:163–186. https://doi.org/10.1016/j.cell.2014.03.001
Kaźmierowska AM et al (2022) Learning about threat from friends and strangers is equally effective: an fMRI study on observational fear conditioning. NeuroImage 263:119648. https://doi.org/10.1016/j.neuroimage.2022.119648
Keum S, Shin HS (2016) Rodent models for studying empathy. Neurobiol Learn Mem 135:22–26. https://doi.org/10.1016/j.nlm.2016.07.022
Keum S, Shin HS (2019a) Genetic factors associated with empathy in humans and mice. Neuropharmacology 159:107514. https://doi.org/10.1016/j.neuropharm.2019.01.029
Keum S, Shin H-S (2019b) Neural basis of observational fear learning: a potential model of affective empathy. Neuron 104:78–86. https://doi.org/10.1016/j.neuron.2019.09.013
Keum S et al (2016) Variability in empathic fear response among 11 inbred strains of mice. Genes Brain Behav 15:231–242. https://doi.org/10.1111/gbb.12278
Keum S et al (2018) A missense variant at the Nrxn3 locus enhances empathy fear in the mouse. Neuron 98:588–601.e585. https://doi.org/10.1016/j.neuron.2018.03.041
Kheirbek MA et al (2013) Differential control of learning and anxiety along the dorsoventral axis of the dentate gyrus. Neuron 77:955–968. https://doi.org/10.1016/j.neuron.2012.12.038
Kim WB, Cho JH (2017) Synaptic targeting of double-projecting ventral CA1 hippocampal neurons to the medial prefrontal cortex and basal amygdala. J Neurosci 37:4868–4882. https://doi.org/10.1523/jneurosci.3579-16.2017
Kim WB, Cho JH (2020) Encoding of contextual fear memory in hippocampal-amygdala circuit. Nat Commun 11:1382. https://doi.org/10.1038/s41467-020-15121-2
Kim JJ, Rison RA, Fanselow MS (1993) Effects of amygdala, hippocampus, and periaqueductal gray lesions on short- and long-term contextual fear. Behav Neurosci 107:1093–1098. https://doi.org/10.1037//0735-7044.107.6.1093
Kim S, Matyas F, Lee S, Acsady L, Shin HS (2012) Lateralization of observational fear learning at the cortical but not thalamic level in mice. Proc Natl Acad Sci USA 109:15497–15501. https://doi.org/10.1073/pnas.1213903109
Kim BS et al (2014) Differential regulation of observational fear and neural oscillations by serotonin and dopamine in the mouse anterior cingulate cortex. Psychopharmacology 231:4371–4381. https://doi.org/10.1007/s00213-014-3581-7
Kim A, Keum S, Shin HS (2019) Observational fear behavior in rodents as a model for empathy. Genes Brain Behav 18:e12521. https://doi.org/10.1111/gbb.12521
Kim S-W, Kim M, Shin H-S (2021) Affective empathy and prosocial behavior in rodents. Curr Opin Neurobiol 68:181–189. https://doi.org/10.1016/j.conb.2021.05.002
Kim S-W et al (2023) Hemispherically lateralized rhythmic oscillations in the cingulate-amygdala circuit drive affective empathy in mice. Neuron 111:418–429.e414. https://doi.org/10.1016/j.neuron.2022.11.001

Kitamura T et al (2009) Adult neurogenesis modulates the hippocampus-dependent period of associative fear memory. Cell 139:814–827. https://doi.org/10.1016/j.cell.2009.10.020
Kitamura T et al (2014) Island cells control temporal association memory. Science 343:896–901. https://doi.org/10.1126/science.1244634
Kitamura T et al (2015) Entorhinal cortical ocean cells encode specific contexts and drive context-specific fear memory. Neuron 87:1317–1331. https://doi.org/10.1016/j.neuron.2015.08.036
Kitamura T et al (2017) Engrams and circuits crucial for systems consolidation of a memory. Science 356:73–78. https://doi.org/10.1126/science.aam6808
Kurtz MM, Gagen E, Rocha NBF, Machado S, Penn DL (2016) Comprehensive treatments for social cognitive deficits in schizophrenia: a critical review and effect-size analysis of controlled studies. Clin Psychol Rev 43:80–89. https://doi.org/10.1016/j.cpr.2015.09.003
LeDoux J (2012) Rethinking the emotional brain. Neuron 73:653–676. https://doi.org/10.1016/j.neuron.2012.02.004
Lidhar NK, Insel N, Dong JY, Takehara-Nishiuchi K (2017) Observational fear learning in degus is correlated with temporal vocalization patterns. Behav Brain Res 332:362–371. https://doi.org/10.1016/j.bbr.2017.06.011
Liu X et al (2012) Optogenetic stimulation of a hippocampal engram activates fear memory recall. Nature 484:381–385. https://doi.org/10.1038/nature11028
Liu L, Ito W, Morozov A (2017) GABAb receptor mediates opposing adaptations of GABA release from two types of prefrontal interneurons after observational fear. Neuropsychopharmacology 42:1272–1283. https://doi.org/10.1038/npp.2016.273
Lyford GL et al (1995) Arc, a growth factor and activity-regulated gene, encodes a novel cytoskeleton-associated protein that is enriched in neuronal dendrites. Neuron 14:433–445. https://doi.org/10.1016/0896-6273(95)90299-6
Maren S, Fanselow MS (1997) Electrolytic lesions of the fimbria/fornix, dorsal hippocampus, or entorhinal cortex produce anterograde deficits in contextual fear conditioning in rats. Neurobiol Learn Mem 67:142–149. https://doi.org/10.1006/nlme.1996.3752
Marin MF et al (2020) Vicarious conditioned fear acquisition and extinction in child-parent dyads. Sci Rep 10:17130. https://doi.org/10.1038/s41598-020-74170-1
Marks WD, Yamamoto N, Kitamura T (2021) Complementary roles of differential medial entorhinal cortex inputs to the hippocampus for the formation and integration of temporal and contextual memory (Systems Neuroscience). Eur J Neurosci 54:6762–6779. https://doi.org/10.1111/ejn.14737
Marks WD, Yokose J, Kitamura T, Ogawa SK (2022) Neuronal ensembles organize activity to generate contextual memory. Front Behav Neurosci 16:805132. https://doi.org/10.3389/fnbeh.2022.805132
Mineka S, Davidson M, Cook M, Keir R (1984) Observational conditioning of snake fear in rhesus monkeys. J Abnorm Psychol 93:355–372. https://doi.org/10.1037//0021-843x.93.4.355
Mobbs D et al (2019) Viewpoints: approaches to defining and investigating fear. Nat Neurosci 22:1205–1216. https://doi.org/10.1038/s41593-019-0456-6
Monfils MH, Agee LA (2019) Insights from social transmission of information in rodents. Genes Brain Behav 18:e12534. https://doi.org/10.1111/gbb.12534
Morelli SA, Rameson LT, Lieberman MD (2014) The neural components of empathy: predicting daily prosocial behavior. Soc Cogn Affect Neurosci 9:39–47. https://doi.org/10.1093/scan/nss088
Morozov A, Ito W (2019) Social modulation of fear: facilitation vs buffering. Genes Brain Behav 18:e12491. https://doi.org/10.1111/gbb.12491
Munksgaard L, DePassille AM, Rushen J, Herskin MS, Kristensen AM (2001) Dairy cows' fear of people: social learning, milk yield and behaviour at milking. Appl Anim Behav Sci 73:15–26. https://doi.org/10.1016/s0168-1591(01)00119-8
Nomura H, Teshirogi C, Nakayama D, Minami M, Ikegaya Y (2019) Prior observation of fear learning enhances subsequent self-experienced fear learning with an overlapping neuronal ensemble in the dorsal hippocampus. Mol Brain 12:21. https://doi.org/10.1186/s13041-019-0443-6

Okuyama T, Kitamura T, Roy DS, Itohara S, Tonegawa S (2016) Ventral CA1 neurons store social memory. Science 353:1536–1541. https://doi.org/10.1126/science.aaf7003
Olsson A, Phelps EA (2007) Social learning of fear. Nat Neurosci 10:1095–1102. https://doi.org/10.1038/nn1968
Olsson A, Nearing KI, Phelps EA (2007) Learning fears by observing others: the neural systems of social fear transmission. Soc Cogn Affect Neurosci 2:3–11. https://doi.org/10.1093/scan/nsm005
Olsson A et al (2016a) Vicarious fear learning depends on empathic appraisals and trait empathy. Psychol Sci 27:25–33. https://doi.org/10.1177/0956797615604124
Olsson A, Kopsida E, Sorjonen K, Savic I (2016b) Testosterone and estrogen impact social evaluations and vicarious emotions: a double-blind placebo-controlled study. Emotion 16:515–523. https://doi.org/10.1037/a0039765
Panksepp JB, Lahvis GP (2011) Rodent empathy and affective neuroscience. Neurosci Biobehav Rev 35:1864–1875. https://doi.org/10.1016/j.neubiorev.2011.05.013
Park J, Ha S, Shin HS, Jeong J (2022) Experience of a hierarchical relationship between a pair of mice specifically influences their affective empathy toward each other. Genes Brain Behav 21:e12810. https://doi.org/10.1111/gbb.12810
Pereira AG, Cruz A, Lima SQ, Moita MA (2012) Silence resulting from the cessation of movement signals danger. Curr Biol 22:R627–R628. https://doi.org/10.1016/j.cub.2012.06.015
Pérez-Manrique A, Gomila A (2018) The comparative study of empathy: sympathetic concern and empathic perspective-taking in non-human animals. Biol Rev 93:248–269. https://doi.org/10.1111/brv.12342
Phillips HL et al (2023) Dorsomedial prefrontal hypoexcitability underlies lost empathy in frontotemporal dementia. Neuron 111:797–806.e796. https://doi.org/10.1016/j.neuron.2022.12.027
Pisansky MT, Hanson LR, Gottesman II, Gewirtz JC (2017a) Oxytocin enhances observational fear in mice. Nat Commun 8:2102. https://doi.org/10.1038/s41467-017-02279-5
Pisansky MT et al (2017b) Mice lacking the chromodomain helicase DNA-binding 5 chromatin remodeler display autism-like characteristics. Transl Psychiatry 7:e1152. https://doi.org/10.1038/tp.2017.111
Plotnik JM, de Waal FBM (2014) Asian elephants (Elephas maximus) reassure others in distress. PeerJ 2:e278. https://doi.org/10.7717/peerj.278
Proverbio AM (2021) Sex differences in the social brain and in social cognition. J Neurosci Res 101(5):730–738. https://doi.org/10.1002/jnr.24787
Quervel-Chaumette M, Faerber V, Faragó T, Marshall-Pescini S, Range F (2016) Investigating empathy-like responding to conspecifics' distress in pet dogs. PLoS One 11:e0152920. https://doi.org/10.1371/journal.pone.0152920
Raviv L et al (2023) Elephants as an animal model for self-domestication. Proc Natl Acad Sci 120:e2208607120. https://doi.org/10.1073/pnas.2208607120
Roberts SA et al (2018) Individual odour signatures that mice learn are shaped by involatile major urinary proteins (MUPs). BMC Biol 16:48. https://doi.org/10.1186/s12915-018-0512-9
Roy DS et al (2016) Memory retrieval by activating engram cells in mouse models of early Alzheimer's disease. Nature 531:508–512. https://doi.org/10.1038/nature17172
Rum Y, Perry A (2020) Empathic accuracy in clinical populations. Front Psych 11:457. https://doi.org/10.3389/fpsyt.2020.00457
Sakaguchi T, Iwasaki S, Okada M, Okamoto K, Ikegaya Y (2018) Ethanol facilitates socially evoked memory recall in mice by recruiting pain-sensitive anterior cingulate cortical neurons. Nat Commun 9:3526. https://doi.org/10.1038/s41467-018-05894-y
Sanders J, Mayford M, Jeste D (2013) Empathic fear responses in mice are triggered by recognition of a shared experience. PLoS One 8:e74609. https://doi.org/10.1371/journal.pone.0074609
Sato N, Tan L, Tate K, Okada M (2015) Rats demonstrate helping behavior toward a soaked conspecific. Anim Cogn 18:1039–1047. https://doi.org/10.1007/s10071-015-0872-2
Sato Y, Hirata S, Kano F (2019) Spontaneous attention and psycho-physiological responses to others' injury in chimpanzees. Anim Cogn 22:807–823. https://doi.org/10.1007/s10071-019-01276-z

Selbing I, Olsson A (2019) Anxious behaviour in a demonstrator affects observational learning. Sci Rep 9:9181. https://doi.org/10.1038/s41598-019-45613-1

Semon RW (1911) Die Mneme als erhaltendes Prinzip im Wechsel des organischen Geschehens, vol 7101. Wilhelm Engelmann

Shi T, Feng S, Shi W, Fu Y, Zhou W (2022a) A modified mouse model for observational fear learning and the influence of social hierarchy. Front Behav Neurosci 16:941288. https://doi.org/10.3389/fnbeh.2022.941288

Shi W, Fu Y, Shi T, Zhou W (2022b) Different synaptic plasticity after physiological and psychological stress in the anterior insular cortex in an observational fear mouse model. Front Synaptic Neurosci 14:851015. https://doi.org/10.3389/fnsyn.2022.851015

Sikich L et al (2021) Intranasal oxytocin in children and adolescents with autism spectrum disorder. N Engl J Med 385:1462–1473. https://doi.org/10.1056/NEJMoa2103583

Silva AJ, Zhou Y, Rogerson T, Shobe J, Balaji J (2009) Molecular and cellular approaches to memory allocation in neural circuits. Science 326:391–395. https://doi.org/10.1126/science.1174519

Silvers JA et al (2021) An exploration of amygdala-prefrontal mechanisms in the intergenerational transmission of learned fear. Dev Sci 24:e13056. https://doi.org/10.1111/desc.13056

Singer T et al (2004) Empathy for pain involves the affective but not sensory components of pain. Science 303:1157–1162. https://doi.org/10.1126/science.1093535

Skversky-Blocq Y, Pine DS, Shechner T (2021) Using a novel paradigm to examine observational fear-learning across development. Depress Anxiety 38(7):731–738. https://doi.org/10.1002/da.23152

Skversky-Blocq Y, Shmuel S, Waters AM, Shechner T (2022) Observational extinction reduces fear and its retention among adolescents and adults. Behav Res Ther 159:104207. https://doi.org/10.1016/j.brat.2022.104207

Smith ML, Asada N, Malenka RC (2021) Anterior cingulate inputs to nucleus accumbens control the social transfer of pain and analgesia. Science 371:153–159. https://doi.org/10.1126/science.abe3040

Song Z, Albers HE (2018) Cross-talk among oxytocin and arginine-vasopressin receptors: relevance for basic and clinical studies of the brain and periphery. Front Neuroendocrinol 51:14–24. https://doi.org/10.1016/j.yfrne.2017.10.004

Sorensen AT et al (2016) A robust activity marking system for exploring active neuronal ensembles. Elife 5:e13918. https://doi.org/10.7554/eLife.13918

Sun X et al (2020) Functionally distinct neuronal ensembles within the memory engram. Cell 181:410–423.e417. https://doi.org/10.1016/j.cell.2020.02.055

Szczepanik M et al (2020) Observational learning of fear in real time procedure. Sci Rep 10:16960. https://doi.org/10.1038/s41598-020-74113-w

Takayama K et al (2022) Autism spectrum disorder model mice induced by prenatal exposure to valproic acid exhibit enhanced empathy-like behavior via oxytocinergic signaling. Biol Pharm Bull 45:1124–1132. https://doi.org/10.1248/bpb.b22-00200

Terranova JI et al (2016) Serotonin and arginine–vasopressin mediate sex differences in the regulation of dominance and aggression by the social brain. Proc Natl Acad Sci 113:13233. https://doi.org/10.1073/pnas.1610446113

Terranova JI, Ferris CF, Albers HE (2017) Sex differences in the regulation of offensive aggression and dominance by arginine-vasopressin. Front Endocrinol (Lausanne) 8:308. https://doi.org/10.3389/fendo.2017.00308

Terranova JI, Ogawa SK, Kitamura T (2019) Adult hippocampal neurogenesis for systems consolidation of memory. Behav Brain Res 372:112035. https://doi.org/10.1016/j.bbr.2019.112035

Terranova JI et al (2022) Hippocampal-amygdala memory circuits govern experience-dependent observational fear. Neuron 110(8):1416–1431.e13. https://doi.org/10.1016/j.neuron.2022.01.019

Terranova JI, Yokose J, Osanai H, Ogawa SK, Kitamura T (2023) Systems consolidation induces multiple memory engrams for a flexible recall strategy in observational fear memory in male mice. Nat Commun 14:3976. https://doi.org/10.1038/s41467-023-39718-5

Tonegawa S, Liu X, Ramirez S, Redondo R (2015) Memory engram cells have come of age. Neuron 87:918–931. https://doi.org/10.1016/j.neuron.2015.08.002

Tonegawa S, Morrissey MD, Kitamura T (2018) The role of engram cells in the systems consolidation of memory. Nat Rev Neurosci 19:485–498. https://doi.org/10.1038/s41583-018-0031-2

Tromp A, Mowry B, Giacomotto J (2021) Neurexins in autism and schizophrenia—a review of patient mutations, mouse models and potential future directions. Mol Psychiatry 26:747–760. https://doi.org/10.1038/s41380-020-00944-8

Twining RC, Vantrease JE, Love S, Padival M, Rosenkranz JA (2017) An intra-amygdala circuit specifically regulates social fear learning. Nat Neurosci 20:459–469. https://doi.org/10.1038/nn.4481

Ueno H et al (2020) Conformity-like behaviour in mice observing the freezing of other mice: a model of empathy. BMC Neurosci 21:19. https://doi.org/10.1186/s12868-020-00566-4

Vaish A, Carpenter M, Tomasello M (2009) Sympathy through affective perspective taking and its relation to prosocial behavior in toddlers. Dev Psychol 45:534–543. https://doi.org/10.1037/a0014322

Van Bourg J, Patterson JE, Wynne CDL (2020) Pet dogs (Canis lupus familiaris) release their trapped and distressed owners: individual variation and evidence of emotional contagion. PLoS One 15:e0231742. https://doi.org/10.1371/journal.pone.0231742

Vaskinn A, Horan WP (2020) Social cognition and schizophrenia: unresolved issues and new challenges in a maturing field of research. Schizophr Bull 46:464–470. https://doi.org/10.1093/schbul/sbaa034

Wang X et al (2022) Empathy, theory of mind, and prosocial behaviors in autistic children. Front Psych 13:844578. https://doi.org/10.3389/fpsyt.2022.844578

Weinreb S, Li F, Kurtz MM (2022) A meta-analysis of social cognitive deficits in schizophrenia: does world region matter? Schizophr Res 243:206–213. https://doi.org/10.1016/j.schres.2022.04.002

Wiltgen BJ, Sanders MJ, Anagnostaras SG, Sage JR, Fanselow MS (2006) Context fear learning in the absence of the hippocampus. J Neurosci 26:5484–5491. https://doi.org/10.1523/JNEUROSCI.2685-05.2006

Wu WY, Cheng Y, Liang KC, Lee RX, Yen CT (2023) Affective mirror and anti-mirror neurons relate to prosocial help in rats. iScience 26:105865. https://doi.org/10.1016/j.isci.2022.105865

Yusufishaq S, Rosenkranz JA (2013) Post-weaning social isolation impairs observational fear conditioning. Behav Brain Res 242:142–149. https://doi.org/10.1016/j.bbr.2012.12.050

Zhang X, Kim J, Tonegawa S (2020) Amygdala reward neurons form and store fear extinction memory. Neuron 105:1077–1093.e1077. https://doi.org/10.1016/j.neuron.2019.12.025

Zhao M-G et al (2023) Major urinary protein 1 acts as a critical molecule in empathy regulation. PREPRINT (Version 1) available at Research Square, 1–35. https://doi.org/10.21203/rs.3.rs-2610178/v1

Zhou C et al (2018) Activation of parvalbumin interneurons in anterior cingulate cortex impairs observational fear. Sci Bull 63:771–778. https://doi.org/10.1016/j.scib.2018.05.030

# Part V
# Beyond the Neurobiology of Engrams

# A Computational Framework for Memory Engrams

**Chiara Gastaldi and Wulfram Gerstner**

**Abstract** Memory engrams in mice brains are potentially related to groups of concept cells in human brains. A single concept cell in human hippocampus responds, for example, not only to different images of the same object or person but also to its name written down in characters. Importantly, a single mental concept (object or person) is represented by several concept cells and each concept cell can respond to more than one concept. Computational work shows how mental concepts can be embedded in recurrent artificial neural networks as memory engrams and how neurons that are shared between different engrams can lead to associations between concepts. Therefore, observations at the level of neurons can be linked to cognitive notions of memory recall and association chains between memory items.

**Keywords** Attractor networks · Hopfield model · Concept cells · Inferior temporal cortex

## Vocabulary Box

| | |
|---|---|
| Activity pattern | A configuration of neurons firing at high rate embedded in a sea of neurons firing at low rate. |
| Assembly | A set of neurons participating in a memory engram with relatively strong connectivity among themselves form an assembly, also called Hebbian assembly. |

C. Gastaldi · W. Gerstner (✉)
Brain Mind Institute – School of Computer and Communication Sciences – School of Life Sciences, EPFL, Lausanne, Switzerland
e-mail: wulfram.gerstner@epfl.ch



J. Gräff, S. Ramirez (eds.), *Engrams*, Advances in Neurobiology 38,
https://doi.org/10.1007/978-3-031-62983-9_13

| | |
|---|---|
| Association | Spontaneous transition between different mental concepts or, in the case of auto-associative memory, between different aspects of the same mental concept. |
| Attractor network | Biological or artificial neural networks with properties such that the network activity converges to certain preferred activity patterns. |
| Concept cell | A single neuron, often in hippocampus or more generally the medium temporal lobe, that responds to the retrieval of a mental concept, independently of the specifics of sensory input |
| Concept | An object such as a person, a place, or an animal. Example: concepts are "my mother," "the Sydney Opera house," and "my dog Max," and it does not matter whether the concept is triggered by an image, a word, or a spontaneous thought. |
| Hopfield model | The Hopfield network is an example of an attractor network where engrams correspond to memory items represented by preferred activity patterns. |
| Memory engram | A set of cells that responds to a specific memory item. In this chapter, we consider an assembly of concept cells as equivalent to an engram. |
| Memory recall or concept retrieval | High activity in assembly of concept cells. |
| Overlap of engrams | Fraction of concept cells shared across engrams. |
| Shared neuron | A neuron that participates in two or more different engrams. |
| Similarity measure, $m^\mu$ | Measure of the correlation between the current state of activity of the memory network and the state of memory recall of a specific concept. |
| Sparse activity | Fraction of neurons in a given brain area that are active during the activation of one engram or one memory item. |

## Associations and Auto-associations in Memory

Humans can memorize thousands of concepts, where the term "concept" is taken in a broad sense encompassing facts ("who is the president of the United States?"), objects of daily life ("imagine a spoon!"), naming of food items ("this is a banana"), meanings of words, episodes of life, persons, family relations, or the layout of your hometown. To check whether your discussion partner remembers a "concept," you may want to give a cue, for example, by asking a question, providing a keyword, or showing a picture.

In computational research, the recall of memories of concepts has been formalized and categorized as either an "associative" or "auto-associative" memory. To understand the idea of "auto-associative" memory, let us consider the example of Fig. 1a, where we see an animal partially hidden behind a tree. We can easily guess that the hidden animal is a horse, even if the visual information is incomplete. Whenever we are able to recall the full memory of a concept starting from a piece of partial information on the same concept, we speak of "auto-associative" memory recall. A rough black-and-white drawing of an apple may enable you to think of the taste and color of an apple and thus helps you to recall the rich concept of "apple" from partial information.

In contrast to auto-associative memory, associative memory (also called association memory) links two separate concepts. The example of Fig. 2b shows Laurel's famous comic character and reminds us of his equally famous comedy partner Hardy. Even though Laurel and Hardy are separate persons and, hence, separate concepts, their repeated and simultaneous appearances on TV have, for most fans of their TV comedies, consolidated the association between the two actors.

Associations are not restricted to concepts of the same type such as two persons. If you visited a famous place together with a good friend, a picture of that place will remind you of the friend. The personally experienced event (the trip in this example) acts as the glue that associates two otherwise unrelated concepts (the friend and the place). This type of learned association, if related to personal experience, is also

**Fig. 1** (**a**) Example of auto-associative memory recall. (**b**) Example of associative or episodic memory recall

called "episodic" memory. Note that, typically, before the trip you could already recognize your friend and had seen pictures of the famous place as separate entities, implying that the memory of the individual concepts was already stored somewhere in your brain; the association between the two concepts was built later. Associative (across mental concepts) and auto-associative (within mental concepts) memory work in parallel and possibly collaborate for concept recall.

## Concept Cells

Memory engrams can be induced in the hippocampal formation of mice by appropriate experimental stimulation paradigms (see other chapters in this book). In a typical experiment, a memory engram is the ensemble of neurons ("assembly") that respond selectively to a context, such as a specific cage or stimulus. While obtaining experimental data in the human brain has ethical and practical limitations, a stream of experimental results hints at a mechanism in the human medium temporal lobe (MTL) that is analogous to the formation of engrams in mice.

Experimentalists found neurons in the human MTL which selectively and consistently respond to stimuli representing specific individuals or places (Quiroga et al. 2005; Ison and Quiroga 2008; Quiroga 2019a). The experiments were conducted in patients suffering from severe treatment-resistant epilepsy, requiring surgical intervention. To determine the location of the epileptic focus in relation to crucial brain areas like those responsible for speech or motor control, electrophysiological recordings are conducted while the patient engages in various tasks. Unlike neurons in the visual cortex that respond specifically to visual stimuli, single neurons in the medial temporal lobe of the human cortex, particularly in the hippocampus, exhibit broader responsiveness to a range of stimuli associated with the same mental concept. For instance, a neuron may respond to both the written word "Sydney Opera" and a picture of the Sydney Opera House (Fig. 2a). This finding allows us to interpret such neurons as being part of an assembly that encodes the mental concept of "Sydney Opera." Other neurons in the same area may respond to the "Tower of Pisa" and yet others to a famous actor. These neurons have been named "concept cells" (Quiroga et al. 2005).

### *Coding for Single Concepts*

Each mental concept is believed to be represented by a group of concept cells (Quiroga et al. 2005; Ison and Quiroga 2008; Quiroga 2019a) which have been hypothesized to form a memory engram (Gastaldi et al. 2021). The group of concept cells simultaneously increase their firing rates when one of the stimuli that trigger concept recall is presented. The estimated fraction, $\gamma$, of MTL neurons involved in representing a given concept is approximately $\gamma = 0.23\%$ (Waydo et al. 2006). Assuming that each memory item is represented by the activation of a fixed, yet

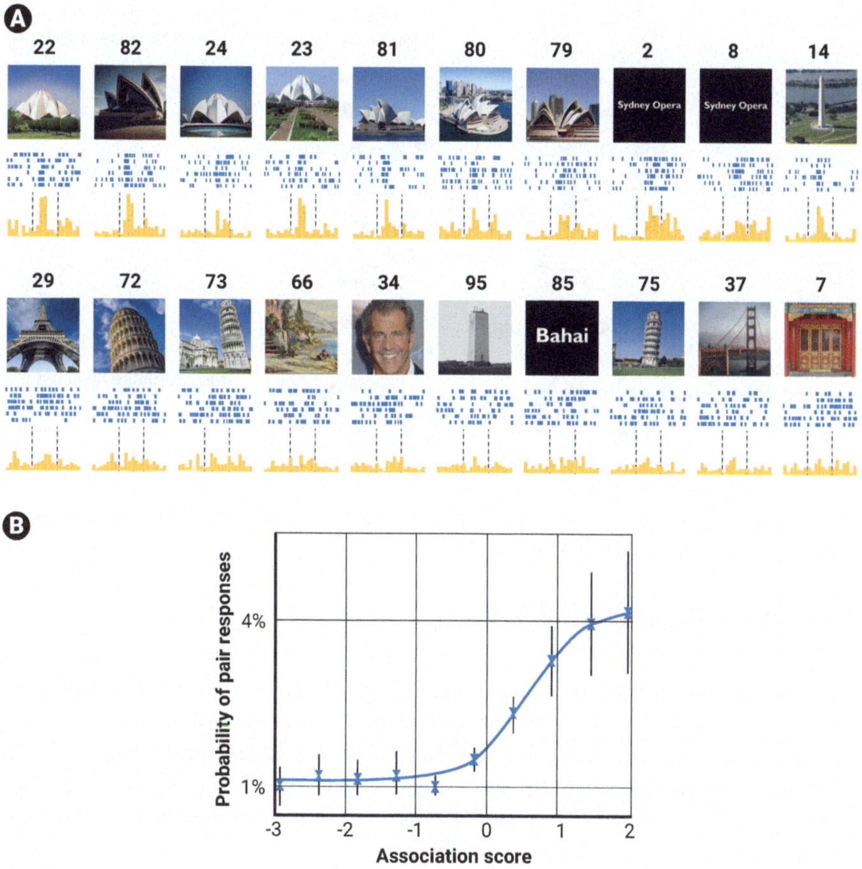


**Fig. 2** (**a**) Recordings from a concept cell selective to the Sydney Opera House. (Taken from Quiroga et al. (2005)) (**b**) Probability of pair response of concept cells. (Adapted from De Falco et al. (2016))

random, subset of active neurons, a single concept is expected to activate $\gamma N$ neurons, while two arbitrary concepts are expected to share $\gamma^2 N$ cells, where $N$ represents the total number of neurons in the relevant brain regions. Suppose that the total number of MTL neurons is $N$ = 200,000. Then, the concept of "Sydney Opera" would be represented by an engram containing 460 neurons. Similarly, the "Tower of Pisa" would also be represented by a different engram of 460 neurons. Since the "Sydney Opera" has no relation to the "Tower of Pisa," we expect from purely statistical arguments that at most one or two neurons are part of both engrams.

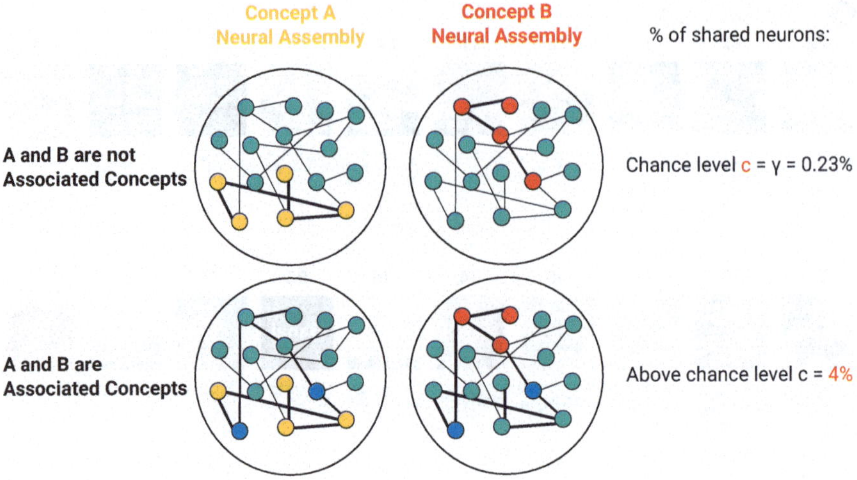


**Fig. 3** Schematic representation of the neural representation of two concepts in the human MTL. Shared memory cells are colored in blue

## *Association Between Concepts*

It is possible to measure the fraction of neurons that respond to more than one concept. It was found that unrelated concepts share less than 1% of neurons, whereas assemblies representing previously associated concepts share approximately 4–5% of neurons (De Falco et al. 2016) (Figs. 2b and 3). Moreover, during the process of building associations between pairs of concepts, individual neurons can become responsive to concepts to which they did not respond initially (Ison et al. 2015). This suggests that an increased proportion of shared neurons facilitates the association between concepts (De Falco et al. 2016; Rey et al. 2018, 2020).

## *Fundamental Questions*

In the presence of shared neurons, the activation of one engram (a place, in the initial example) may trigger the activation of another engram (such as a friend). This raises three fundamental issues which we address in this computational chapter:

First, for the brain to function effectively as a memory network, it must retain the ability to recall the two associated concepts separately (e.g., recalling the place without necessarily thinking of the friend). However, if the concepts share too many neurons, it becomes increasingly likely that the two concepts cannot be distinguished and instead merge into a broader engram counting a larger number of neurons. In the section "How Can Shared Concept Cells Encode Associations?," we address the following question: What is the maximum allowable fraction, $c_{max}$, of neurons shared between two assemblies before the possibility of separate memory recall breaks down?

Second, if mental concepts are related to engram cells, then associations between concepts should cause activation patterns of engrams cells, but how? The activation of a first concept can enhance the recall of a second, associated, concept that activates in parallel to the first one; or it may trigger a sequential activation of first, second, and potentially more concepts in the form of an association chain (as observed in free memory recall tasks (Romani et al. 2013; Recanatesi et al. 2015, 2017; Naim et al. 2020)). If these transitions from one concept to the next are caused by neurons that are shared between engrams, we must pose a second question: is there a minimum fraction of shared neurons, $c_{min}$, required to facilitate a reliable activation of the associated concepts?

## Neuronal Assemblies as Memory Engrams

Memory engrams are mathematically linked to neural assemblies which play a central role in the network algorithm for memory retrieval that we will discuss in the section "Generalised Hopfield Model." Neuronal assemblies (Hebb 1949) are subnetworks of strongly connected neurons that represent an abstract concept. The assembly as a subgroup of strongly connected neurons has been an influential theoretical notion, introduced by Donald Hebb (1949). Moreover, Hebb suggested that the strong connections within the assembly could be the consequence of synaptic plasticity (Hebb 1949). Finding real assemblies in the brain is technically challenging since neurons belonging to an assembly do not have to be neighbors but can be widely distributed across one, or even several, brain areas. If we define an assembly as a group of neurons that (i) respond simultaneously to a group of stimuli related to each other and (ii) have strong connectivity among themselves, then the second condition is hard to assess experimentally whereas the first condition has an obvious link to memory engrams.

The concept cells described above are primary candidates of neurons forming memory assemblies. We hypothesize that the simultaneous activation of concept cells by stimuli related to the concept is the result of strong recurrent connectivity within the assembly. Moreover, the fact that some, but not all, of the neurons show responses that persist after the end of the stimulus presentation (Quiroga et al. 2005; Ison et al. 2015; De Falco et al. 2016; Rey et al. 2018; Quiroga 2012) is an indirect indication that information is held in some form of working memory and could be a sign of strong interconnectivity. In this chapter, we link the Hebbian notion of neural "assembly" with the notion of "engram": We assume that the set of neurons participating in an engram form a Hebbian assembly and have relatively strong connectivity amongst themselves. Furthermore, we assume that the groups of concept cells participating in the same concept are the human equivalent of memory engrams in mice. However, it is worth emphasizing that, as of today, these are hypotheses and we cannot exclude other explanations.

## Modeling Memory Networks

### *Attractor Networks*

"Attractor neural networks" have been widely used to model memory systems in recurrent neuronal networks such as the CA3 area of the hippocampus (Hopfield 1982; Weisbuch and Fogelman-Soulie 1985; Amit 1992; Kanter and Sompolinsky 1987; Tsodyks and Feigel'man 1988). In attractor networks, each memory item is encoded as a memory engram (Tonegawa et al. 2015; Josselyn and Tonegawa 2020) consisting of a fixed random subset of neurons. The joint activity of a large fraction of neurons that are part of the same engram represents the memory item and indicates that the memorized "concept" is recalled.

An attractor network is a recurrent dynamical network, that evolves toward a stable state, called a "fixed point" of the network dynamics. In computational neuroscience, attractor networks are built such that the memory engrams are the fixed points of the network dynamics. If a stimulus activates a subset of neurons within an engram, the interactions of neurons within the network activate other neurons of the same engram and suppress the activity of other neurons that are not part of the engram. These interactions assure that the memory network has the auto-associative property introduced at the beginning of this chapter: the full concept is recalled (a large fraction of engram neurons are active together) triggered by partial information (the initial cue that stimulated a subset of neurons). From a birds-eye perspective, the network activity is "attracted" toward a state of memory recall.

### *Bio-plausibility of Attractor Memory Networks*

Animal studies have provided evidence of attractor dynamics in the CA3 area of the hippocampus (Rennó-Costa et al. 2014; Wills et al. 2005). Since concept cells have been found in the human hippocampus and its surroundings (Quiroga et al. 2005; Ison and Quiroga 2008; Quiroga 2019a), attractor dynamics in the hippocampus is a likely candidate to describe the activity of concept cells in humans—or engram cells in mice.

Attractor memory networks exhibit two key functional properties: (i) the ability to retrieve memories when presented with partial cues and (ii) the capacity to sustain activity even after the stimulus is no longer present. Traditionally, the analysis of attractor networks using the replica (Amit et al. 1987) or cavity methods (Mezard et al. 1987; Shamir and Sompolinsky 2000) has drawn criticism due to the unrealistic assumption of symmetric connections. However, the approach presented in this chapter is based on dynamic systems arguments (Shiino and Fukai 1992) and enables a straightforward generalization to the case of asymmetric connectivity. Moreover, in the original model of Hopfield (1984), each memory engram involved 50% of neurons, whereas in the human hippocampus, the fraction g of neurons involved in a single memory engram is at most 1% ("sparse" engrams) (Waydo et al. 2006; De Falco et al. 2016). This number is estimated indirectly from the probability that an experimentalist would find a neuron responding to a, say, famous person known to the patient if an electrode is placed randomly in a given brain area, knowing the number of presented stimuli and the number of neurons from which the experimentalist recorded (Waydo et al. 2006; De Falco et al. 2016). Modern attractor network theory does not rely on symmetric connectivity and is characterized by sparse (Tsodyks and Feigel'man 1988) memory engrams and random (Amit and Brunel 1997) connectivity. Therefore, modern attractor networks have emerged as promising models for understanding biological memory.

## *Generalized Hopfield Model*

We work with an attractor neural network, made of N neurons, which has stored P memory engrams. Each engram $\mu$ (with $\mu = 1, \ldots, P$), is represented by a string of random binary variables $\xi_i^\mu \in \{0,1\}$.

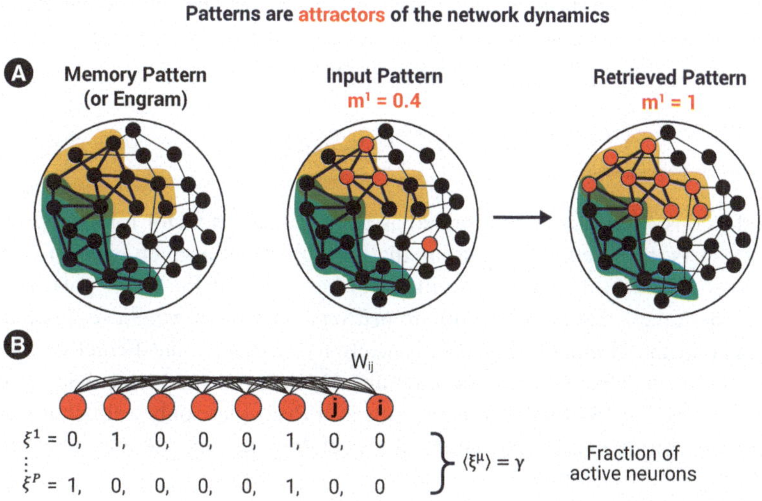


**Fig. 4** (**a**) Memory recall in a generalized Hopfield network. (**b**) Schematics of the memory engrams as represented into the generalized Hopfield model

For example, $\xi_i^3 = 1$ and $\xi_i^4 = 1$ indicate that neuron $i$ is part of both memory engrams 3 and 4; similarly, $\xi_j^3 = 0$ and $\xi_j^4 = 1$ indicates that neuron $j$ is not part of the memory engram 3, but part of engram 4 (Fig. 4b—redraw the figure with less math).

The state of each neuron $i$ is characterized by its firing rate $r_i$ (with $i = 1, \ldots, N$). The change $dr_i/dt$ of the firing rate of neuron $i$ is driven by the total input $h_i$:

$$\tau \frac{dr_i}{dt} = -r_i + \phi(h_i), \tag{1}$$

where $\phi(h_i) = r_{\max}/\{1 + \exp[-b(h - h_0)]\}$ is the frequency-current ($f$-$I$) curve (also called transfer function), characterized by the firing threshold $h_0$, the maximal steepness $b$, and the maximal firing rate $r_{\max}$.

The total input driving the neuron i is

$$h_i(t) = \sum_{j=1}^{N} \omega_{ij} r_j(t) + I_i(t), \tag{2}$$

where the synaptic weights $w_{ij}$, which can be interpreted as the strength of signal transmission from neuron $j$ to neuron $i$, is related to the amplitude of the (excitatory or inhibitory) postsynaptic current.

We assume the learning process has already happened in the past so we consider the weights as fixed. The engrams $\xi_i^\mu$ are encoded in synaptic weights $w_{ij}$ as follows: the basic idea is that two neurons $i$ and $j$ that participate in the same engram (e.g., engram 4) have strong excitatory connections, in both directions; moreover, if neuron $i$ participates in engram 3, but neuron $j$ does not, then the connections between the two neurons are weakly inhibitory. However, since the connection from neuron $j$ to neuron $i$ cannot be both excitatory and inhibitory at the same time, we take the average by summing over the contributions of all engrams:

$$\sum_a^b \omega_{ij} = A \sum_{\mu=1}^{P} \left(\xi_i^\mu - \gamma\right)\left(\xi_j^\mu - \gamma\right). \tag{3}$$

Here, the constant $A$ can be interpreted as a normalization factor for averaging and $\gamma = 0.0023$ is the fraction of neurons that participate in a given engram in MTL (Waydo et al. 2006). If the sum on the right-hand side of Eq. (3) is positive, then neuron $j$ has an excitatory connection to neuron $i$. The set of weights defined in Eq. (3), is called the Hopfield-Tsodyks connectivity (Tsodyks and Feigel'man 1988; Pereira and Brunel 2018) for sparse engrams. By default, we assume that two unrelated concepts (e.g., Sydney Opera and Tower of Pisa) share only a small number of neurons that correspond to the statistical expectation, i.e., the memory engrams are statistically independent.

Mathematical analysis, as well as simulations of the model, show that with the connections defined in Eq. (3) and the firing rate changes defined in Eq. (1), the network has (at least) $P$ stable stationary states where each state corresponds to a configuration where nearly all neurons that should be active in engram $\mu$ are indeed active and all others are inactive. Since these configurations are stable states, the network dynamics are "attracted" toward these special configurations—and these states correspond exactly to the retrieval of a stored concept. In the stationary state where engram $\mu$ is retrieved, neuronal firing rates have a fixed value $r_i$ which is high if $\xi_i^\mu = 1$ and low if $\xi_i^\mu = 0$. Attraction means that when the configurations of firing rates across the network are similar to one of the stored memories $\mu$, then the attractor dynamics drive the network to recall memory $\mu$, by showing persistent activity of all those neurons that belong to the assembly of concept $\mu$.

Importantly, it is possible to measure how close (or how far) is the network state is to retrieve one of the stored engrams, $\mu$, thanks to the similarity measures $m^\mu$:

$$m^\mu(t) = \frac{A}{r_{\max}} \sum_{i=1}^{N} \left(\xi_i^\mu - \gamma\right) r_i(t). \tag{4}$$

The similarity measures the correlation between the firing rates $\{rj(t)\}$, with $j = 1, \ldots,$ $N$, and the stored engrams $\xi_i^\mu$ such that if memory concept $\mu$ is retrieved, then $m^\mu \sim 1$ (schematics in Fig. 4a), and, if no memory is recalled (resting state), then $m^\mu \sim 0$ for all $\mu$. The similarity of the network activity with a stored memory changes over time as the network state, given by the firing rates $r_i(t)$ of the individual neurons, changes.

## *Theory, Computer Simulations, and Low-Dimensional Factors*

To answer the fundamental questions above, we consider a memory network with $P$ memories stored and then focus on memories 1 and 2. By default their engram share neurons at chance level (memories 1 and 2 are not associated); alternatively, to implement associations between the two memories, we also consider the case that the fraction of shared neurons is above chance. While the network dynamics can also be analyzed mathematically (Gastaldi et al. 2021), we focus here on computer simulations, which can be done at two different levels of detail.

First, at the detailed level, we explicitly simulate a network of $N$ neurons. This involves integrating numerically for each neuron the Eqs. (1), (2), and (3). In other words, each neuron and each synaptic connection is modeled explicitly.

Second, at a coarser level, we use a common trick from the literature called the mean-field approach (Hopfield 1982; Weisbuch and Fogelman-Soulie 1985; Amit 1992; Kanter and Sompolinsky 1987; Tsodyks and Feigel'man 1988). Such an approach assumes that the network is very large and that neurons belonging to the same memory engram behave similarly. It is then possible to fully describe the network dynamics using the similarity measures $m^\mu$. Each similarity measure $m^\mu$ can be viewed as a single, network-wide variable that captures the global state of the network in terms of its similarity with memory $\mu$. Since we are only interested in the retrieval of concepts $\mu = 1$ and 2, we can assume the similarity of the present network state with other memories $\mu > 2$ to be close to zero: we will refer to these nonactivated memories as "background engrams." Under these assumptions, it is possible to derive dynamical mean-field equations that fully describe the network dynamics through the similarity variables $m^1$ and $m^2$. The details of such derivation are beyond the aim of this chapter, and we refer to other texts to deepen the topic (Amit and Brunel 1997; Amit et al. 1985, 1987; Mezard et al. 1987; Shamir and Sompolinsky 2000; Shiino and Fukai 1992); however, we will illustrate the main results of this theoretical approach in what follows. The main take-home message is that the mean-field approach allows one to predict the collective behavior of a very large network using only a few equations and without the need to implement the dynamics of every single neuron in the network. The spirit of the mean-field approach is similar to that of identifying "factors" (DePasquale et al. 2016) or low-dimensional neural manifolds (Gallego et al. 2017) in experimental data.

## How Can Shared Concept Cells Encode Associations?

As suggested in the section "Association Between Concepts," the creation of episodic associations between different concepts (such as a person and a place) might be caused by common neurons shared across the corresponding memory engrams (Ison et al. 2015). Drawing inspiration from these experiments, we artificially introduce shared neurons in the generalized Hopfield model to create pairwise associations between multiple concepts. We refer to "overlapping engrams" when the number of shared concept cells exceeds the expected number of $\gamma^2 N$ cells shared by chance (see Fig. 5). We check by computer simulations whether increasing the overlap between two engrams causes a measurable increase in association performance.

Let us imagine gradually increasing the fraction of shared neurons between the memory engrams 1 and 2. At the lowest end, the two memories are not associated, since cell assemblies 1 and 2 share only a small fraction of neurons corresponding to chance level. It is well known, that in this case, each memory engram generates a

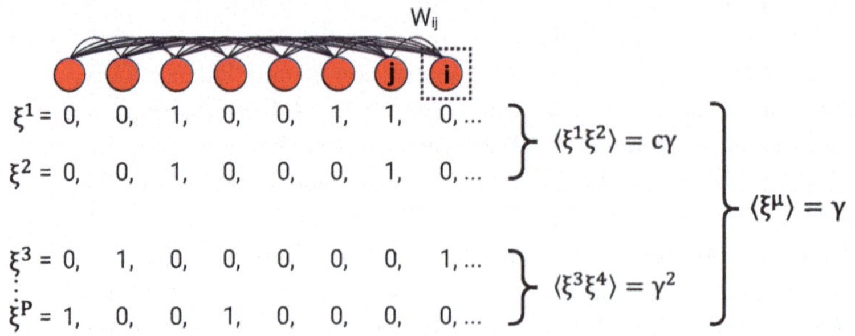


**Fig. 5** Schematic representation of memory engrams

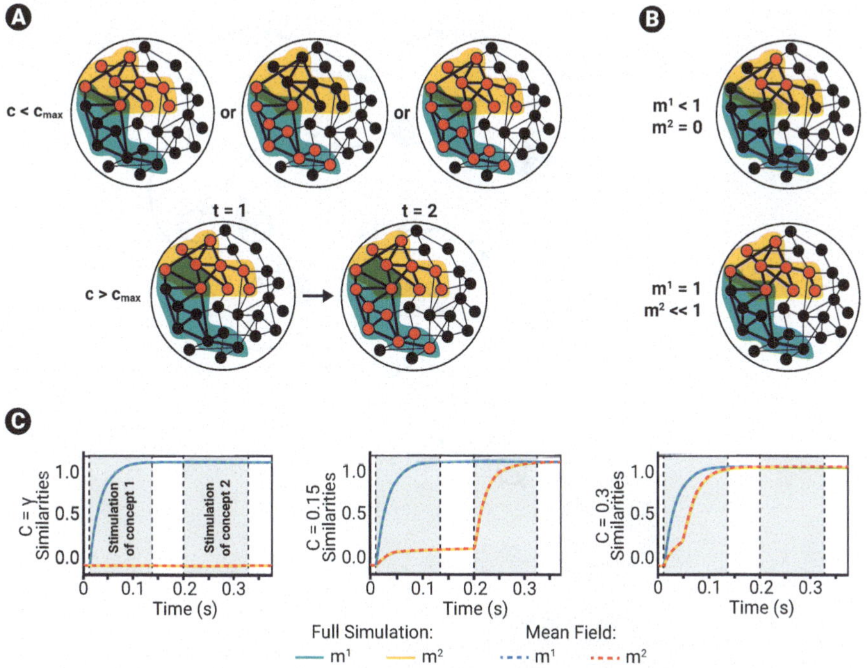


**Fig. 6** Overlapping concepts can be retrieved separately and jointly. (Adapted from Gastaldi et al. (2021)) **(a) Top figure: when the fraction of shared neurons is lower than the maximal value, the following states are stable network states: retrieval of concept 1, retrieval of concept 2, and retrieval of concepts 1 and 2 simultaneously. Bottom figure: on the contrary, when the fraction of shared neurons is above the maximal value, the retrieval of concept 1 triggers to the retrieval of concept 2: in this case the simultaneous retrieval state is the only stable retrieval state of the network, beside the trivial state in which no neuron is active. (b) Example of the retrieval of concept 1 from a partial information. Some neurons in concept 1 are activated leading to the activation of the rest of memory engram 1, in an auto-associative fashion. (c) Simulation of an attractor network in which two concepts are stored, different levels of overlap: no overlap on the left, some overlap in the center and large overlap on the right. In all three cases, a first a strong stimulation is given to concept 1 and later a weak stimulation is given to concept 2**

separate attractive fixed point of the network dynamics (Tsodyks and Feigel'man 1988), indicating that the two corresponding concepts can be retrieved separately. However, experimental data reports that, for associated concepts, the fraction of shared neurons $c \times$ 4–5% (De Falco et al. 2016) is much larger than chance level $\gamma$ ~ 0.23%. Let's now consider the case in which memories 1 and 2 share more neurons than by chance, i.e., the fraction c of shared neurons is larger than $\gamma$. In the upper limit case of a large fraction of shared neurons $c$ close to 1, the two memory engrams share all neurons, and it is clearly impossible to retrieve one memory without the other. In other words, the two memories are indistinguishable so that there are no longer two memories, but only a single, larger one.

**Fig. 7** Memory recall task

Computer simulations of a network of $N = 10{,}000$ interacting neurons indicate that, if one of two engrams that share concept cells is stimulated for 120 ms, then the similarity of the network activity with this engram increases to a value close to one, indicating that the memory has been recalled (Fig. 6c middle) while the second memory is only weakly activated quantified by a small, but non-zero similarity. However, if the fraction of shared neurons is above a maximally allowed fraction $c_{\max}$, then the second memory always gets activated even before it is stimulated (Fig. 6c bottom) indicating that associations are so strong that the two concepts have been merged. Hence, the fraction of neurons shared between two engrams should be above chance but remain relatively low so as to guarantee that concepts (e.g., your friend and "Sydney Opera") can remain separated if needed but can also be recalled jointly if so desired.

## *Association Chains*

The notion of common neurons shared among memory engrams has also been proposed as the foundation for recalling a list of memorized words. In earlier research (Romani et al. 2013; Recanatesi et al. 2015, 2017; Naim et al. 2020; Katkov et al.

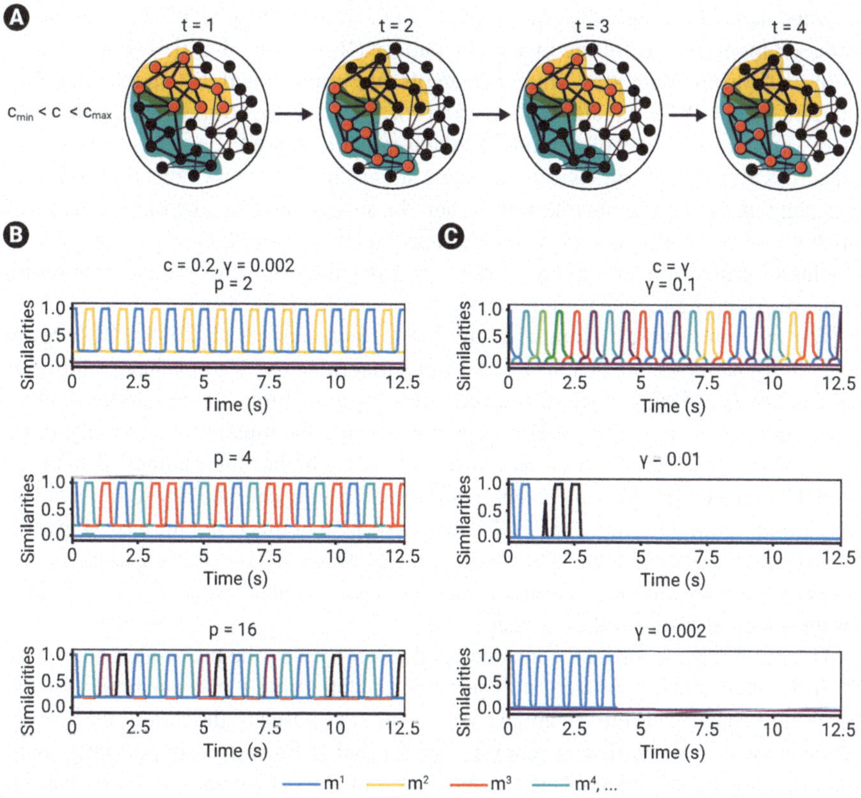


**Fig. 8** Chain of associations requires shared concepts cells. (Adapted from Gastaldi et al. (2021)) **(a) Schematic representation. (b) Network simulations**

2015), Romani and Tsodyks analyzed human behavioral experiments, during which the subjects had some time to memorize a list of words and then to freely recall as many words as possible (Fig. 7). They noticed that the subjects doing better at the task were those who associated the words in the list in small groups, and they seemed to recall the words following those personal associations. The same scientists also proposed a computational explanation (Romani et al. 2013; Recanatesi et al. 2015, 2017; Naim et al. 2020) of the experimental results, using a generalized Hopfield model similar to that introduced above, but with two additional components that we now add to our model.

Firstly, we introduce global inhibitory feedback that is periodically modulated in strength mimicking hippocampal oscillatory activity. These oscillations serve as a clock signal, triggering transitions between overlapping concepts. Secondly, we introduce an adaptation current, $\theta_i(t)$, to each neuron $i$, preventing the network state from immediately reverting to the previous concept. With this expanded model, the network state transitions from one concept to the next (Fig. 8a). These transitions are repeated, but eventually, the network state returns to one of the previously retrieved memories, resulting in a cyclic pattern (Romani et al. 2013) (Fig. 8a).

In network simulations where concepts are represented by sparse memory engrams ($\gamma = 0.2\%$), we allow a subgroup of $p = 2$, 4, or 16 memory engrams to share a fraction $c = 20\%$ of neurons. As the number of shared concept cells is identical across all concept pairs within the same subgroup, the order of recalled concepts is dependent on the initial condition. When the subgroup of overlapping engrams is small ($p = 2$ or 4), all memory items are successfully retrieved. However, in the case of a larger group of overlapping engrams ($p = 16$), the cycle closes once a subgroup of the overlapping memory engrams has been recalled (Fig. 8b).

The previous studies (Romani et al. 2013; Recanatesi et al. 2015, 2017; Naim et al. 2020; Katkov et al. 2015) have assumed that memory engrams have a substantial fraction ($\gamma = 10\%$) of active neurons, allowing transitions to rely on the chance-based number of shared units. However, considering the much lower sparsity value in the MTL ($\gamma = 0.23\%$), it is natural to question whether the number of neurons shared by chance ($c = \gamma$) is sufficient to induce a sequence of memory retrievals. The simulations from (Gastaldi et al. 2021) demonstrate that this is not the case (Fig. 8c). In a memory network with a realistic level of sparsity ($\gamma \sim 0.2\%$), associations between memory engrams require a fraction of shared neurons above chance level for the successful retrieval of concept chains.

The same conclusions are confirmed and reinforced by the mean-field approach. With the theoretical approach, it is possible to determine the lower bound of the fraction of shared neurons, denoted as $c_{0\min}$, corresponding to the minimum overlap between two engrams that is required for a reliable transition. Importantly, with suitable choices of neuronal and network parameters, association chains can be achieved for the values of $\gamma$ and $c$ as observed in the human MTL. This suggests that, in principle, associations can be implemented as sequences of transitions if the number of shared neurons exceeds $c_{0\min}$.

In summary, overlaps between memory engrams are necessary to encode associations. More precisely, the fraction of concept cells shared across concepts must significantly exceed chance level, to account for the phenomenon of free memory recall as a chain of associations in recurrent networks like the CA3 region of the human brain. Notably, in these networks, each memory engram is represented by only a small fraction of neurons.

## Conclusions

This chapter bridges experimental observations and theories from four distinct fields: (1) experimental investigations of concept cells in the human MTL, (2) empirical studies on memory engrams in mice, (3) the theory of association chains for free memory recall, and (4) the classic theory of attractor neural networks. Our theory assumes that concepts are represented by the activation of sparse subgroups of neurons participating in the same engram. If engrams represent the memorized concepts, then associations between concepts can be implemented by an increased fraction of concept cells shared across engrams. This increased fraction must surpass the chance level but remain below a certain maximal threshold to enable reliable encoding of associations. Experimental evidence indicates a 4–5% overlap between memory engrams in the human MTL (De Falco et al. 2016), and these numbers lie in the range of values that are supported by simulations of association chains (Gastaldi et al. 2021).

Association chains could form the basis of a "stream of thought" where the direction of transitions from one concept to the next is based on learned associations. In large networks with sparse coding levels ($\gamma \sim 0.23\%$), neurons shared by chance are not enough to reliably induce the retrieval of a chain of concepts. Sequential memory retrieval is possible only for overlaps larger than chance, potentially representing associations learned during real-life episodes. Potentially, the fraction of shared concept cells could increase due to Hebbian learning if two concepts repeatedly occur simultaneously. However, the existence of a maximal fraction of shared neurons highlights the need for Hebbian learning to operate in conjunction with an intrinsic control mechanism to prevent the undesired merging of distinct concepts.

Overall, our computational approach shows that the creation and retrieval of memory engrams is amenable to mathematical analysis. Here, we focused on the challenges and advantages of overlapping memory engrams, and for this, we had to rely on data from human MTL, but future experiments in mice might bring further evidence on how overlapping engrams can build or reinforce associations between memory concepts.

## Other Applications of Hopfield and Generalized Hopfield Networks

### *Other Areas: ITC*

The area CA3 of the hippocampus is not the only one that can be modeled with attractor neural networks. Indeed, this model was first introduced with the intention of modeling cortical areas. The inferior temporal cortex (ITC) has often been taken as an example area to be modeled with generalized Hopfield networks (Amit 1992; Kanter and Sompolinsky 1987; Tsodyks and Feigel'man 1988; Tonegawa et al. 2015; Josselyn and Tonegawa 2020; Rennó-Costa et al. 2014; Wills et al. 2005; Hopfield 1984; Khona and Fiete 2022).

The inferotemporal cortex (IT) is a region of the brain that plays a crucial role in processing visual stimuli related to objects within our field of vision. It is responsible for extracting complex visual features and attributes, enabling us to recognize and identify objects.

One of the key functions of the IT cortex is its involvement in memory and memory recall processes. Once we have seen and processed an object, the IT cortex stores representations of its visual features and characteristics as memory engrams. These memory engrams are subsequently utilized during memory recall to recognize and identify the object when encountered again.

We can represent the objects memorized in the ITC as attractor states within the network. Each attractor state corresponds to a particular object or category, and when the network is presented with a visual input, it undergoes dynamics that converge to the nearest attractor state, thus recognizing the object in the input.

## *Nonbinary Engrams*

Above we have considered the so-called "binary" memory engrams, where a neuron either participates in the memory engram or it does not. This is a modeling simplification, but experimental results do not exclude other types of memory engrams. Indeed, in the concept cells experiments, a neuron (Quiroga et al. 2005; Ison et al. 2015; De Falco et al. 2016; Rey et al. 2018; Quiroga 2012) is considered to be a concept cell if its firing rate is deviating of at least 3 or 5 (depending on the experiment) standard deviations from the firing rate distribution of all other recorded neurons during the stimulus presentation. This means that the criterion of assigning concept cells is binary but not necessary as the concept cells responds, which might follow a different distribution.

It is possible to define the generalized Hopfield model for nonbinary firing rate engrams; for example, in Pereira and Brunel (2018), they proposed a model for Gaussian firing rate engrams. In this case, retrieving a memory means that the neurons' firing rate follows the same Gaussian distribution of the equivalent memory engram.

## *Spiking Hebbian Networks*

So far, we have considered the firing rate of neurons as the only relevant parameter of the neuron. This is an approximation, but ideally, we would like to achieve the same type of memory neural network with more realistic spiking neurons (Wilson and Cowan 1972; Muscinelli et al. 2019; Zenke et al. 2015).

Moreover, in the generalized Hopfield model, synaptic weights can be both positive and negative, allowing for bidirectional signaling. However, experimental observations have revealed a phenomenon known as Dale's law. According to Dale's law, all connections originating from the same presynaptic neuron have the same sign, either excitatory or inhibitory. This empirical finding has led to the primary classification of neurons into two categories: excitatory neurons, which promote the firing of postsynaptic neurons, and inhibitory neurons, which suppress or inhibit the firing of postsynaptic neurons.

It is possible to create a more detailed and realistic computational spiking neuron model in which Dale's law is respected, yet it preserves the same dynamical behavior of a generalized Hopfield network. In a neural network scenario, there exists a population of excitatory spiking neurons that interact with two distinct populations of inhibitory spiking neurons. The encoding of memory engrams occurs through the formation of Hebbian assemblies within the excitatory population. All neurons in the network follow an integrate-and-fire behavior.

According to theoretical predictions, the first inhibitory population is expected to be activated to a degree where the gain function (as shown in the left inset) exhibits an approximately linear response. This linear activation range ensures effective regulation of excitatory neuron activity and helps maintain a stable network state.

On the other hand, the activation of the second inhibitory population occurs when the total input to the network surpasses a specific threshold value (as depicted in the right inset). This threshold-based activation mechanism serves as a control mechanism to prevent excessive excitation or maintain the stability of the network by suppressing neuronal firing when necessary.

Overall, this architecture and activation scheme involving excitatory and inhibitory populations contribute to the dynamic regulation of neural activity and play a crucial role in shaping the network's behavior and information processing capabilities.

## Bibliography


Akrami A, Russo E, Treves A (2012) Lateral thinking, from the Hopfield model to cortical dynamics. Brain Res 1434:4–16. https://doi.org/10.1016/j.brainres.2011.07.030. PMID: 21839426

Amit DJ (1992) Modeling brain function: the world of attractor neural networks. Cambridge University Press

Amit DJ, Brunel N (1997) Model of global spontaneous activity and local structured activity during delay periods in the cerebral cortex. Cereb Cortex 7(3):237–252. PMID: 9143444

Amit DJ, Gutfreund H, Sompolinsky H (1985) Storing infinite numbers of patterns in a spin-glass model of neural networks. Phys Rev Lett 55(14):1530. https://doi.org/10.1103/PhysRevLett.55.1530. PMID: 10031847

Amit DJ, Gutfreund H, Sompolinsky H (1987) Information storage in neural networks with low levels of activity. Phys Rev A 35(5):2293. https://doi.org/10.1103/PhysRevA.35.2293. PMID: 9898407

Andersen P, Morris R, Amaral D, Bliss T, O'Keefe J (2006) The hippocampus book. Oxford University Press

De Falco E, Ison MJ, Fried I, Quiroga RQ (2016) Long-term coding of personal and universal associations underlying the memory web in the human brain. Nat Commun 7:13408. https://doi.org/10.1038/ncomms13408. PMID: 27845773

DePasquale B, Churchland MM, Abbott LF (2016) Using firing-rate dynamics to train recurrent networks of spiking model neurons. arXiv preprint arXiv:1601.07620. https://doi.org/10.48550/arXiv.1601.07620

Gallego JA, Perich MG, Miller LE, Solla SA (2017) Neural manifolds for the control of movement. Neuron 94(5):978–984

Gastaldi C, Schwalger T, De Falco E, Quiroga RQ, Gerstner W (2021) When shared concept cells support associations: theory of overlapping memory engrams. PLoS Comput Biol 17(12):e1009691

Guzman SJ, Schlögl A, Frotscher M, Jonas P (2016) Synaptic mechanisms of pattern completion in the hippocampal CA3 network. Science 353(6304):1117–1123. https://doi.org/10.1126/science.aaf1836. PMID: 27609885

Hebb DO (1949) The organization of behavior. Wiley, New York

Hopfield JJ (1982) Neural networks and physical systems with emergent collective computational abilities. Proc Natl Acad Sci 79(8):2554–2558. https://doi.org/10.1073/pnas.79.8.2554. PMID: 6953413

Hopfield JJ (1984) Neurons with graded response have collective computational properties like those of two-state neurons. Proc Natl Acad Sci 81(10):3088–3092. https://doi.org/10.1073/pnas.81.10.3088. PMID: 6587342

Ison MJ, Quiroga RQ (2008) Selectivity and invariance for visual object perception. Front Biosci 13:4889–4903. https://doi.org/10.2741/3048. PMID: 18508554

Ison MJ, Quiroga RQ, Fried I (2015) Rapid encoding of new memories by individual neurons in the human brain. Neuron 87(1):220–230. https://doi.org/10.1016/j.neuron.2015.06.016. PMID: 26139375

Josselyn SA, Tonegawa S (2020) Memory engrams: recalling the past and imagining the future. Science 367(6473):eaaw4325. https://doi.org/10.1126/science.aaw4325. PMID: 31896692

Kanter I, Sompolinsky H (1987) Associative recall of memory without errors. Phys Rev A 35(1):380. https://doi.org/10.1103/PhysRevA.35.380. PMID: 9897963

Katkov M, Romani S, Tsodyks M (2015) Effects of long-term representations on free recall of unrelated words. Learn Mem 22(2):101–108. https://doi.org/10.1101/lm.035238.114. PMID: 25593296

Khona M, Fiete IR (2022) Attractor and integrator networks in the brain. Nat Rev Neurosci 23:744–766. https://doi.org/10.1038/s41583-022-00642-0

Mezard M, Parisi G, Virasoro M (1987) Spin glass theory and beyond: an introduction to the replica method and its applications, vol 9. World Scientific Publishing Company

Muscinelli SP, Gerstner W, Schwalger T (2019) How single neuron properties shape chaotic dynamics and signal transmission in random neural networks. PLoS Comput Biol 15(6):e1007122. https://doi.org/10.1371/journal.pcbi.1007122. PMID: 31181063

Naim M, Katkov M, Romani S, Tsodyks M (2020) Fundamental law of memory recall. Phys Rev Lett 124(1):018101. https://doi.org/10.1103/PhysRevLett.124.018101. PMID: 31976719

Pereira U, Brunel N (2018) Attractor dynamics in networks with learning rules inferred from in vivo data. Neuron 99(1):227–238.e4. https://doi.org/10.1016/j.neuron.2018.05.038. PMID: 29909997

Podlaski WF, Agnes EJ, Vogels TP (2020) Context-modular memory networks support high-capacity, flexible, and robust associative memories. BioRxiv. https://doi.org/10.1101/2020.01.08.898528

Quiroga RQ (2012) Concept cells: the building blocks of declarative memory functions. Nat Rev Neurosci 13(8):587–597. https://doi.org/10.1038/nrn3251. PMID: 22760181

Quiroga RQ (2019a) Neural representations across species. Science 363(6434):1388–1389. https://doi.org/10.1126/science.aaw8829. PMID: 30923208

Quiroga RQ (2019b) Plugging in to human memory: advantages, challenges, and insights from human single-neuron recordings. Cell 179(5):1015–1032. https://doi.org/10.1016/j.cell.2019.10.016

Quiroga RQ, Reddy L, Kreiman G, Koch C, Fried I (2005) Invariant visual representation by single neurons in the human brain. Nature 435(7045):1102. https://doi.org/10.1038/nature03687. PMID: 15973409

Recanatesi S, Katkov M, Romani S, Tsodyks M (2015) Neural network model of memory retrieval. Front Comput Neurosci 9:149. https://doi.org/10.3389/fncom.2015.00149. PMID: 26732491

Recanatesi S, Katkov M, Tsodyks M (2017) Memory states and transitions between them in attractor neural networks. Neural Comput 29(10):2684–2711. https://doi.org/10.1162/neco_a_00998. PMID: 28777725

Rennó-Costa C, Lisman JE, Verschure PFMJ (2014) A signature of attractor dynamics in the CA3 region of the hippocampus. PLoS Comput Biol 10(5):e1003641. https://doi.org/10.1371/journal.pcbi.1003641. PMID: 24854425
Rey HG, De Falco E, Ison MJ, Valentin A, Alarcon G, Selway R, Richardson MP, Quiroga RQ (2018) Encoding of long-term associations through neural unitization in the human medial temporal lobe. Nat Commun 9(1):1–13. https://doi.org/10.1038/s41467-018-06870-2. PMID: 30348996
Rey HG, Gori B, Chaure FJ, Collavini S, Blenkmann AO, Seoane P, Seoane E, Kochen S, Quiroga RQ (2020) Single neuron coding of identity in the human hippocampal formation. Curr Biol 30(6):1152–1159.e3. https://doi.org/10.1016/j.cub.2020.01.035. PMID: 32142694
Romani S, Pinkoviezky I, Rubin A, Tsodyks M (2013) Scaling laws of associative memory retrieval. Neural Comput 25(10):2523–2544. https://doi.org/10.1162/NECO_a_00499. PMID: 23777521
Russo E, Treves A (2012) Cortical free-association dynamics: distinct phases of a latching network. Phys Rev E 85(5):051920. https://doi.org/10.1103/PhysRevE.85.051920. PMID: 23004800
Russo E, Namboodiri VMK, Treves A, Kropff E (2008) Free association transitions in models of cortical latching dynamics. New J Phys 10(1):015008. https://doi.org/10.1088/1367-2630/10/1/015008
Shamir M, Sompolinsky H (2000) Thouless-Anderson-Palmer equations for neural networks. Phys Rev E 61(2):1839. https://doi.org/10.1103/PhysRevE.61.1839. PMID: 11046469
Shiino M, Fukai T (1992) Self-consistent signal-to-noise analysis and its application to analogue neural networks with asymmetric connections. J Phys A Math Gen 25(7):L375. https://doi.org/10.1088/0305-4470/25/7/017
Tonegawa S, Pignatelli M, Roy DS, Ryan TJ (2015) Memory engram storage and retrieval. Curr Opin Neurobiol 35:101–109. https://doi.org/10.1016/j.conb.2015.07.009. PMID: 26280931
Tsodyks MV, Feigel'man MV (1988) The enhanced storage capacity in neural networks with low activity level. Europhys Lett 6(2):101. https://doi.org/10.1209/0295-5075/6/2/002
Waydo S, Kraskov A, Quiroga RQ, Fried I, Koch C (2006) Sparse representation in the human medial temporal lobe. J Neurosci 26(40):10232–10234. https://doi.org/10.1523/JNEUROSCI.2101-06.2006. PMID: 17021178
Weisbuch G, Fogelman-Soulie F (1985) Scaling laws for the attractors of Hopfield networks. J Phys Lett 46(14):623–630. https://doi.org/10.1051/jphyslet:019850046014062300
Wills TJ, Lever C, Cacucci F, Burgess N, O'Keefe J (2005) Attractor dynamics in the hippocampal representation of the local environment. Science 308(5723):873–876. https://doi.org/10.1126/science.1108905. PMID: 15879220
Wilson HR, Cowan JD (1972) Excitatory and inhibitory interactions in localized populations of model neurons. Biophys J 12(1):1–24. https://doi.org/10.1016/S0006-3495(72)86068-5. PMID: 4332108
Zenke F, Agnes EJ, Gerstner W (2015) Diverse synaptic plasticity mechanisms orchestrated to form and retrieve memories in spiking neural networks. Nat Commun 6(1):6922

# Engram Studies: A Call for Historical, Philosophical, and Sociological Approaches

**Luca Chiapperino and Francesco Panese**

**Abstract** In this chapter, we identify three distinct avenues of research on the philosophical, historical, and sociopolitical dimensions of engram research. First, we single out the need to refine philosophical understandings of memory within neuroscientific research on the engram. Specifically, we question the place of constructivist and preservationist philosophical claims on memory in the formulation of the engram concept and its operationalization in contemporary neuroscience research. Second, we delve into the received historiography of the engram claiming its disappearance after Richard Semon's (1859–1918) coinage of the concept. Differently from this view, we underline that Semon's legacy is still largely undocumented: Unknown are the ways the engram circulated within studies of organic memory as well as the role Semon's ideas had in specific national contexts of research in neurosciences. Finally, another research gap on the engram concerns a socio-anthropological documentation of the factual and normative resources this research offers to think about memory in healthcare and society. Representations of memory in this research, experimental strategies of intervention into the engram, as well as their translational potential for neurodegenerative (e.g., Alzheimer's disease) and psychiatric (e.g., post-traumatic stress disorder) conditions have not yet received scrutiny notwithstanding their obvious social and political relevance.

All these knowledge gaps combined call for a strong commitment towards interdisciplinarity to align the ambitions of a foundational neuroscience of the engram with a socially responsible circulation of this knowledge. What role can the facts, metaphors, and interventional strategies of engram research play in the wider society? With what implications for philosophical questions at the foundation of memory, which have accompanied its study from antiquity? And what can neuro- and social scientists do jointly to shape the social and political framings of engram research?

L. Chiapperino (✉) · F. Panese
STS Lab, Institute of Social Sciences, Faculty of Social and Political Sciences, University of Lausanne, Lausanne, Switzerland
e-mail: Luca.Chiapperino@unil.ch

 
J. Gräff, S. Ramirez (eds.), *Engrams*, Advances in Neurobiology 38,
https://doi.org/10.1007/978-3-031-62983-9_14

**Keywords** Engram · Preservationism · Constructivism · Richard Semon · Science and technology studies

## Introduction

Engram studies open novel and important perspectives into neuroscientific research on memory. As the present book testifies, there is a wealth of evidence that points to the existence of the long-sought memory engram and the engram cells (see also Josselyn and Tonegawa 2020). Yet, little attention has been devoted to the humanistic and social dimensions of this research. Although "excitement" defines the scientists' assessment of this research (Albo and Gräff 2019, 198), contemporary engram studies have received little critical scrutiny as to their philosophical, historical, and sociopolitical dimensions.

In this chapter, we identify three distinct avenues of social sciences and humanities research on the foundations, framings, and interpretations of contemporary engram studies. Far from being exhaustive, our exercise wishes to open engram studies to a pluralistic research agenda—and, notably, a social-scientific one. To this purpose, we single out three knowledge gaps on the philosophical, historical, and sociopolitical dimensions of engram research. These, we argue, call for (i) refining philosophical understandings of memory grounded on neuroscientific research on the engram; (ii) questioning the historiography of the engram claiming its disappearance after Richard Semon's (1859–1918) coinage of the concept; and (iii) multiplying social studies of engram research dissecting the factual and normative resources it offers to the understanding of memory in healthcare and the wider society.

## The Foundations of Engram Theory: Philosophical Issues

It is often argued that the idea of memory as trace left on the mind/soul goes back to Plato's Theaetetus. In this dialogue, Plato accounts for memory and mind as substances molded by perception and thought (Plato 2004, 191d–e). Yet, as noted by Felipe De Brigard (2014), the first argument about the nature of memory as trace (rather than just impression) is owed to Aristotle and his De Memoria et Reminiscentia. There, the Stagirite grapples with the philosophical nature of memory traces: These require an ontological continuity between the past (what is remembered) and the present (the act of remembering). As one of us has shown elsewhere (Panese 2018), the concept of the "trace" was also central in the seventeenth-century theories of "animal spirits"—inaugurated by Descartes and pursued by Malebranche in particular. With due differences, their respective theories were metaphorical accounts of the relationship between the body and the mind. Metaphorical, in the sense of philosopher Hans Blumenberg: rather than being developed from experimentation, these theories were heuristic devices "invested with [the] pragmatic function" of making intelligible phenomena whose empirical and conceptual foundations proved "resistant" to investigation (Blumenberg 2010, 4–5). In other words, while memory escaped empirical objectivation, the metaphor of "trace" functioned as a powerful explanatory resource: It allowed these thinkers to fine-tune a great spectrum of phenomena happening as, for instance, conscience and thinking, but also as biological change and physiology. Far from being an aporia, the quest for the material alterations that constitute learning, memory, and retrieval has been a generative difficulty for many centuries: The challenge of offering a convincing articulation of a philosophy of the subject and a physiology of the brain.

Richard Semon's physical theory of human memory epitomizes this generative difficulty (Semon 1921). In the words of philosopher Bertrand Russell, Semon managed to bring this long-standing question "in various directions [that] are interesting and ingenious" (Russell 2005, 67). His concept of engram can in fact be positioned both in continuity with and in contrast to these classical ideas. As argued by philosopher John Sutton (2007), his engram theory does not merely offer another iteration of these long-standing views of memory as stored information kept aside for future use. Certainly, Semon's The Mneme calls for a strong empirical emphasis on the substrate holding learned information in the nervous system and the body writ large. What he calls the "engraphic action" of a stimulus is a "permanent record [that] has been written or engraved on the irritable substance" (Semon 1921, 24). The engram denotes in Semon's work a specific "permanent change"; the sum of the engrams, an "engram-store"; and the phenomena spurred by the existence of engrams in an organism "mnemic phenomena" (ibid., 24). However, Sutton argues, Semon offers a more articulate view of learning as change in the material substrates of the body and the brain (which, in Sutton's view, makes him akin to Descartes more than Aristotle or Plato). In fact, Semon's theory was not just an iteration of views of memory as repository.[1] Rather, his theory has made it possible to problematize the phenomenon of memory through an approach that can be defined connectionist; that is, one that emphasizes the "interactive relations between self, body, mind, and coursing nervous fluids" (Sutton 2007, 1).

In this section, we argue that contemporary neuroscientific literature reproduces this ambiguity in Semon's work when narrating its theoretical relevance. Yet, we argue that it matters for the neurosciences of memory whether Semon is considered as the unsung "tragic hero" of engram research (Josselyn et al. 2017, 4649) for any or both of the following stipulations: On the one hand, for having established the idea that certain physical changes in the brain—which consist in the storage and use of information—make the engram; or, on the other hand, for having anticipated many current memory concepts by emphasizing that learning (engraphy) and

[1] Nor, as we shall see in the next section, his theory was just another versions of popular (at his time) reductionist views of "organic memory"—that is, the view that postulates that experiences and "cultural change", as well as their "effects on individual learning are to be understood in exactly the same terms as the effects of evolutionary history" on the organism's adaptation (Sutton 2007, 188).

retrieval (ecphory) are not static and change with engram use. Establishing more precisely where the "amazing insightfulness" (Tonegawa et al. 2015, 918) of Semon's work lies has consequences, we conclude, for the philosophical foundations of contemporary memory studies and the experimental priorities of this neuroscientific research.

The engram is often described as the mediating trace of memory which explains how retention of past information instructs future experiences in ways that would not have been possible if the retained information was absent. In philosophical terms, this would squarely position the notion among the so-called "preservationist" views about the nature of memory (Michaelian et al. 2018). This is the traditional "storehouse" view of memory, which postulates that past events produce persistent traces that are preserved in memory within a specific storage place and play a causal role in the act of remembering. As pointed out by Jonathan Najenson (2021), preservationist views can vary but are in general strongly committed to two postulates. On the one hand, memory lies in the traces: These have to be localized and persistent in mediating the act of remembering. On the other hand, preservationist commit to an informational view of memory: Memory traces carry a unique information related to the original event.

While comforting and omnipresent it is however far from clear whether this preservationist reading of the engram captures the insight on memory offered by contemporary engram studies (e.g., Tonegawa et al. 2015; Khalaf and Gräff 2016; Josselyn and Tonegawa 2020). Observational studies have confirmed that there exists an overlap between the population of neurons involved in the act of learning and in the act of retrieving this information. As argued by influential engram researchers Sheena A. Josselyn and Susumu Tonegawa (2020, 11), there exists little doubt that the engram is no longer the "elusive" trace of memory and that these overlaps among populations of neurons "exceed chance levels." At the same time, it is far from clear whether these populations are clearly delimited and can be easily localized: The correspondence between neurons learning and retrieving information, these scientists argue, is "relatively low (roughly 10–40% depending on the study)" (Josselyn and Tonegawa 2020, 11). This lack of overlap means therefore that considering the engram a memory trace is a rough approximation. A given memory is rather supported by (what could be called) an "engram complex," made up of functionally connected cells, which are not necessarily localized in the same place, but rather dispersed across multiple brain regions. Under this reading, connectivity and plasticity, dynamic, and structural connections produce the acts of learning and remembering more than the localized inscription of any given memory trace (Khalaf and Gräff 2016).

Circling back to philosophical theories of memory, this view of the engram as neural dynamics challenges the interpretation of engram theory as preservationist; that is, as the long-standing theoretical framework that sees memory as delimited, stored information. In this vein, the engram would rather fit with "constructivist" views, which see memory as the capacity for constructing the representation of past events from general, stored information (De Brigard 2014). According to Felipe De

Brigard, evidence of the engram complex—as opposed to the engram trace—suggests that memories are rather reconstructed from neural structures that are not specific to the learned experience but rather partake into other mental capacities too. Thus, a crucial difference is that on the constructivist side, there exists no one-to-one correspondence between the physical changes that persist in the nervous system after an event, the entity we call a memory as well as its informational content. The engram, in this view, is a dispositional property intrinsic to neural networks, which can reactivate, refine, and reconsolidate patterns of relation encoded during learning events.

Further research is needed, complementing the few remarkable exceptions to this rule (e.g. Najenson 2021), to dissect these philosophical issues in contemporary engram studies. Primarily, a deeper historical-epistemological understanding of Semon's law of engraphy is needed to ascertain the place of these constructivist and preservationist claims in his original formulation of the engram concept. But, more importantly, additional scrutiny is required for the contemporary idea that remembering involves both a fixed, stable engram, and an enduring trace that is plastic, dynamic, and open to reconsolidation. Either one of these two options has been a guiding framework of philosophical theories of neurosciences: So, how to reckon with this conflicting view of the engram offered by neuroscientists today? What bits of engram knowledge are in turn a challenge, a revision, or a deeper appreciation of Semon's engram theory? And what does resolving conflicts among preservationist and constructivist views mean for engram research? What is the bearing of these alternative philosophical views of memory on the ways neuroscientists speak about "trace," "storage," or even "information" in experimental research? These questions call for a strong commitment toward analyses of incommensurability and convergence of constructivist and preservationist views of memory. Discussions on the philosophical foundations of memory—which have accompanied its study from antiquity and played out in Semon's work too—remain, in other words, a topical issue.

## A Gap in the Historiography of the Engram

Neuroscientists insist on owing the term "engram" to Richard Semon and his study of memory in The Mneme (Semon 1921). This widespread historiography of the concept rests upon three interrelated claims. First, one that attributes to him the intuition that memories-as-information are held in an identifiable substrate within in the nervous system. Semon's groundbreaking idea is that of a medium operating memory, that is, a neural substrate and process that physically records, preserves, and retrieves learned information. Second, neuroscientists take Semon's engram as having postulated these ideas long before the conditions to investigate them empirically were reunited (Josselyn et al. 2017). This explains why the few attempts at locating the engram before contemporary neurosciences, such as the one by Karl

Lashley in the 1950s, have failed to deliver on the task.[2] Finally, this received historiography of the engram rests upon the idea that contemporary engram research is a vindication of Semon's ideas, which have been "largely overlooked or dismissed during his lifetime" and beyond (Josselyn and Tonegawa 2020, 1). As we mentioned in the previous section, he is the "overlooked" (Josselyn et al. 2017, 4648) and "tragic" (ibid., 4649) hero of contemporary engram research.

However, besides few notable exceptions (e.g., Logan 2015), little historical work has analyzed extensively the legacy and circulation of Semon's ideas in research attempting to locate and characterize the process of memory-making (cf. Schacter 1982). While it is certainly true that Semon's ideas encountered strong resistance, there exists a gap in the historical studies of his influence on the search for the engram in neurosciences. As we learn from Cheryl A. Logan (2015), Semon's views on psychological/physiological memory were part of a broader theoretical and experimental approach to organic memory. Organic memory consisted of the idea that actions, experiences, or injuries that affected one generation could not just become memories but become hereditary and engrained (as memory traces) in the biological makeup of succeeding generations (Otis 1994). Often associated with Jean-Baptiste Lamarck and the inheritance of acquired characteristics, these views linking somatic, psychological, and hereditarian processes of memory-making were largely discarded at the dawn of the twentieth century. Logan argues that Semon's The Mneme was, if anything, rather considered an authoritative and promising attempt at vindicating the view linking memory-as-learning and memory-as-inheritance. We could add that prominent philosopher Bertrand Russell—a contemporary reader of Semon's The Mneme—went as far as defining him the "best writer on mnemic phenomena known" in his The Analysis of Mind (Russell 2005, 64).

Why, then, the received view, among contemporary neuroscientists, about Semon's work being ignored in his time? What his critics found wanting was his concept of organic memory "*as applied to heredity*" (Logan 2015, 409 emphasis added). Semon's idea that the activation and stimulation of engram complexes has an adaptive role, in an evolutionary sense, was in fact quickly dismissed as obsolete. Due to the emergence of dominant views of hard heredity, inspired by the work of August Weismann (1889; see Winther 2001), the strong commitment of Semon's views of memory to heredity simply became an unacceptable theoretical view. But only few questioned instead the regulatory theory that engram complexes are subject to states of relative plasticity that modify mnemic traces: Excitation and reaction, engraphy, and ecphory per se were not the source of peer resistance. In fact,

[2] Karl Spencer Lashley (1890–1958) was a neuropsychologist who pioneered experimental research on the localization, functioning, and neuropathology of memory. He famously published, in 1950, an article (Lashley 1950) entitled *In Search of the Engram* building upon his 30 years of research on memory. The article offers a series of damning conclusions on the state of the field at that time. Of note, Lashley's paper emphasizes the lack of any demonstrable localization of a memory trace and the need to consider memory as distributed across brain regions. While essential for memory function, Lashley underscore that these regions and their parts are, however, functionally equivalent—thus suggesting also no direct connection between these parts and specific memories (see Bruce 2001).

Logan concludes, the "engram" concept did not really disappear. During a historical period (1890–1940) of several scientific disciplines taking shape, the engram rather moved away from a strong commitment and engagement with organic memory (i.e., away from an intimate link with the study of heredity). Semon's ideas influenced neurophysiologists and psychiatrists, especially in Switzerland and "outside [of] the geographic and national mainstream of the original debates" (Logan 2015, 419). He was a close friend of the influential Swiss psychiatrist Auguste Forel (1848–1931), who not only defined The Mneme "epoch-making work" but also considered Semon's attempt at bridging biology, physiology, anatomy, and psychology a desirable integrative approach to scientific activity (Schacter 1982, 139). Similarly, Constantin von Monakow (1853–1930) conducted neuropathological research in Zurich that was heavily influenced by Semon's ideas. Seeking "a synthesis of psychiatry, neurology, and biology as the foundation for a 'mnemic biopsychology'" (Logan 2015, 419), von Monakow elaborated Semon's concepts into a veritable brain science that studied the effects of pathological injuries to the human brain, its evolution and development. As put concisely by Logan, although "[r]ejected in one field, the concept's value was transformative for another" (Logan 2015, 419).

There is, in other words, a long way before the thesis about the disappearance of Semon's engram gets settled by more extensive historical studies. Paraphrasing Michel Morange on the historiography of molecular biology, the disappearance of the engram is a "vision [that] is too common not to deserve closer examination" (Morange 2018, 15). Still unknown are the ways the engram circulated beyond his rejection as part of theories and studies of organic memory. Also unknown are the ways these uses of Semon's ideas remorphed his intuitions into the nascent discipline of neuropathological sciences. Or, another gap is about the existence of national morphings of his views (e.g., the case of Swiss neuropsychiatric research mentioned by Logan) that kept the engram geographically and intellectually close to specific contexts of research in neurosciences. Thanks to the pioneering work of scholars such as Cheryl A. Logan, it becomes clearer that the engram was alive and well also when it was not part of mainstream debates in heredity or physiology. More historical studies are needed to complement the received historiography that postulates the disappearance of the engram throughout the twentieth century (cf. Schacter 1982).

## Representing and Intervening: Social Studies of Engram Manipulation

Contemporary neurosciences of memory leverage a specific set of experimental strategies to characterize engrams and intervene into them (Josselyn and Tonegawa 2020). Besides observational studies, which illustrate that specific populations of neurons are activated by memory function (e.g., learning and retrieval), current memory research rests also upon distinct interventional study designs to manipulate engrams (i.e., loss-of-function, gain-of-function, and mimicry studies). Among the biological mechanisms used to establish, update, and consolidate memories, a lot of attention has gone into epigenetic modifications. While epigenetics lacks a clear-cut definition for both life (Horsthemke 2022) and social/human scientists (Stotz and Griffiths 2016), its methods, experimental approaches, and biological mechanisms are often taken to drive biological plasticity and mediate gene-environment interactions. An extensive literature exists on how the epigenome shapes patterns of behavior, cognitive skills, personality traits, or mental health (Sandi and Haller 2015). It is therefore no surprise that epigenetic modifications have been invested with much promise in research manipulating the synaptic plasticity required to memory formation and persistence (Khalaf and Gräff 2016). Memory has long been associated to epigenetics in a broader sense than merely somatic or psychological memory (e.g., Holliday 1994, 2006). This research—which connects back to the relationship between lifelong memory and heredity found in Semon's work (see section "A Gap in the Historiography of the Engram")—postulates the epigenome can offer not just a molecular basis to understand the establishment and endurance of lifelong memories, but it may be also implicated in the transgenerational transmission of experiences and exposures (Bohacek and Mansuy 2015). Epigenetics offers therefore to engram research a repertoire of knowledge and methods to both give meaning to and intervene into the molecular events that govern the ways nurture (i.e., experiences and the social environment) shapes nature (i.e., the body, the brain, and the genome)—hereby including the ways the organism can learn and remember (e.g., through engram formation) across different temporal frames (i.e., over the life course and, possibly, multiple generations).

In the words of anthropologist Paul Rabinow (Rabinow 1996, 93), "representing and intervening" in the context of experimental life sciences is not a neutral process. Rather, this practice produces a connection between "knowledge and power," between ways of understanding biological processes and ways of collectively addressing them. These two discourses—one factual, the other political—are in fact "built in, from the start, as simultaneous goals and means" (Rabinow 1996, 93). And while Rabinow's work fleshed out these issues in the case of the nascent genomics of the Human Genome Project, his intuitions can extend to research on the engram and call for developing similar studies of these research practices. As Rabinow argued, the key for exploring the social and political possibilities built into scientific research is the mapping of the specific repertoires of intervention into these biological processes practiced by molecular biologists. These approaches offer in fact a snapshot of the ways nature can became "culture understood as practice" (1996, 99); that is, how these biological processes can become amenable to expert practice, analysis and techno-scientific interventions. What, then, are the opportunities that interventional research on the engram (e.g., the one based on epigenetics) offer to the politics and policies around health matters related to memory (e.g., traumatic memories) in our societies?

Critical scholarship in social studies of science and technology (STS) has already devoted extensive attention to the opportunities and challenges raised by epigenetic research (Niewöhner 2011; Chiapperino and Testa 2016; Kenney and Müller 2017;

Chiapperino 2021). As one of us argued elsewhere (Chiapperino 2023, 21), the "way political views and technoscientific strategies of intervention (e.g., pharmacological, environmental, behavioural, clinical, populational)" intersect in current epigenetic research is far from uncontroversial. The most widely used approaches to the manipulation of epigenetic mechanisms (including those involved in engram formation and consolidation) (see Gulmez Karaca et al. 2020; Odell et al. 2020) consist of specific classes of drugs known to act on the epigenetic machinery, such as DNA methyltransferases (DNMTs) and histone deacetylases (HDACs) inhibitors (Szyf 2009; Day 2014). To this drug-based approach, recent technological developments in the field offer the addition of epigenetic editing techniques that promise to leverage the same mechanisms with more predictable and targeted approaches than so-called epi-drugs (Policarpi et al. 2021). However, while being practical and convenient to experimentation, these approaches run the risk of advancing a problematic biomedical mode of intervention into the epigenome (Chiapperino 2023). There exists, in fact, a relatively developed set of complementary evidence, which suggests that the material/social environment(s) can have clear-cut (beneficial) effects on the epigenome and health. Enriched environments—alternatively operationalized as improved cognitive, physical, or relational stimulation—have shown potential for the reversal of the epigenetic predispositions to disease (McCreary and Metz 2016; see also Chiapperino 2021). This insight goes beyond animal studies too and resonates with human evidence of the effects of psychological therapy on the epigenome (Vinkers et al. 2021) as well as studies of psychiatric resilience in social epigenetics (Brody et al. 2016; Smeeth et al. 2021). From the perspective of those studying the biological and sociological causes of mental illness through epigenetic methods, an exclusive focus on pharmacological interventions risks therefore missing on the importance and potential actionability of social determinants of mental health. Rather, a holistic suite of interventions is needed to address the biological, psychological, and social components of epigenetic differences. The recognition that our epigenome is affected by complex bio-psycho-social processes calls therefore for expanding knowledge of the interventional strategies of the field beyond pharmacology: What can targeted interventions in the social environment do for therapy and for psychiatric resilience? And how can psychosocial and/or structural interventions affecting the epigenome and health be made relevant for a policy translation of the field (Chiapperino 2021, 2023)?

These features of epigenetics are at play also in the interventional studies on engram cells. A distinct strand of engram research leverages epigenetic knowledge and molecular technologies (e.g., drugs, epigenetic editing) to manipulate these cells (see Khalaf and Gräff (2016; see Fuentes-Ramos et al. 2021 for a review of this research). But without simplistically extrapolating from a general trend in epigenetics, the existence of this critical literature on its biomedical modes of interventions should be taken as an opportunity to scrutinize and anticipate the normative dimensions of engram research. What are the hidden normative assumptions behind study designs directed at the manipulation of engrams? How are specific views of memory in this research made relevant for society writ large? With what expectations about who should act to correct memory processes gone wrong, or the illness conditions associated with it? In the preparation of this chapter, we could not identify any piece of literature nor life-social sciences collaboration around the practical (and political) repertoire built into contemporary neurosciences of memory. Experimental strategies of intervention into the engram, scientists' discourses about this research's translational potential for neurodegenerative (e.g., Alzheimer's disease) and psychiatric (e.g. post-traumatic stress disorder) conditions, as well as public translations and circulations of memory studies have not yet been investigated to the extent of other examples of cognate epigenetic research in the last few years. It is likely that the sociopolitical dimensions of engram research will exhibit their own specificities when closely scrutinized from a social sciences and humanities perspective. However, it is worth underlining the urgence of a question: What should neuro- and social scientists do to shape the social and political framings of engram research in society?

## Conclusion

Research on the engram is developing rapidly (Josselyn and Tonegawa 2020). In the face of such advancements, we are grateful for the opportunity to draft a preliminary agenda for the much-needed philosophical, historical, and sociological studies of this field. The integration of this chapter in a book by life scientists is a testament to the recognition that, far from being only a concern of social and humanistic scholarship, critical perspectives on the foundations, framings, and interpretations of engram research are on the agenda of neuroscientists too. Social scientists and humanities scholars can only look at this shared reflexivity, and the opportunities it offers, with undiminished critical spirit and a renewed optimism for the future of social-life sciences conversations around neurosciences (see Fitzgerald and Callard (2015) and Callard and Fitzgerald (2015) for a comprehensive critical uptake of this relationship). However, much more work is needed.

To start this conversation, we have identified three main research gaps that could contribute to critical scrutiny of engram research and the development of its future agenda. First, we have highlighted the need to take seriously the impact of contemporary engram research on philosophical conceptions of memory. Closer attention should be paid to the incommensurability of constructivist and preservationist claims in this research as well as to the role of recent evidence pointing to synaptic plasticity and dynamic engram complexes for the understanding of human memory (cf. Najenson 2021). Are engrams dropping in and out of specific brain regions the vindication of constructivist conceptions of memory? Or is a middle-ground convergence of both constructivist and preservationist views of memory the way forward? Second, we have questioned the received historiography of the engram that emphasizes the disappearance of Semon's ideas throughout the twentieth century. Building upon the little historical literature on his work (e.g., Logan 2015), we have distinguished between the resistances Semon's ideas encountered as theory of psychological memory and, as part of existing debates on heredity, the inheritance of acquired characters and organic memory. While it is unquestionable that the latter were overshadowed by emerging views of hard heredity at the beginning of the twentieth century, it is far from clear whether the engram actually disappeared from neuropathological research on memory and brain. Further historical research is needed to map the specific contexts of research that Semon's engram may have inspired during its alleged disappearance. Finally, we also examined the relationship between contemporary engram research and its social circulation. We underlined the lack of studies dissecting the relationship between neurosciences of memory and public health discourses, or healthcare intervention strategies into pathologies related to memory (e.g., neurodegenerative and psychiatric disorders). To interrogate the role that the facts, metaphors, and interventional strategies of engram research play in the wider society, critical social studies of these scientific practices are much needed. How is memory being understood and valued in this research? What are the political undertones of distinct interventional strategies into the biological mechanisms of memory-making, such as epigenetic modifications? What aspects of the environment, as psychosocial relations or social structures, can be critically nurtured in experimentation to point to diverse approaches to healthcare intervention into the neuropathologies of memory (see Chiapperino 2024)?

But besides critical scrutiny, engram studies call for an integration of social/life sciences perspectives within collaborative approaches to the nature of memory and experience. As argued by philosopher John Dewey (1920, 6), the fact that "memory is not a remembering of actual facts, but is association, suggestion, dramatic fancy" only calls for rejecting simplistic views of experience and the organisms as passive receptacles of external stimuli. Contemporary neurosciences of memory are reminiscent of these—and others—old ideas, without however necessarily being commemorative of the deep philosophical, historical, and social implications they entail. If anything, therefore, this chapter should be read as a call to shy away from simply bringing in social and humanistic scrutiny ex post to reflect on nominal responsibility of research, public communication, or policy controversies spurred by engram research (Owen et al. 2012). Taking inspiration from existing collaborative scholarship in STS (Niewöhner 2016; Greenhough et al. 2020), our goal is to add on the existing research agenda of engram research the following question: How can this research be driven *also* by philosophical, historical and sociological questions and concerns?

We foresee that this question will play a central role in shaping the future of engram research. The time is ripe for a closer humanistic attention to the experimental evidence accumulating on the neurophysiology of memory and the possibilities to manipulate it. It is crucial that novel links between social scientists, humanities scholars, and life scientists are leveraged to inform, structure, and make sense of this research. This, we hope, will not only expand the disciplinary composition of engram studies but also align the ambition of a foundational approach to the neurosciences of memory with a socially responsible circulation of this knowledge.

## References

Albo Z, Gräff J (2019) Engram excitement. Neuron 101(2):198–200. https://doi.org/10.1016/j.neuron.2018.12.033
Blumenberg H (2010) Paradigms for a metaphorology. Cornell University Press
Bohacek J, Mansuy IM (2015) Molecular insights into transgenerational non-genetic inheritance of acquired behaviours. Nat Rev Genet 16(11):641–652. https://doi.org/10.1038/nrg3964
Brody GH, Tianyi Y, Beach SRH (2016) Resilience to adversity and the early origins of disease. Dev Psychopathol 28(4 Pt 2):1347–1365. https://doi.org/10.1017/S0954579416000894
Bruce D (2001) Fifty years since Lashley's in search of the engram: refutations and conjectures. J Hist Neurosci 10(3):308–318. https://doi.org/10.1076/jhin.10.3.308.9086
Callard F, Fitzgerald D (2015) Rethinking interdisciplinarity across the social sciences and neurosciences. Palgrave Macmillan UK, Basingstoke/New York. https://www.palgrave.com/gp/book/9781137407955
Chiapperino L (2021) Environmental enrichment: an experiment in biosocial intervention. BioSocieties 16(1):41–69. https://doi.org/10.1057/s41292-019-00181-5
Chiapperino L (2023) The turn towards "the biosocial" in epigenetics: ontological, epistemic and socio-political considerations. In: Giroux É, Merlin F, Fayet Y (eds) Integrative approaches in environmental health and exposome research: epistemological and practical issues. Springer International Publishing, Cham, pp 9–33. https://doi.org/10.1007/978-3-031-28432-8_2
Chiapperino L (2024) Enacting biosocial complexity: Stress epigenetic biomarkers and the tools of postgenomics. Soc Stud Sci (Online first). https://doi.org/10.1177/03063127231222613
Chiapperino L, Testa G (2016) The epigenomic self in personalised medicine: between responsibility and empowerment. In: Meloni M, Williams SJ, Martin P (eds) Biosocial matters: rethinking the sociology-biology relations in the twenty-first century, Sociological review monograph, vol 616. Wiley-Blackwell, Chichester/Malden
Day JJ (2014) New approaches to manipulating the epigenome. Dialogues Clin Neurosci 16(3):345–357
De Brigard F (2014) The nature of memory traces. Philos Compass 9(6):402–414. https://doi.org/10.1111/phc3.12133
Dewey J (1920) Reconstruction in philosophy, 1st edn. Henry Holt and Company
Fitzgerald D, Callard F (2015) Social science and neuroscience beyond interdisciplinarity: experimental entanglements. Theory Cult Soc 32(1):3–32. https://doi.org/10.1177/0263276414537319
Fuentes-Ramos M, Alaiz-Noya M, Barco A (2021) Transcriptome and epigenome analysis of engram cells: next-generation sequencing technologies in memory research. Neurosci Biobehav Rev 127:865–875. https://doi.org/10.1016/j.neubiorev.2021.06.010
Greenhough B, Read CJ, Lorimer J, Lezaun J, McLeod C, Benezra A, Bloomfield S et al (2020) Setting the agenda for social science research on the human microbiome. Palgrave Commun 6(1):1–11. https://doi.org/10.1057/s41599-020-0388-5
Gulmez Karaca K, Kupke J, Brito DVC, Zeuch B, Thome C, Weichenhan D, Lutsik P, Plass C, Oliveira AMM (2020) Neuronal ensemble-specific DNA methylation strengthens engram stability. Nat Commun 11(1):639. https://doi.org/10.1038/s41467-020-14498-4
Holliday R (1994) Epigenetics: an overview. Dev Genet 15(6):453–457. https://doi.org/10.1002/dvg.1020150602
Holliday R (2006) Epigenetics: a historical overview. Epigenetics 1(2):76–80
Horsthemke B (2022) A critical appraisal of clinical epigenetics. Clin Epigenetics 14(1):95. https://doi.org/10.1186/s13148-022-01315-6
Josselyn SA, Tonegawa S (2020) Memory engrams: recalling the past and imagining the future. Science 367(6473):eaaw4325. https://doi.org/10.1126/science.aaw4325
Josselyn SA, Köhler S, Frankland PW (2017) Heroes of the engram. J Neurosci 37(18):4647–4657. https://doi.org/10.1523/JNEUROSCI.0056-17.2017

Kenney M, Müller R (2017) Of rats and women: narratives of motherhood in environmental epigenetics. BioSocieties 12(1):23–46. https://doi.org/10.1057/s41292-016-0002-7
Khalaf O, Gräff J (2016) Structural, synaptic, and epigenetic dynamics of enduring memories. Neural Plast 2016:1–11. https://doi.org/10.1155/2016/3425908
Lashley KS (1950) In search of the engram. In: Physiological mechanisms in animal behavior, Society's symposium IV. Academic Press, Oxford, pp 454–482
Logan CA (2015) Engrams and biological regulation: what was "wrong" with organic memory? Mem Stud 8(4):407–421. https://doi.org/10.1177/1750698015575189
McCreary JK, Metz GAS (2016) Environmental enrichment as an intervention for adverse health outcomes of prenatal stress. Environ Epigenet 2(3):dvw013. https://doi.org/10.1093/eep/dvw013
Michaelian K, Debus D, Perrin D (eds) (2018) New directions in the philosophy of memory, 1st edn. Routledge, New York
Morange M (2018) The historiography of molecular biology. In: Dietrich M, Borrello M, Harman O (eds) Handbook of the historiography of biology, Historiographies of science. Springer International Publishing, Cham, pp 1–20. https://doi.org/10.1007/978-3-319-74456-8_11-1
Najenson J (2021) What have we learned about the engram? Synthese 199(3):9581–9601. https://doi.org/10.1007/s11229-021-03216-2
Niewöhner J (2011) Epigenetics: embedded bodies and the molecularisation of biography and milieu. BioSocieties 6(3):279–298. https://doi.org/10.1057/biosoc.2011.4
Niewöhner J (2016) Co-laborative anthropology: crafting reflexivities experimentally. In: Jouhki J, Steel T (eds) Ethnological interpretation and analysis: towards a transparent research process. Ethnos, Helsinki, pp 81–125
Odell SC, Taki F, Klein SL, Chen RJ, Levine OB, Skelly MJ, Nabila A et al (2020) Epigenomically bistable regions across neuron-specific genes govern neuron eligibility to a coding ensemble in the hippocampus. Cell Rep 31(12):107789. https://doi.org/10.1016/j.celrep.2020.107789
Otis L (1994) Organic memory: history and the body in the late nineteenth and early twentieth centuries. University of Nebraska Press, Lincoln
Owen R, Macnaghten P, Stilgoe J (2012) Responsible research and innovation: from science in society to science for society, with society. Sci Public Policy 39(6):751–760. https://doi.org/10.1093/scipol/scs093
Panese F (2018) Sur les traces des esprits animaux. Lecture anachronique d'une métaphore psychophysiologique entre Nicolas de Malebranche et les sciences du cerveau contemporaines. In: Louis-Courvoisier M, Kleiman-Lafon S (eds) Les esprits animaux (XVIème – XXIe siècles). Littérature, histoire, philosophie. Epistémocritique, Genève, pp 113–128. https://epistemocritique.org/sur-les-traces-des-esprits-animaux-lecture-anachronique-dune-metaphore-psychophysiologique-entre-nicolas-de-malebranche-et-les-sciences-du-cerveau-contemporaines/
Plato (2004) Plato: Theaetetus (trans: Sachs J), 1st edn. Focus, Newburyport
Policarpi C, Dabin J, Hackett JA (2021) Epigenetic editing: dissecting chromatin function in context. Bioessays 43(5):e2000316. https://doi.org/10.1002/bies.202000316
Rabinow P (1996) Essays on the anthropology of reason, Princeton studies in culture/power/history. Princeton University Press, Princeton
Russell B (2005) The analysis of mind. Routledge, London
Sandi C, Haller J (2015) Stress and the social brain: behavioural effects and neurobiological mechanisms. Nat Rev Neurosci 16(5):290–304. https://doi.org/10.1038/nrn3918
Schacter DL (1982) Stranger behind the engram: theories of memory and the psychology of science, 1st US-1st printing edn. Lawrence Erlbaum Associates Inc, Hillsdale
Semon RW (1921) The Mneme. G. Allen & Unwin Limited
Smeeth D, Beck S, Karam EG, Pluess M (2021) The role of epigenetics in psychological resilience. Lancet Psychiatry 8(7):620–629. https://doi.org/10.1016/S2215-0366(20)30515-0
Stotz K, Griffiths P (2016) Epigenetics: ambiguities and implications. Hist Philos Life Sci 38(4):22. https://doi.org/10.1007/s40656-016-0121-2

Sutton J (2007) Philosophy and memory traces: descartes to connectionism, 1st edn. Cambridge University Press

Szyf M (2009) Epigenetics, DNA methylation, and chromatin modifying drugs. Annu Rev Pharmacol Toxicol 49(1):243–263. https://doi.org/10.1146/annurev-pharmtox-061008-103102

Tonegawa S, Liu X, Ramirez S, Redondo R (2015) Memory engram cells have come of age. Neuron 87(5):918–931. https://doi.org/10.1016/j.neuron.2015.08.002

Vinkers CH, Geuze E, van Rooij SJH, Kennis M, Schür RR, Nispeling DM, Smith AK et al (2021) Successful treatment of post-traumatic stress disorder reverses DNA methylation marks. Mol Psychiatry 26(4):1264–1271. https://doi.org/10.1038/s41380-019-0549-3

Weismann A (1889) Essays upon heredity and kindred biological problems (trans: Poulton EB, Shipley AE), vol 1. Clarendon Press

Winther RG (2001) August Weismann on germ-plasm variation. J Hist Biol 34(3):517–555

# If Engrams Are the Answer, What Is the Question?

**Fionn M. O'Sullivan and Tomás J. Ryan**

**Abstract** Engram labelling and manipulation methodologies are now a staple of contemporary neuroscientific practice, giving the impression that the physical basis of engrams has been discovered. Despite enormous progress, engrams have not been clearly identified, and it is unclear what they should look like. There is an epistemic bias in engram neuroscience toward characterizing biological changes while neglecting the development of theory. However, the tools of engram biology are exciting precisely because they are not just an incremental step forward in understanding the mechanisms of plasticity and learning but because they can be leveraged to inform theory on one of the fundamental mysteries in neuroscience—how and in what format the brain stores information. We do not propose such a theory here, as we first require an appreciation for what is lacking. We outline a selection of issues in four sections from theoretical biology and philosophy that engram biology and systems neuroscience generally should engage with in order to construct useful future theoretical frameworks. Specifically, what is it that engrams are supposed to explain? How do the different building blocks of the brain-wide

This is a preprint of the following chapter: O'Sullivan & Ryan, "If engrams are the answer, what is the question?" published in *Engram Biology*, edited by S. Ramirez & J. Gräff, 2023, Springer Nature, reproduced with permission of Springer Nature.

F. M. O'Sullivan
School of Biochemistry and Immunology, Trinity College Dublin, Dublin, Ireland

Trinity College Institute of Neuroscience, Trinity College Dublin, Dublin, Ireland

T. J. Ryan (✉)
School of Biochemistry and Immunology, Trinity College Dublin, Dublin, Ireland

Trinity College Institute of Neuroscience, Trinity College Dublin, Dublin, Ireland

Florey Institute of Neuroscience and Mental Health, Melbourne Brain Centre, University of Melbourne, Melbourne, VIC, Australia

Child & Brain Development Program, Canadian Institute for Advanced Research (CIFAR), Toronto, ON, Canada
e-mail: tomas.ryan@tcd.ie



J. Gräff, S. Ramirez (eds.), *Engrams*, Advances in Neurobiology 38,
https://doi.org/10.1007/978-3-031-62983-9_15

engram come together? What exactly are these component parts? And what information do they carry, if they carry anything at all? Asking these questions is not purely the privilege of philosophy but a key to informing scientific hypotheses that make the most of the experimental tools at our disposal. The risk for not engaging with these issues is high. Without a theory of what engrams are, what they do, and the wider computational processes they fit into, we may never know when they have been found.

**Keywords** Ecphory · Representation · Instinct · Recall · Information

## Introduction

An outside observer might be forgiven for thinking that hidden in memory engram biology papers, or implicitly known among the systems neuroscience community, there is a clear understanding of what engrams are and how they relate to memory. Memory is defined at a behavioral level as changes resulting from past experience and may also be defined as reconstructed internal representations (Dudai 2007). In contrast, engram biology as a research program is dedicated to defining the biological basis of memory. Engrams are not memories, instead they are what provide "the necessary (physical) conditions for a memory to emerge" (Moskovitch 2007). Although not always, within neuroscience, the terms memory trace and engram are used synonymously, with the term engram introduced in Semon's biological theory of memory (1904/1921, 1909/1923). Inspired by Semon, in contemporary neuroscience, engrams are operationalized as whatever physical or chemical changes meet four criteria: these changes are elicited by learning and persist, they can be reactivated ("ecphory"), that content is preserved from encoding to retrieval, and that between these processes the changes are not permanently active but dormant (Josselyn et al. 2015). Because cell ensembles can be experimentally labelled through co-opting activity-dependent immediate-early genes (IEGs) to express fluorescent and other proteins in those neurons activated during a particular learning experience, and because reactivating these specific labelled cell ensembles (but not others) often reproduces the original learned behavior, and their silencing or ablation fails to reproduce that exact same behavior, scientists have claimed these cells fulfil the function of the memory trace or engram and have appropriated and operationalized Semon's theory to help make sense of these experimental findings (Josselyn et al. 2015; Tonegawa et al. 2015a, b; Josselyn and Tonegawa 2020).

However, for all the pragmatic contemporary usage of the word, the engram remains an open-ended scientific construct. Semon's definition, restated in Tonegawa et al. (2015a) as the "enduring physical and/or chemical changes that were elicited by learning and underlie the newly formed memory associations" is not tied to any specific biological level. It is telling that we must reach back over 100 years for a conceptual framework to operationalize for current use in neuroscience.[1] In other words, although Semon's framework has been validated, it has not been elaborated, and there are no fleshed out alternative biological theories that would create healthy competition in the field. In general, the memory engram field remains undertheorized (Robins 2023). Is this something that can be remedied by more advanced experimentation alone, or is new theory also needed?

For some, the concept of an engram may serve as a working abstract placeholder because the engrams that the scientific construct refers to have not been truly found, and the abstraction affords the elbow room required to maneuver while figuring out how engrams are realized in biological tissue. Despite experimental limitations regarding the temporal resolution of the cell-tagging protocol, differences in endogenous IEG function, controlling for off-target effects, and detecting exact differences between engram and non-engram cells, referring to the existence of "engram cells" is a pragmatic step forward based on progressive behavioral evidence that cannot be dismissed. Examining the cells involved in learning and recall did not a priori suggest that we would find cells labelled at encoding being reactivated at recall or that these overlapping cells would be functional at the level of behavior. Rather, this was demonstrated by empirical investigation (Reijmers et al. 2007, Liu et al. 2012; Garner et al. 2012). Meanwhile, the understanding of "how information is stored in an engram" is acknowledged as the major outstanding overarching question in the engram field (Queenan et al. 2017; Josselyn and Tonegawa 2020) and cognitive science generally (Poeppel and Idsardi 2022).

For others, not only have engrams not been satisfactorily found, they may never be because the concept is not well defined (Hardt and Sossin 2020), and there are views of memory that do not place prime importance on the existence of engrams (Schacter and Addis 2007). If their existence can neither be proven or disproven, their scientific utility is questionable. Additionally, it is unclear from either theory or empirical work what the information content is and how it is supposed to persist in the form or format of an engram. The language used in discussing existing engram theory can hide this lack of knowledge. To say that physical and/or chemical changes "underlie" memory is either a truism or insinuates that we have a clear theory for what the enduring changes are and how they do this underlying.[2] We have neither. Performing observational, gain-of-function, loss-of-function, and mimicry experiments has established that there is a replicable and specific linking between these unique cell ensembles and behavior. However, when the theory being tested is ambiguous, such studies are limited in their power to explain what engrams are or how this specific linking works (Krakauer et al. 2017; Tonegawa et al. 2015b). There are now several putative levels for where the persistent changes that function as engrams may be found. These include (i) synaptic weight plasticity (Martin et al.

---

[1] This is not to disregard the original theory's insightfulness, proper historical accreditation, or the very real sociological, ideological, and experimental limitations that might have previously discouraged further theorizing about the nature of the engram during the twentieth century (Schacter 2001).

[2] The use of such "filler-verbs" is a kind of interpretative (mal)practice not unique to engram biology but endemic to all disciplines of contemporary neuroscience (Krakauer et al. 2017).

2000; Kandel 2001), (ii) intracellular and molecular mechanisms such as protein or epigenetic modifications (Sacktor 2011; Bédécarrats et al. 2018; Campbell and Wood 2019; Gershman 2023), or (iii) that the relevant mechanism is supracellular, such as changes in the structural connectivity of the connectome (Chklovskii et al. 2004; Ryan et al. 2021) or non-neuronal structures, like perineuronal nets (Tsien 2013), astrocytes and oligodendrocytes (Kol and Goshen 2021), or microglia (Wang et al. 2020). However, none of these proposals explain the how—how such changes are the appropriate mechanism capable of supporting recall of a given behavior over any other.[3] This discrepancy—between the establishment of a linking between labelled neuronal ensembles (or any other level) and behavior, but a dearth of understanding of how this works—is a gauntlet that needs to be picked up by a new theoretical framework. However, our objective here is not to outline a concrete "engram theory." Developing new theories first requires understanding what our current framework is missing, including whole types of theory. This chapter therefore discusses current ambiguities in order to identify those areas any future theories must pay attention to.

Section "Levels, Questions, and Explanations" addresses how there are in fact multiple things-to-be-explained when we are concerned with the biological basis of memory, and apparently competing explanations may not actually be in conflict. Section "What Flavor of Model Are We Using in Engram Biology?" singles out the dominant theory of engrams as operationalized in contemporary neuroscience, focusing on the limitations raised by its assumed modular architecture. Attention is paid to questioning the assumption that differences at the level of cell ensembles actually correspond to differences in the "type of information stored" by said ensembles. Section "If Engrams Are Vehicles, What Are They Vehicles of?" argues this conflation occurs due to failure to dissociate questions about information content from questions about vehicles of content, and emphasizes the need for more careful terminology regarding differing conceptions of what the engram should be. Section "Information and Coding" is devoted to considering the limitations of theories of information in neuroscience and biology. For us, one of the most exciting aspects of engram technology is its potential to inform a new conception of information in the brain. Finally, in the "Discussion" section we summarize the implications of this approach for helping to guide experimental research.

## Levels, Questions, and Explanations

### *If Engram Is the Answer, What Was the Question?*

The major theoretical hurdle faced by engram biology is a lack of clarity regarding what the problems to be solved are or not spelling them out in sufficient detail. The engram is assumed to be a good explanation, but what exactly are engrams posited

---

[3] Our explanation may involve integration across many of these levels, but a multi-level explanation must still account for the *how*.

to explain? The existence of something capable of functioning as an engram is usually invoked in order to explain the general phenomena of learning and memory. However, "learning and memory" is not a coherent, singular phenomenon, and when the phenomenon we want to explain is too general, multiple competing explanations will appear equally adequate. There are different questions that need to be parsed. In this section, three different methods for doing so are presented based on (i) biological organization, (ii) question-type, and (iii) question-subdivision.

## *Dividing Explanations Based on Levels of Biological Organization*

Many biological processes and computations involve changes across multiple organismal levels. The changes studied in learning and memory research cover several orders of magnitude of various nested biological processes (Josselyn et al. 2015; Craver 2007, ch.5), from molecular to systems physiology. However, we are not sure which levels of organization are most relevant for theorizing about how the brain can be said to store information. It may be useful to draw analogies with similar problems in other areas of neuroscience. In their discussion comparing the relevance of neuron-and-circuit-level versus population-manifold-level descriptions when it comes to explaining cognitive phenomena, Barack and Krakauer (2021) illustrate the importance of choosing the best level of explanation with analogy to different answers to the question "why did the window break?" The first-level explainer (we denote here as $E_1$) should be "because the ball was thrown at it" and not explanations based on the physiology of arm-throwing or the Si-O amorphous molecular structure of glass. These second- and third-level explainers ($E_2$ and $E_3$) are defined by their ability to explain $E_1$ but that does not make $E_2$ and $E_3$ better answers to the original question.[4]

If we are seeking to explain the biological basis of memory, we are not seeking to explain "the engram." Engrams may be posited as a first-level explanation ($E_1$) for the biological basis of memory or may be a given step ($E_n$) along the way, say $E_5$. Alternatively, the engram may refer to a given chain of processes, with no one privileged level. Since we do not know what the engram is, our focus shifts to picking out the next level of explanation down ($E_{n-1}$) to that which might support the engram. Thus, when a biological mechanism or process is conjectured as an explanation, care should be taken to ascertain whether this is an explanation for the biological basis of memory or actually the-thing-that-explains-the-thing (etc.) that explains the biological basis of memory.[5]

---

[4] If the question is not fixed but we vary it, then what is $E_2$ to one question may become the $E_1$ for another.

[5] The pitfall to avoid is to take, say, $E_{20}$ for $E_2$.

Although this seems like a straightforward pitfall to avoid, it is not when it comes to memory. This is because the problem to which memory is a solution to, that of dealing with dynamically changing external conditions that have some statistical dependencies, is not a problem that manifests once at a single scale.[6] It is evolutionarily universal, and solutions or adaptations have manifested across multiple levels of biological organization and phylogeny (unlike, say, speech production or motor imagery). Bacteria have been said to store information in protein post-translational modifications (Sterling and Laughlin 2015; Gershman 2023) and slime molds are capable of anticipating temperature changes (Saigusa et al. 2008). Should any biological mechanism of "storing information" by modifying some structure be considered an engram? Avoiding a situation of "endless engrams" all the way down, in which vastly different processes all stake claim to the same name, implying some nontrivial equivalence across all, is preferable to us.[7,8] Granted, different biological processes across the evolutionary continuum and in parallel across multiple scales within multicellular organisms involve the creation of specific changes could be said to function as engrams that are relevant for various kinds of phenotypic plasticity.[9] However, the question to be addressed in order to explain behavioral memory is which aspect of the problem is each mechanism or process a solution to (see Fig. 1) and how these different information storage mechanisms are hierarchically scaffolded together (Simon 2002; Hoffmeyer 2008). Experimental data without theory does not reveal this.[10]

Given these general considerations, competing proposals for the location of "the engram" suddenly no longer seem to be in competition at all. In recognizing that synapses (or synaptic strength at least) may not be the site of the lasting changes that count as an engram (Chen et al. 2014; Ryan et al. 2015), two theories, one that goes down a level of biological organization from synapses to nucleotides and protein conformational states (Gallistel 2017; Gershman 2023), and one that goes up a level

---

[6] This question-answer dialectic of an environment posing problems to which an organism evolves "solutions" is imperfect (see Lewontin 1983) but illustrative for our purposes here.

[7] There may be nothing that unites them other than uncertainty reduction very generally defined.

[8] Biological processes are nested and previous solutions are repurposed, but this does not make a solution at one-level universal or that it will be used for addressing the same problem in future evolution (Gould and Vrba 1982; Kauffman 2019). Solutions to general evolutionary problems can be convergent in some cases, but they can also be divergent. When the same problem is faced at different levels of organization, even within the organism, divergent solutions are found. Both we and our immune cells have to "navigate." We use a musculoskeletal system and the algorithm of walking. Immune cells use cytoskeletal rearrangements and roll along tissue with surface proteins. Like the general problem of navigation, it is sensible to assume the similarly general problem of learning and the solution of storing changes also has many levels.

[9] For example, DNA, immune memory, and external memory (Donald 1991; Clark and Chalmers 1998).

[10] Although we might claim that the level of learning and memory we study in systems neuroscience is tethered to whole-organism behavioral parameters, and all experimental data is collapsed relative to this, the changes we choose to look at, stain, sequence, or quantify are not determined by the data but by hypotheses of where to look.

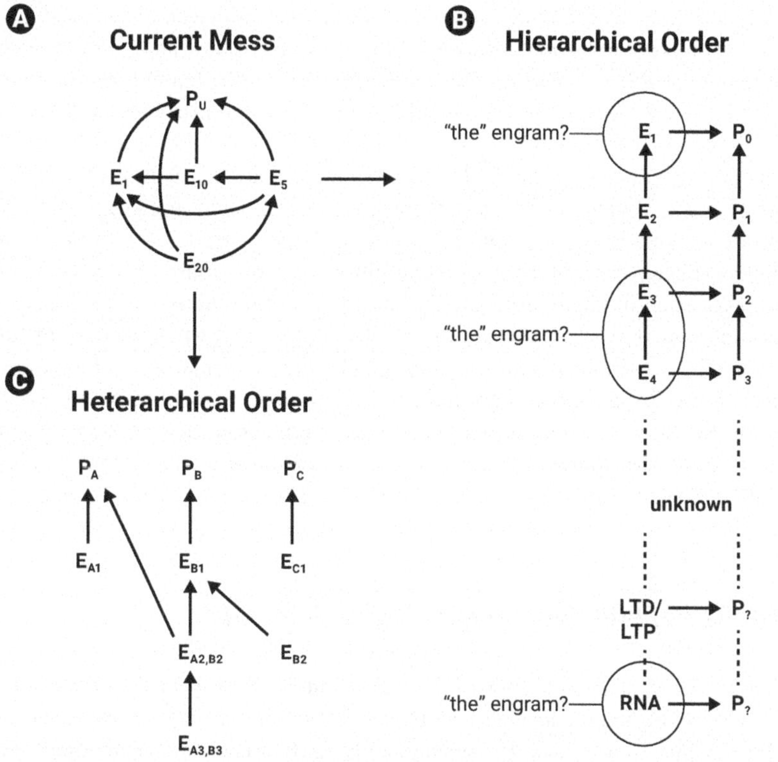


**Fig. 1** (**a**) When the problem to which engrams are posited to solve is articulated over-generally as a kind of universal problem, $P_U$, multiple biological mechanisms may seem to qualify as equally adequate explanations. Either many biological mechanisms get to be engrams or their conglomeration is an engram, neither of which is illuminative. In moving from (**a**) to (**b**) or (**c**) by unfolding $P_U$ into a series of more specific problems that have to be solved, the biological mechanisms can be parsed and scaffolded into a more coherent theory, whether hierarchical or heterarchical. We may choose to define the engram as a particular causal motif (Ross 2021) or as that mechanism which offers the greatest control in terms of causal manipulation of memory content, or as a chain of biological mechanisms, none of which alone qualify as an engram. Specifying problems discretely rarely works in practice and here only illustrates a conceptual point. LTP: long-term potentiation; LTD: long-term depression

to structural connectivity (Chklovskii et al. 2004; Tonegawa et al. 2015b; Ryan et al. 2021), have seen a resurgence or been proposed (for a detailed assessment, see Colaço and Najenson 2023). However, the spatial scales of causal influence implicated at each level may differ dramatically. It may be completely accurate to claim that there is information in neurons, that it is molecular and that stable distributed molecular states could function as engrams, but these may be engrams for the cell informing itself about, say, cytoskeletal rearrangements, and not engrams directly for whole-organism behavior that is retrievable over millisecond timescales.[11] The former is unlikely to be an $E_1$ for the biological basis of how an organism such as a mammal is capable of creating an episodic memory. It may feature in a future theory of the biological basis of episodic memory at $E_{10}$, say, and may indeed be necessary, but $E_{10}$ alone is insufficient.

In contrast, it is also possible that the engram will not be the $E_1$ for the biological basis of episodic memory but will require it being situated in a larger context or a chain of explanations. A "manifold account of the engram," for example, might include explanations linking the disinhibition of a previously latent connectivity pattern with neural population activity taking on a different dynamical trajectory in a certain context (Langdon et al. 2023). The connectivity pattern may in turn be explained by cytoskeletal rearrangements and other molecular level changes, just as motor neuron firing can at a certain level be said to implement throwing a ball at a window. Spelling the problem(s) out at different gradations of complexity will facilitate selecting the appropriate type of engram or information storage mechanism posited in a given hierarchy.

## *Dividing Explanations Based on Question-Type*

So far we have discussed choosing explanations with respect to a fixed question type (and by varying the level at which the question manifests), but we can also vary the type of question we ask. For example, classic heuristics like Tinbergen's (1963) "Four Questions" for any given behavior differentiates how-questions (the development of a behavior within an individual lifespan, and its mechanism of function) from why-questions (the evolutionary history of the behavior and what adaptation it affords). We used a why-question in asking why the window broke, but if we ask "how did the window break?" then explanations for the brittleness of glass and velocity of the ball are indeed the relevant form of explanation. It could be argued that when it comes to the memory relevant for behavior, the what (the behavior we are studying) is learning about the social and physical environment, the why is recall in the future for adaptive behavior, and the how is the engram and its expression. However, there is no singular answer to the how-question, because how to form an engram versus how to use an engram we presume are different processes. Formation and reactivation may therefore require different levels of description. For example, an explanation for how an engram is used may require situating structural changes in terms of their downstream effects (e.g., funneling network activity into a subspace regardless of where the network starts at the time of retrieval, but which preserves some high-level invariant properties with the landscape at the time of encoding), before jumping to explaining behavior.

[11] Stable molecular states may be $E_{n+2}$ explanations for cytoskeletal rearrangements, $E_{n+1}$, which are explanations for $E_n$, but this does not make $E_{n+2}$ an explanation for $E_n$.

### *Dividing Explanations Based on Subdividing the Problem*

Section "Dividing Explanations Based on Levels of Biological Organization" introduced the problem of potentially many different processes being collapsed into an unknown hierarchical relevance with respect to a particular observed behavior. Section "Dividing Explanations Based on Question-Type" demonstrated different kinds of questions. Drawing on both concerns, this section focuses on taking the general problem of "dealing with dynamically changing external conditions…" and decomposing it into subproblems. For example, we might focus on the problem of efficient information storage (Sterling and Laughlin 2015), or the problem of making stable changes accessible (Ryan and Frankland 2022), or the problem of whether a given type of engram is computationally content-addressable or address-addressable (Gallistel 2017), to speculate on a few. The most relevant level description may differ for each, and there may be multiple versions of each problem across scales. Deciding how to divide a problem like biological information storage into subproblems is also not something that can be approached solely as an engineering problem but requires consideration of the phylogenetic history of the organism and its ancestral environments (Cisek 2019).

Failure to decompose the problem hinders inferring which changes in what process are most relevant. If all we are looking for is structured variation in a particular behavioral parameter, there is an endless array of variables across the scales of biological organization that we can intervene with that will produce such effects. This risk, as raised in Jonas and Kording's (2017) problem of inferring the logic of a microprocessor using standard neuroscientific gain-of-function and loss-of-function causal interventional techniques, is that we will take the wrong level of explanation to be $E_1$ just because manipulation at that level resulted in structured outcomes in behavioral parameters. Structured variation in behavioral outcomes due to protein synthesis inhibition (e.g., no memory consolidation, see Ryan et al. 2015) does not necessarily mean that our $E_1$ for how an engram instantiates a particular form of information will involve describing a given protein's function.

## What Flavor of Model Are We Using in Engram Biology?

### *Current Engram Theory Is Modular and Compositional*

The existence of multiple levels of engrams aside, the dominant sense of the engram that has emerged in contemporary systems neuroscience research is that the engram is somehow instantiated as cellular ensembles and a brain-wide distributed complex (Josselyn et al. 2015; Tonegawa et al. 2015a; Josselyn and Tonegawa 2020; Wheeler et al. 2013; Vetere et al. 2017; Roy et al. 2022). But exactly what kind of model is being proposed? Must the engram necessarily be a global and nondecomposable unitary entity or can discrete "sub-engrams" have fixed identities while still making up a "super-engram" or "engrome" and is this the most appropriate way to carve up the system? Having a clearer understanding of what assumptions this model entails will aid in refining or replacing the existing paradigm.

The terms engram cells, engram cell pathways, engram components, and engram complexes were defined and proposed in Tonegawa et al. (2015a) in a contemporary attempt to operationalize Semon's engram theory in light of the last century of advances in biology. In short, there are ensembles of cells (engram cells) that instantiate different components of a much larger, distributed engram (the engram complex) joined together by connectivity pathways between these engram cell subpopulations (Josselyn et al. 2015; Josselyn and Tonegawa 2020). Each of these different sub-engrams or engram building-blocks contribute different pieces of information, such as hippocampal DG and CA3 for contextual information and CA1 for contextual and temporal information (MacDonald et al. 2011; Tonegawa et al. 2015a; Roy et al. 2022). A further level of complexity is added by the existence of inhibitory neuronal ensembles ("inhibitory engrams") contributing to broader engram complexes which have been proposed as a general feature of brain function (Barron et al. 2017; Koolschijn et al. 2019; Sun et al. 2020; Nambu et al. 2022). The picture that emerges is one with a strong sense of modularity and compositionality, where parts of specialized function can be viewed as plug-and-playable modules, participating in a larger distributed engram which in sum we presume has sufficient capacity and richness of stored information to reconstitute a memory.

If explicitly articulated, a modular or compositional theory may serve as a flexible and helpful architecture to explore. However, there are different ways of treating the parts of any modular theory. An essentialist approach explains the behavior of a system by appealing to the intrinsic power and casual abilities that inhere within its parts. These elements have necessary properties that make them what they are. In contrast, a relational approach explains the behavior of a system by appealing to the unique way the parts are configured with respect to each other (Buzsáki 2006). Another way of describing these positions is as follows: what confers on the trace the properties that it has, the intrinsic qualities of trace, or the system in which it is embedded? We explore the differing consequences of essentialist and relational approaches to the basic modular theory of engram biology across different levels of granularity.[12]

## *Relational and Essentialist Conceptions of "the Engram"*

We start with the foundational construct of the engram first before talking of cell-ensembles and ensemble-complexes. An essentialist approach views the engram as or distributed in the constellation of cells that were activated during learning and

[12] We do not have to limit ourselves to two positions. See Laplane (2016) for a fourfold classification of "stemness" in stem cell research. Additionally, we also do not focus on hypothesis that the engram is molecular in this section in the interest of space. However, there is nothing that precludes imagining differences in relational or essentialist approaches at a molecular level.

whose reactivation produces a behavior indicative of recall. Any cells that were active during learning that did not get incorporated into the ensemble, or neurons not active during learning but were recruited later, are an unexplained nuisance. Information is instantiated in a fixed, solid entity, and any alterations will change or corrupt the information preserved. So-called "representational drift" (Rule et al. 2019) might therefore be taken to be a threat to the integrity of the stored information.

A relational approach to defining the engram does not deny that these neurons must play an important part but attempts to appreciate their role in a wider context. We summarize this attitude with the maxim it's not what the changes are, it's what the changes do. For example, Guskjolen and Cembrowski (2023) recommend that the construct of the engram must expand to include both nonneuronal cells and also those neurons that are actively silenced during learning and must be re-silenced for retrieval. In other words, knowing which cells are inactive is just as important as those that are active for information storage and retrieval, because plasticity in engram cells and their connectivity patterns will change the local connectome around them. Unlike artificial networks with scalar activations, excitatory/inhibitory balance is an essential principle of nervous system function, where computationally inhibition can be understood as transiently modifying the morphology of other neurons (Buzsáki 2006). Whether we consider a necessary unity between excitatory and inhibitory cells to be required for something to count as an engram, or we treat excitatory and inhibitory ensembles to be dissociable structures that cooperate, it is clear that any theory of engrams must account for the relevant context that makes an engram possible. Just as a gene is understood not merely as the sequence of base pairs that code for a protein but a region that includes enhancers, promotors, introns, the presence and absence of general and specific transcription factors, so too is there ample room to speculate about the neurobiological context that makes something an engram or makes it the engram that it is. This sets us up to consider engrams not as monolithic entities or "data structures" of one format, such as a singular "information molecule" or singular kind of information (e.g., binary states) realized in several different biomolecules or larger structures, but rather complex (that is, non-decomposable) informational units that may have many necessary pieces, none of which on their own are sufficient to count as a piece of latent information. But what exactly are these "pieces of information"? Do differences at a cellular, ensemble, or complex level explain the differences in the information content held by one engram versus another?

## *Ensembles Are Conflated with Pieces of Information*

A common assumption, introduced in section "Current Engram Theory Is Modular and Compositional", is that different cell-ensembles each instantiate different pieces of information to a brain-wide engram. This encompasses any statements to the effect that one ensemble is said to provide spatial context, another temporal, perhaps a valence component from the amygdala and the various sensory modality engram components from sensory areas, and so forth. However, it is a strong assumption that differences at the ensemble-level are the right level of explanation for differences in the "type of information content" the hypothetical engram is posited as preserving. Relating these is a potential category error (Ryle 1949) and cannot be assumed. It is not yet clear that cell populations in different areas with different functions somehow neatly instantiate the informational components that make up the full engram. It is a promising hypothesis but must be articulated as a hypothesis and not an assumption, as it still lacks experimental evidence. There are at least two items this hypothesis must be able to account for, (i) Intra-ensemble differences: What is it about one ensemble of a given type that differentiates it from another of the same type? For example, temporal engram component X versus temporal engram component Y in CA1. (ii) Inter-ensemble differences: What is it about one type of ensemble that makes it different from another type? A dentate gyrus engram ensemble differs from a medial prefrontal cortex engram ensemble, but in what way are they the same and in what ways do they differ? Is it region alone? Ideally, we want a taxonomy of engram types that avoids the mistake of grouping heterogeneous things under a unifying construct (lumping error), while also avoiding unnecessary fragmentation of engram-ensemble types into as many as there are possible ensembles (splitting error). The essentialist-relational distinction can be used to draw out questions from these ambiguities.

## *Essentialist and Relational Views of Ensemble and Complex Function*

At an inter-ensemble level, there are three essentialist ways of explaining what makes one type of ensemble instantiate the information that it does. First, perhaps it is the type of changes undergone (whether intracellular or membrane localized) that confers the informational specificity of that engrams function (1). Given two approximately identical ensembles in different locations, the unique changes undergone determine, for example, whether the engram is a fear engram, because the cells underwent "fear changes." The second is that all engram cells undergo the same kinds of changes, but it is the cells in which the changes occur that confers informational specificity (2). Since the region does not itself confer properties, either it is something about the intrinsic properties of the cells in that region (i), the connectivity of the cells in that region (ii), or a combination of both (iii). The third possibility meets (1) and (2) somewhere in the middle: There are different kinds of changes, but location matters too (3). For example, Shpokayte et al. (2022) demonstrated topographic segregation (with some overlap) between ventral hippocampal engram cells for aversive and appetitive experiences, as well as different transcriptomic profiles for each population. Whatever the specific changes are in each case, they denote either the positive or negative aspect of the experience. Regardless of which account of how the type of information is determined at the level of the engram ensemble, an essentialist explanation for how these ensembles work together is straightforward—their place in the brain-wide engram-complex is merely additive.[13] Failure to recruit one ensemble still results in a functional memory, minus, say, knowledge of when an experience happened.

A relational approach would argue that helpful as the three distinctions are, these changes are not explanations for why one ensemble instantiates a given type of latent information over any other. The changes as described do not explain how it is that this ensemble can store a "spatial type of information," while a given other ensemble is capable of "positive valence information storage." We want to know what confers this stored informational capacity and type, and if changes at the level of the ensemble even are the most relevant level for determining the type of latent information. A relational approach would also be open to the possibility that information is not determined by persistent changes at the level of cell ensembles, but only at the level of the brain-wide engram when activated. In contrast to an essentialist view of engram components as fully formed cogs that are simply "called up" by context with appropriate syntactic synchrony, a relational approach might argue that each component is mutually determined or has a strong dependence on the other components present based on how they are arranged within the connectome (Vetere et al. 2017).[14] This view informs how we test questions about whether or not the cell ensemble is the best level to locate discrete storage arising, whether there is a minimal number of ensembles needed,[15] and whether or not ensemble parsing and their syntactic activation order (Buzsáki 2010) must be taken into consideration for determining informational specificity at a brain-wide level for successful recall.

## Cutting Across the Levels

To conclude section "What Flavor of Model Are We Using in Engram Biology?", the struggle to flesh out the details of the modular framework go hand in hand with a lack of clear hypotheses regarding the basic concept of the engram itself. It is likely that the way we have divided the problems faced at each level is misplaced. Some problems may apply across all levels and to solve the problem at one level requires solving them across all.

[13] Ensemble-type$_1$ + Ensemble-type$_2$ + Ensemble-type$_3$ = successful retrieval of memory to guide behavior.

[14] In linguistics and philosophy, this is called semantic inferentialism or conceptual role semantics. In fusional but especially polysynthetic languages, morphemes (the units of meaning) when split apart often do not make sense on their own, thus communicating involves using sentence-words. A missing morpheme is easily guessed (pattern completion).

[15] Say a protein may be functional only with eight subunits of different isoforms. If only seven subunits are transcribed, the protein is not coherent, either it does not form or forms something else with a different function.

## If Engrams Are Vehicles, What Are They Vehicles of?

### *Representations Versus Neural Representations*

In this section, we introduce a common if controversial distinction made in philosophy between the vehicle and content of mental representations (Dennet 1978; Hurley 1998). Content or semantic content is what a representation is about, such as a particular scene or scent in the case of an immediate perceptual representation and/or an indirect uncoupled conceptual representation such as an idea, abstract concept, or mental map that we use to stand in and flexibly guide decisions in the absence of direct experience. The vehicle is often taken to be a physical (e.g., neural) part or process (Shea 2018; Burnston 2021). There is much debate over what conditions must be satisfied for something to count as a representation (Ramsey 2007; Krakauer 2022), and whether they map neatly to discrete vehicles or not. However, this debate on representations is typically agnostic as to neural implementation. Thus, both senses of representation just outlined (immediate and uncoupled) are very different senses of representation than when neuroscientists refer to an activity pattern as a "neural representation" of a face or tuning curves or firing rates as representing some features of a stimulus (e.g., Chang and Tsao 2017). This is a softer concept (that of encoding), and one we will assess in section "Information and Coding". For now we need only to note that neural representations are unlikely to qualify as representations in the sense outlined above. Instead, they may be candidate vehicles of representations or recording-apparatus-relative proxies of vehicles at best.[16] Nonetheless, employing the vehicle/content distinction to engrams is instructive, as it can help us separate out at least three categories of engram concepts with their own subtypes. These three are engram-as-content, engram-as-vehicle, and engram-as-ensemble.

### *Engrams-as-Content*

As similarly outlined in Gershman (2023) and Najenson (2021), understanding the biological basis of memory encompasses both understanding memory content and how it is instantiated by some physical vehicle. The first conception is that the engram refers only to the content. Although this is not the most common conception of the engram, it is worth addressing. Within this, there are two senses of content that need to be distinguished—representational and nonrepresentational content. Conceiving of engrams as representations includes viewing them as downsampled

[16] There are more pragmatic accounts of what it means for something to be a representation in practice, under which the following discussion may qualify as treating engrams as representations (Cao 2022). We do not adopt this pragmatism here for the purposes of avoiding conflation between memories and engrams.

and/or "encoded" versions of memories (Moskovitch 2007).[17] However, memories (at ecphory) may also be referred to as representations (Dudai 2007). This is a view of memory representations as "the same thing" transforming from latent trace to active and back again, flip-flopping between two states. While we colloquially refer to the contents of an image and the contents of a file using the same word contents, engrams were never envisioned to be of the same stuff as memory, hence their separation from ecphory (Robins 2018). The second sense of engrams-as-content is that engrams are not representations with rich semantic content comparable to human concepts but some form of simple informational content locally for the neural system itself. For example, this includes the idea that an engram (or a hippocampal engram) is a particular index for constructing a given memory (Goode et al. 2020), meaning the informational content of these engrams is addressed. For example, this includes the idea that an engram is a particular index for constructing a given memory. Talking of information rather than representation suggests that there is a difference in format between engram and memory so large it may require a whole other way of conceiving of engrams other than as the "encoded version" of a memory. This is an idea we will return to but requires clarification.

## *Engrams-as-Vehicles*

As per Semon and recent definitions, the engram is envisioned as an implementational construct that concerns physical and chemical biological changes. However, we often fail to distinguish between the conception of engram-as-content and engram-as-vehicle. There are at least three possible senses of engram-as-vehicle, based on what the engram is a vehicle of or for. First is that engrams are not memories but the neural vehicles of memories. However, whatever the neural implementation of memory is, this concerns what happens at time of recall. The engram is a separate neural implementation which persists even when a memory is not being created. Engrams, therefore, do not have to be the vehicles of memories.[18] Given this, we may propose a second though similar option—that engrams are not the vehicles of memory but are vehicles instantiating a stored, reformatted, and downsampled encoding of memory. This does not abet the problems outlined with the engram-as-representation view above. Finally, there is the conception that engrams are not vehicles of representations but vehicles of a simpler informational content. The content may not be seen as semantic in the representational sense of being about or standing in for things outside the brain, rather the content of the vehicle is what it does: For an engram theory that places primacy on connectivity, it may be shifting the probability of a given neural activity pattern being expressed through

[17] This differs from the sense of representations outlined in the previous paragraph, which pertains to first-person deliberative use of representations (Krakauer 2022).

[18] This conflation may come from using "memory" to refer ambiguously to both memory and memory traces.

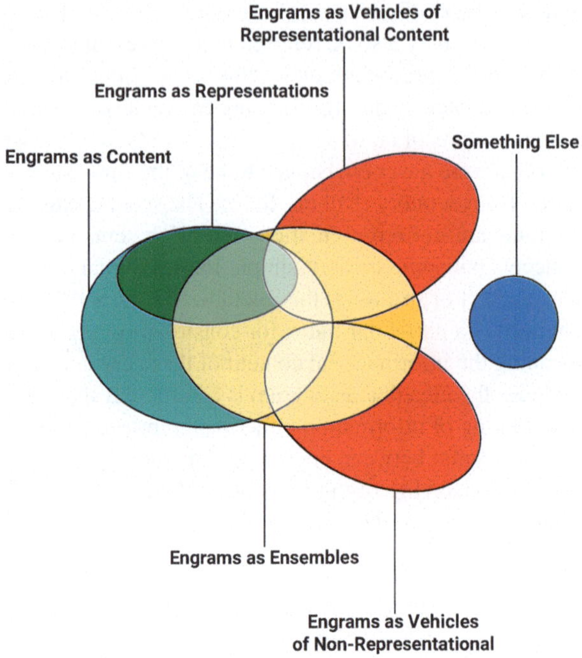


**Fig. 2** Different overlapping conceptions of what engrams should be

altering the pathways that neural activity is capable of taking. At this point in time, it may no longer be fruitful to look at the engram from separate vehicle and content perspectives or talk of content at all (Fig. 2). This is the view we are sympathetic to and wish to see clarified further in the context of engram biology.

## *Engrams-as-Ensembles*

The last category of engram views is that engrams have already been discovered, and they are unique sparse ensembles that we can partially label and manipulate. These cell-ensembles may or may not qualify as vehicles and could overlap the previous section (see Fig. 2). However, if these ensembles are not the vehicles of stored information, then these engrams do not fulfil the objective of explaining the biological basis of memory, and we will need to posit some other construct that fulfils this role. If these ensembles are indeed the vehicles, then we need to be able to bring to the foreground those most relevant changes and relegate those not relevant but still picked up by the tagging protocol. However, current experimental methods have not revealed to us what these most relevant changes are (are they molecular changes, connectivity changes, or something else?). Although adopting the engrams-are-ensembles definition may seem pragmatic, it gives the illusion that the research program of engram biology has finished answering its fundamental question and moved on to simply looking for applications to behavioral neuroscience. Although the existing techniques of engram biology are powerful and will undoubtedly lead to more applications where a true understanding of the biological basis of memory is not essential (e.g., in understanding the biology of forgetting, addiction, trauma, delusions, etc.), if we want the engram to explain the specificity of memory, then we must admit that satisfying Semon's criteria is only a beginning and is not alone sufficient (Ortega-de San Luis and Ryan 2022). In other words, we have not yet isolated the engram. Rather, these labelled ensembles may encapsulate, be adjacent to, or partially instantiate the "enduring physical/chemical changes" that constitute the engram.[19]

## *A Vehicle-Information Distinction*

These distinctions are controversial[20] and coarsely wielded, but entertaining these differences opens a space for considering other possibilities. If engrams do not implement storage of representations, what do they have, or what can they be said to function as? Applying the vehicle-content distinction dissociates different senses of the word engram, but this is not its only utility. Even if engrams do not have representational content, there may be utility in employing a modified vehicle-information distinction, whereby "information…persists via a memory trace, which is the vehicle for information preservation" (Mace 2021). There are two avenues leveraging this distinction that helps us consider. Just as we presume that not anything can count as a representational vehicle, we presume that not everything is competent to function as an engram. What would it take for a vehicle to function as preserving information and what criteria would exclude a biological structure from qualifying as competent (e.g., scar tissue preserves persistent changes despite cellular turnover)? This means further refining the four criteria for what it takes for something to count as an engram (see Josselyn et al. 2015). Second, this distinction makes clear an asymmetrical bias in engram biology research toward characterizing the vehicle of stored information. Discoveries about the vehicle, even figuring out the mechanism that makes something an engram, does not necessarily tell us anything about what information is stored. As will be discussed in section "Information and Coding", we have a poor understanding of what this latent information may be. At the very least, we can use the vehicle-information distinction to check ourselves, asking "do the questions addressed in this study focus solely on the vehicle, or do they also tell us exactly what information is stored?"

---

[19] Different IEG expression profiles do not always overlap, leading to the confusion that two disjunct labelled populations may both be called engrams. At best, differences in IEG function may be for implementing preservation of different types or layers of information that ultimately comprise the engram (Nambu et al. 2022).

[20] To distinguish between latent information and vehicles is not to separate them and treat them as separate entities but to look at the same phenomena from different perspectives. Nonetheless, it creates a dualistic divide that is not always desirable (Mitchell 2023).

This has precedent in biology, as a clear vehicle-information distinction is seen in genetics. A gene may be given a physicochemical description as a string of nucleic acids with certain molecular properties or an informational description as a symbolic sequence that has a particular meaning in the context of the cell. It can be interpreted through specific cellular processes into a functional peptide or nucleic acid. DNA/genes lead a "double life" as matter and semiotic symbol (Gazzaniga 2018; Pattee 1987/2012). In the case of genetics, an entirely new way of thinking about biology had to be created, that of thinking in terms of a symbolic code system: The eventual cracking of the DNA triplet codon to tRNA amino acid allocation proved that the mapping is arbitrary and therefore code-like, with redundancy. This shift required a level of abstraction that turned genetics from a form of molecular biology into a kind of biological information science.

Important for our discussion is that information does not fall out of physical knowledge of the vehicle. The physical form of two genes with the same base-pair length may be near-identical, but what the genes code for may differ radically (a plant enzyme or a SARS-CoV-2 cell surface protein). The meaning can be divergent even from the same information in cases of differential splicing and post-translational modifications. Whether we are looking at DNA with the fine resolution of an electron microscope or a chromosomal karyotype, we are not able to see this genetic information. We are characterizing its vehicle. When we look at a labelled cell ensemble in a brain tissue slice, we are looking only at one side of the engram. Distinguishing between engram and non- or other-engram cells is not looking at stored information. Double-blinded to a tissue slice of labelled cells, we could not say "this is a fear engram" or "this is an episodic memory from this place and time." Again, knowledge of the vehicle of information is not enough to tell us what this engram is for (Eichenbaum 2016). In the case of engram neuroscience, we may even have the resolution needed but cannot from the forest pick out (i) what the vehicle is or (ii) how it affords the capacity for the specificity of memory and therefore behavior. These are two different targets of explanation; analogously, explaining the structure of DNA is not the same as explaining the role of DNA in heredity and evolution. Regarding (i), identifying the vehicle is not a binary YES/NO achievement, since there are levels of precision. When Avery et al. (1944) and Hershey and Chase (1952) discovered DNA to be the molecule responsible for transmitting genetic information and not protein, they had identified the vehicle in a coarse sense. This did not tell us what the genetic information was or reveal which changes in the physical vehicle were most relevant for storing information. When we find that unique labelled cell-ensembles are capable of reinstating a behavior, we may (at best) have identified engrams in a similarly coarse sense. Additionally, we would not say that a gene stores representations of proteins, organelles, or organs. Rather there is a stable molecule (DNA) that constrains the creation of functional molecules in a highly reproducible way using an arbitrary code-like mapping. Genes don't store representations of proteins or organelles, but the triplet code allows for an emergent link between a chemical structure and a biological outcome, through a conserved symbolic relationship between tRNAs and amino acidsrole (Pattee 1987/2012) in constraining the creation of functional molecules in a highly reproducible way during the lifetime of an organism. This informational description, that of a one-dimensional sequence and code, is an abstract and imperfect model we create to understand the semiotic logic implemented in biological systems at a cellular level. Similarly, we are not suggesting that we try to "find the information in the brain" in a literal sense, rather we are arguing that engram biology needs a framework that makes a similarly abstract leap to help us understand whatever constructs serve as the biological basis of memory. We feel a vehicle-information distinction is important because it illustrates how engram biology is focused exclusively on the vehicle side of the engram and how there is a dark side of the moon to engram biology that is rarely discussed. With this in mind, we now focus on how we might understand this information.

## Information and Coding

### *The Necessity of a Theory of Information*

Talk of information-processing is ubiquitous throughout neuroscience, but what is meant by information is rarely specified. This same state of affairs applies in engram biology, where understanding the biological basis of memory is often framed in terms of discovering how information is stored plasticity allows information to be stored in the brain and manages to persist through time (Josselyn and Tonegawa 2020). But what exactly is this information? The use of information concepts in biology is controversial (Godfrey-Smith and Sterelny 2016). However, in engram biology, there are behavioral and other phenomena that may be difficult to explain without invoking the existence of stored information. For example, various forms of memory learned during the infantile amnesia window are not lost but rather inaccessible and can be artificially retrieved (Guskjolen et al. 2018; Power et al. 2023), and other apparently lost memories, including in Alzheimer's disease models, can also be recalled (Ryan et al. 2015; Roy et al. 2016; Perusini et al. 2017; Abdou et al. 2018; Poll et al. 2020; Lei et al. 2022; Bolsius et al. 2023; Autore et al. 2023; O'Leary et al. 2023). Accounting for these phenomena without positing latent information, or another concept that performs a related function, is problematic. To the extent that it is helpful to posit the existence of information and use information-concepts, a stronger theory of what this information is needed (Robins 2023). We therefore survey two different existing general conceptions of information and conclude that neither are completely satisfactory for use in engram biology. Instead, the techniques of engram biology mandate creating a new conception of information for use in neuroscience.

## Classical Information Theory and Coding Metaphors

The first general class of views on information are inflected by classical information theory and computer science (Stone 2022). Information is treated as something entering the brain from an external source (e.g., a stimulus) whereupon it is "encoded" (Roediger 2007) and is passed around between neurons, perhaps traveling in interspike intervals as a rate or combinatorial code (Gallistel 2017). Learning here is the acquisition and consolidation of relevant information extracted from "sense data," with stored information held in something like a weight matrix (Luo 2020). However, classical information theory is not a theory of information content but of quantifying how information can be optimally transmitted. It explicitly excludes accounting for the semantics, or meaning, of what is communicated (whether termed a message, signal, or information).[21] Neuroscientific theories about how information is coded, the computational and therefore metabolic benefits of using coding principles (Friston 2010), inherit this same problem. Although the details of such views can differ, as not everyone might agree that the brain computes using bits (Sterling and Laughlin 2015), or uses a read-write computing architecture (Gallistel and King 2009), they do not provide an account for what the information being coded is. This leaves much to be desired, as neuroscientists often care about the meaning of what brain signals are communicating.

The limitations of this heavily abused coding metaphor are summarized in Brette (2018). Principally, when we look for a correlation between any external variable, whether sensory or higher order task structure, and neural activity (e.g., $Ca^{2+}$ events or blood oxygenation level-dependent patterns as proximate measurements), we have assumed the existence of information, but only in our capacity as a third-person observer fixing a relation between two variables, one inside and one outside the brain. The brain does not have this third-person perspective, and so this cannot be assumed to be the information stored in engrams. Instead, for any future engram theory seeking inspiration from computer science, it may be more fruitful to direct attention to computer hardware engineering principles.

## Information Is Relational and Momentary

A second view, one that tries to accommodate the meaning of information, is that information is not a noun, property, or thing that can ever be localized but is inherently relational or contextual. There is nothing in DNA. Nucleic acids like DNA and RNA do not carry information by themselves. Outside the context of the cell, they are merely polynucleotides. There is only information for a given system in context.

---

[21] In telecommunications and computer science, this is not a problem, as it is humans who design the hardware and determine the arbitrary conventions for how variables should be encoded for other humans to decode them.

Neurotensin, interleukin-17, insulin, and serotonin only have biological meaning or unique consequences in the context of signaling and are differentially responded to by cells on the basis of the receptors they express. The meaning of a signal has physical constraints, such as molecular mass and hydrophilicity, but these constraints do not necessarily determine the meaning, which is fixed by evolutionary convention and may be modified by learning. Information in biology is not something that inheres to any physical pattern, it is information only in relation to or for some system in context (Brette 2018; Buzsáki 2006; Deacon 2021). This means that information can never be something held or stored but is being continually created. In the context of theories of memory, this may fit with a view of information only existing at the time of recall (Schacter and Addis 2007). Under this view, the function of the engram (if posited) is not to store information but to participate in the process of reproducing information meaningful for action in the relevant context. There is none of this information in the unique pattern of persistent changes that make an engram. Although viewing recall as a process of forming transiently existing information using engrams fits with Semon's account of ecphory, this view still leaves the biological basis of memory to be explained. This perspective needs much more refinement it is limited to live processes unfolding from moment to moment and therefore does not provide a clear way of thinking about information that is not live, dormant, or potential. This is orthogonal to the objectives of engram biology, which seeks to explain (i) a mechanism for how organisms appear to possess dormant information at all and (ii) explain how it is that engrams have the specific potential they have for eliciting a given behavior. In other words, how it is that the reaction of a specific ensemble, rather than any other, exerts such strong behavioral specificity.

## *What's Next?*

The computationalist view of information-processing and the information as relational and created view both have appealing characteristics, but neither are adequate for understanding engrams. It is unclear what form of theory would be adequate. This is not a question that would have been asked two decades ago, before the development of engram technology. It is a new problem. Nonetheless, there are several issues that any theory of engrams ought to be able to deal with: (i) arbitrariness, (ii) coding strength and universality, and (iii) reading.

### **Arbitrariness**

Although decoding the meaning of any purported information in the brain is not possible if an external variable is being used as a reference point by a third-person observer, this does not mean that internal codes within the nervous system, rather than between the nervous system and world, are not employed. A key characteristic of a code is the use of a symbolic or arbitrary relation, such as between a triplet codon "standing for" a particular amino acid (but unlike between the triplet codon of DNA and a triplet of RNA, which is a relation based on similarity or resemblance). Understanding the origins and degree of arbitrariness will be important to deciphering any internal code. For example, the taste or meaning of a chemical compound is not determined by the intrinsic qualities of the chemical compound but is fixed instead by anatomical wiring, namely, whether the neuron expressing the relevant taste receptor is part of a labeled line to sweet or bitter cortex (Peng et al. 2015; Wang et al. 2018). Thus, to decode the meaning requires knowledge of evolutionary and developmental history of the organism. If we cannot infer anything about the properties of a chemical compound at a sensory level based on the neuronal activity it elicits, "reading out" what mnemonic features a given ensemble or complex would be capable of reinstating (without activating the ensemble and looking at neural activity) is considerably more conceptually challenging. There is the possibility that the cells that comprise a given engram are entirely arbitrary. We want to be able to say why these cells, rather than those cells, should be involved. For example, if allocation is always stochastic, then there is no other reason why these cells, rather than those cells, should be involved. At least in the hippocampus, silencing the specific neurons previously labeled during conditioning while being exposed to the same conditioning paradigm simply results in a new and nonoverlapping engram to form (Tanaka et al. 2014). Therefore, an engram is constrained, but not determined by anatomy. However, if engrams are not determined by their anatomy, this does not mean that connectome-level analysis is not relevant. Instead, the kinds of invariants that may form the basis of an internal code may be structural topological invariants, rather than geometric properties of the relevant network.[22] The degree of arbitrariness may also vary regionally and phylogenetically.

## Coding Strength and Universality

Related to arbitrariness is how strong the internal code is, namely, what degree of determination over recall and ultimately behavior a given engram is capable of exerting. Cao and Rathkopf (2019) suggest that nervous systems learn their own internal modest codes. This means that there is no one-to-one correspondence between neural signal and function. The function is indeterminate or nondisjunct. For example, Dorst et al. (2023) demonstrated that the activation of the same ensemble while varying the size of the environment results in different behaviors. Thus, it is unlikely there will be a one-ensemble-one-behavior relationship. The causal contribution of a latent structure may vary as it confluences with the online contextual information of the retrieval environment. In the case of instincts, these may involve

[22] There is an important distinction to be made here between structural topology and topological properties of high dimensional data.

internal codes that do not have to be learned and may be at the stronger end of the spectrum (Ryan et al. 2021).

Closely tied to the concept of coding strength is coding universality, that is, whether there is any conservation in internal codes between individuals (e.g., how conserved one animal's internal code for a simple specific associative fear memory is relative to another's). If there is no selection pressure for there to be any level of similarity, then even if it were possible to "read out" any features based off a labelled ensemble in one animal, this knowledge would not be transferable. However, knowing the internal code may not be necessary to test this hypothesis. As discussed in section "A Vehicle-Information Distinction", before the "hereditary factors" that explained the specificity to heredity were known, transformation experiments were performed. RNA transfer experiments are one kind of transformation experiment and have been successful in Aplysia (Bédécarrats et al. 2018) but have yet to be tested in model organisms with larger nervous systems. Even if we do not think that RNA is the source of the "enduring changes" that comprise the engram in mammalian nervous systems, surrogate transformation experiments can still be performed. If a we hypothesize that the topology of the connectome is where the enduring changes are held (Ortega-de San Luis et al. 2023), then based on the topological characteristics of animals that have learned the behavior, experimental manipulation that implements that same topological invariance should change the behavior in a way that a topological theory predicts, but simple anatomy would not.

## Reading

Finally, any new theory of information must combat the reading problem. It is important to ask how these engram vehicles have their informational contents read. Here is where the standard coding metaphor can creep back in. Gallistel and King (2009) argue that the brain uses von-Neumann addressable read-write mechanisms where information is written into cells in the changes they undergo and that it is then read by a reader mechanism that knows this code. Others deem it too awkward to apply a classical computer architecture onto the brain. Godfrey-Smith (2014) draws the distinction between two kinds of biological information. Genetic information does indeed have a read mechanism. It has a defined coding-relationship for how to go from the latent informational structure, the gene, to a protein, but no symmetrical write mechanism. Instead mutations, duplications, and deletions occur during the course of cell division and evolution. Therefore, there is an evolve-read mechanism. The brain conversely has no clear reader(s), rather a write-only (Donohoe 2010) or rather write-activate system: "Evolved neural machinery has the function of introducing marks or traces into the brain as a result of experience, but these marks have useful effects on behaviour without being read" (Godfrey-Smith 2014). While there may be a kind of distributed reading that occurs, locating the processes doing this without turning them into readers is notoriously difficult (Cao 2012), although attempts to consider partial and multiple observers within a system have been proposed (Kolchinsky and Wolpert 2018). Again, should it prove less cumbersome to think without the concepts of information, storage, and codes, a robust theory of how memories can be created out of accumulated changes that explains the specificity in behavior from reactivating labelled cell ensembles is still required.

## Discussion

For all the methodological advances in recent years in characterizing the biological basis of memory, conceptual advances in understanding whether or not there are memory traces and what their precise functions are, or what they might look like have not kept pace. There is a multiplicity of unclear views, areas of consideration missing, and much yet to be articulated and clarified before we have a formal theory. On a practical level, how does the above analysis help? First and most straightforward, one can ask "which sense of engram have I been using? Multiple? None of those discussed?" This piece has not aimed to be prescriptive regarding which sense of the engram should be used. In response, a classification system or typology could be constructed of the different mechanisms or processes involved and the different types and levels of engrams. This approach would wrestle effort away from debating what "the engram" is toward the unique reasons competing classification systems use to unify similarities and explicate differences among trace-supporting mechanisms. Additionally, it prioritizes the need for work in a variety of model organisms before theorizing about what traces are to be found in humans. Even if the term engram becomes reserved exclusively for the results of a given form of learning, the theory must be able to differentiate other forms of information-storage mechanisms and explain how these trace types differ. Along with putative classification systems, we also want to see a vibrant future for the contemporary systems neuroscience of memory, where different modular frameworks with clearly staked hypotheses as to which functional units (such as cells, ensembles, and complexes) best explain observed informational specificity, will compete to falsify each other in a healthily adversarial way. Our analysis also enables us to better appraise the explanatory remit of experimental approaches. We can ask "what is this study aiming to address and what do these results inform? Do they further our understanding primarily of the implementational vehicle aspect of the engram, or do they inform the more abstract, informational questions of the engram?" We also emphasize the size and shape of the explanatory gap we face in engram biology and systems neuroscience generally and the nature of some of the questions we will need to engage with and move beyond in order to develop a deeper understanding. Finally, what is at stake here for engram biology and systems neuroscience? We believe it is our understanding of how the brain is capable of storing anything. Here, engram biology can pursue two paths. As a subdiscipline of neuroscience coming into its own, it can attempt to help resolve the existing conceptual confusions of the wider field by articulating an engram theory within this worldview using existing, poorly defined concepts. Alternatively, the techniques and ambitions of engram neuroscience can be used as an inside job, to challenge dominant metaphors and attempt articulating a wholly different theory of what there is to be found in the brain. Importantly, if after pursuing this approach "the engram" remains even more ethereally elusive or seems to have been transformed or assimilated into different biological processes leaving in its place an assemblage of new and distinct problems to be explained, we would consider this exciting progress.

**Acknowledgments** The authors would like to thank Caitlin Mace, Clara Ortega-de San Luis, and Paul Conway for their helpful comments on a draft of this manuscript and the members of the Ryan Lab and the Trinity College Dublin Representations: Past, Present, and Future group for the useful discussions.

## References

Abdou K, Shehata M, Choko K, Nishizono H, Matsuo M, Muramatsu SI, Inokuchi K (2018) Synapse-specific representation of the identity of overlapping memory engrams. Science 360(6394):1227–1231. https://doi.org/10.1126/science.aat3810

Autore L, O'Leary JD, Ortega-de San Luis C, Ryan TJ (2023) Adaptive expression of engrams by retroactive interference. Cell Rep 42(8):112999. https://doi.org/10.1016/j.celrep.2023.112999

Avery OT, MacLeod CM, McCarty M (1944) Studies on the chemical nature of the substance inducing transformation of pneumococcal types: induction of transformation by a deoxyribonucleic acid fraction isolated from pneumococcus type III. J Exp Med 79(2):137–158. https://doi.org/10.1084/jem.79.2.137

Barack DL, Krakauer JW (2021) Two views on the cognitive brain. Nat Rev Neurosci 22:359–371. https://doi.org/10.1038/s41583-021-00448-6

Barron HC, Vogels TP, Behrens TE, Ramaswami M (2017) Inhibitory engrams in perception and memory. Proc Natl Acad Sci USA 114(26):6666–6674. https://doi.org/10.1073/pnas.1701812114

Bédécarrats A, Chen S, Pearce K, Cai D, Glanzman DL (2018) RNA from trained Aplysia can induce an epigenetic engram for long-term sensitization in untrained Aplysia. eNeuro 5(3):ENEURO.0038-18.2018. https://doi.org/10.1523/ENEURO.0038-18.2018

Bolsius YG, Heckman PRA, Paraciani C, Wilhelm S, Raven F, Meijer EL, Kas MJH, Ramirez S, Meerlo P, Havekes R (2023) Recovering object-location memories after sleep deprivation-induced amnesia. Curr Biol 33(2):298–308.e5. https://doi.org/10.1016/j.cub.2022.12.006

Brette R (2018) Is coding a relevant metaphor for the brain? Behav Brain Sci 42:e215. https://doi.org/10.1017/S0140525X19000049

Burnston DC (2021) Contents, vehicles, and complex data analysis in neuroscience. Synthese 199(1–2):1617–1639

Buzsáki G (2006) Rhythms of the brain. Oxford University Press

Buzsáki G (2010) Neural syntax: cell assemblies, synapsembles, and readers. Neuron 68(3):362–385. https://doi.org/10.1016/j.neuron.2010.09.023

Campbell RR, Wood MA (2019) How the epigenome integrates information and reshapes the synapse. Nat Rev Neurosci 20(3):133–147. https://doi.org/10.1038/s41583-019-0121-9

Cao R (2012) A teleosemantic approach to information in the brain. Biol Philos 27:49–71

Cao R (2022) Putting representations to use. Synthese 200(2):151

Cao R, Rathkopf C (2019) Modest and immodest neural codes: can there be modest codes? Behav Brain Sci 42:e221. https://doi.org/10.1017/S0140525X19001420

Chang L, Tsao DY (2017) The code for facial identity in the primate brain. Cell 169(6):1013–1028.e14. https://doi.org/10.1016/j.cell.2017.05.011

Chen S, Cai D, Pearce K, Sun PY-W, Roberts AC, Glanzman DL (2014) Reinstatement of long-term memory following erasure of its behavioral and synaptic expression in Aplysia. eLife 3:e03896. https://doi.org/10.7554/eLife.03896

Chklovskii DB, Mel BW, Svoboda K (2004) Cortical rewiring and information storage. Nature 431(7010):782–788. https://doi.org/10.1038/nature03012

Cisek P (2019) Resynthesizing behavior through phylogenetic refinement. Atten Percept Psychophys 81(7):2265–2287. https://doi.org/10.3758/s13414-019-01760-1

Clark A, Chalmers D (1998) The extended mind. Analysis 58(1):7–19

Colaço D, Najenson J (2023) Where memory resides: is there a rivalry between molecular and synaptic models of memory? Philos Sci:1–11. https://doi.org/10.1017/psa.2023.126

Craver C (2007) Explaining the brain. Oxford University Press

Deacon TW (2021) How molecules became signs. Biosemiotics 14:537–559. https://doi.org/10.1007/s12304-021-09453-9

Dennet DC (1978) Toward a cognitive theory of consciousness. In: Savage CW (ed) Minnesota studies in the philosophy of science, volume 9: perception and cognitions, issues in the foundation of psychology. University of Minnesota Press, Minneapolis, pp 201–228

Donald M (1991) Origins of the modern mind. Harvard University Press

Donohoe J (2010) Man as machine. Review of memory and the computational brain, by C. R. Gallistel and A. P. King. Behav Philos 38:83–101

Dorst K, Senne RA, Diep AH, de Boer AR, Suthard RL, Leblanc H, Ruesch EA, Skelton S, McKissick OP, Bladon JH, Ramirez S (2023) Hippocampal engrams generate flexible behavioral responses and brain-wide network states. bioRxiv. 2023.02.24.529744. https://doi.org/10.1101/2023.02.24.529744

Dudai Y (2007) Memory: it's all about representations. In: Roediger HL, Dudai Y, Fitzpatrick SM (eds) Science of memory: concepts. Oxford University Press, pp 17–21

Eichenbaum H (2016) Still searching for the engram. Learn Behav 44(3):209–222. https://doi.org/10.3758/s13420-016-0218-1

Friston K (2010) The free-energy principle: a unified brain theory? Nat Rev Neurosci 11(2):127–138. https://doi.org/10.1038/nrn2787

Gallistel CR (2017) The coding question. Trends Cogn Sci 21(7):498–508. https://doi.org/10.1016/j.tics.2017.04.012

Gallistel CR, King AP (2009) Memory and the computational brain: why cognitive science will transform neuroscience. Wiley-Blackwell

Garner AR, Rowland DC, Hwang SY, Baumgaertel K, Roth BL, Kentros C, Mayford M (2012) Generation of a synthetic memory trace. Science 335(6075):1513–1516. https://doi.org/10.1126/science.1214985

Gazzaniga M (2018) The consciousness instinct. Farrar, Straus and Giroux

Gershman SJ (2023) The molecular memory code and synaptic plasticity: a synthesis. Biosystems 224:104825

Godfrey-Smith P (2014) Sender-receiver systems within and between organisms. Philos Sci 81(5):866–878. https://doi.org/10.1086/677686

Godfrey-Smith P, Sterelny K (2016) Biological information. In: Zalta EN (ed) The stanford encyclopedia of philosophy. Stanford University (Summer 2016 ed) https://plato.stanford.edu/archives/sum2016/entries/information-biological/

Goode TD, Tanaka KZ, Sahay A, McHugh TJ (2020) An integrated index: engrams, place cells, and hippocampal memory. Neuron 107(5):805–820. https://doi.org/10.1016/j.neuron.2020.07.011

Gould S, Vrba E (1982) Exaptation—a missing term in the science of form. Paleobiology 8(1):4–15. https://doi.org/10.1017/S0094837300004310

Guskjolen A, Kenney JW, de la Parra J, Yeung BA, Josselyn SA, Frankland PW (2018) Recovery of "Lost" Infant Memories in Mice. Current biology 28(14):2283–2290.e3. https://doi.org/10.1016/j.cub.2018.05.059

Guskjolen A, Cembrowski MS (2023) Engram neurons: encoding, consolidation, retrieval, and forgetting of memory. Mol Psychiatry 28:3207–3219. https://doi.org/10.1038/s41380-023-02137-5
Hardt O, Sossin WS (2020) Terminological and epistemological issues in current memory research. Front Mol Neurosci 12:336. https://doi.org/10.3389/fnmol.2019.00336
Hershey A, Chase M (1952) Independent functions of viral protein and nucleic acid in growth of bacteriophage. J Gen Physiol 36(1):39–56. https://doi.org/10.1085/jgp.36.1.39
Hoffmeyer J (2008) Semiotic scaffolding of living systems. In: Barbieri M (ed) Introduction to biosemiotics. Springer, Dordrecht
Hurley SL (1998) Consciousness in action. Harvard University Press
Jonas E, Kording KP (2017) Could a neuroscientist understand a microprocessor? PLoS Comput Biol 13(1):e1005268. https://doi.org/10.1371/journal.pcbi.1005268
Josselyn SA, Tonegawa S (2020) Memory engrams: recalling the past and imagining the future. Science 367(6473):eaaw4325. https://doi.org/10.1126/science.aaw4325
Josselyn SA, Köhler S, Frankland PW (2015) Finding the engram. Nat Rev Neurosci 16(9):521–534. https://doi.org/10.1038/nrn4000
Kandel ER (2001) The molecular biology of memory storage: a dialogue between genes and synapses. Science 294(5544):1030–1038. https://doi.org/10.1126/science.1067020
Kauffman S (2019) A world beyond physics. Oxford University Press
Kol A, Goshen I (2021) The memory orchestra: the role of astrocytes and oligodendrocytes in parallel to neurons. Curr Opin Neurobiol 67:131–137. https://doi.org/10.1016/j.conb.2020.10.022
Kolchinsky A, Wolpert DH (2018) Semantic information, autonomous agency and non-equilibrium statistical physics. Interface Focus 8(6):20180041. https://doi.org/10.1098/rsfs.2018.0041
Koolschijn RS, Emir UE, Pantelides AC, Nili H, Behrens TEJ, Barron HC (2019) The hippocampus and neocortical inhibitory engrams protect against memory interference. Neuron 101(3):528–541.e6. https://doi.org/10.1016/j.neuron.2018.11.042
Krakauer JW (2022) Representation in cognitive science by Nicholas Shea: but is it thinking? The philosophy of representation meets systems neuroscience. Stud Hist Phil Sci 92:267–269. https://doi.org/10.1016/j.shpsa.2021.05.014
Krakauer JW, Ghazanfar AA, Gomez-Marin A, MacIver MA, Poeppel D (2017) Neuroscience needs behavior: correcting a reductionist bias. Neuron 93(3):480–490. https://doi.org/10.1016/j.neuron.2016.12.041
Langdon C, Genkin M, Engel TA (2023) A unifying perspective on neural manifolds and circuits for cognition. Nat Rev Neurosci 24(6):363–377. https://doi.org/10.1038/s41583-023-00693-x
Laplane L (2016) Cancer stem cells: philosophy and therapies. Harvard University Press
Lei B, Lv L, Hu S, Tang Y, Zhong Y (2022) Social experiences switch states of memory engrams through regulating hippocampal Rac1 activity. Proc Natl Acad Sci USA 119(15):e2116844119. https://doi.org/10.1073/pnas.2116844119
Lewontin RC (1983) The organism as the subject and object of evolution. In: Levins R, Lewontin RC (eds) The dialectical biologist. Harvard University Press
Liu X, Ramirez S, Pang PT, Puryear CB, Govindarajan A, Deisseroth K, Tonegawa S (2012) Optogenetic stimulation of a hippocampal engram activates fear memory recall. Nature 484(7394):381–385. https://doi.org/10.1038/nature11028
Luo L (2020) Principles of neurobiology, 2nd edn. Taylor & Francis
MacDonald CJ, Lepage KQ, Eden UT, Eichenbaum H (2011) Hippocampal "time cells" bridge the gap in memory for discontiguous events. Neuron 71(4):737–749. https://doi.org/10.1016/j.neuron.2011.07.012
Mace CB (2021) Casting light on the search for engrams: on the reductionism-mechanism debate (Order No. 28644486). Available from ProQuest Dissertations & Theses A&I; Publicly Available Content Database (2597766427)
Martin SJ, Grimwood PD, Morris RG (2000) Synaptic plasticity and memory: an evaluation of the hypothesis. Annu Rev Neurosci 23:649–711. https://doi.org/10.1146/annurev.neuro.23.1.649

Mitchell K (2023) The origins of meaning – from pragmatic control signals to semantic representations. PsyArXiv:1–16. https://doi.org/10.31234/osf.io/dfkrv

Moskovitch M (2007) Memory: why the engram is elusive. In: Roediger HL, Dudai Y, Fitzpatrick SM (eds) Science of memory: concepts. Oxford University Press, pp 17–21

Najenson J (2021) What have we learned about the engram? Synthese 199:9581–9601. https://doi.org/10.1007/s11229-021-03216-2

Nambu MF, Lin YJ, Reuschenbach J, Tanaka KZ (2022) What does engram encode?: heterogeneous memory engrams for different aspects of experience. Curr Opin Neurobiol 75:102568. https://doi.org/10.1016/j.conb.2022.102568

O'Leary J, Bruckner R, Autore L, Ryan TJ (2023) Natural forgetting reversibly modulates engram expression. eLife 12:RP92860. https://doi.org/10.7554/eLife.92860.1

Ortega-de San Luis C, Ryan TJ (2022) Understanding the physical basis of memory: molecular mechanisms of the engram. J Biol Chem 298(5):101866. https://doi.org/10.1016/j.jbc.2022.101866

Ortega-de San Luis C, Pezzoli M, Urrieta E, Ryan TJ (2023) Engram cell connectivity as a mechanism for information encoding and memory function. Curr Biol 33:5368. https://doi.org/10.1016/j.cub.2023.10.074

Pattee HH (2012) Cell psychology: an evolutionary approach to the symbol-matter problem. In: Pattee HH, Rączaszek-Leonardi J (eds) Laws, language and life: Howard Pattee's classic papers on the physics of symbols with contemporary commentary. Springer, Dordrecht. (Original work published 1987)

Peng Y, Gillis-Smith S, Jin H, Tränkner D, Ryba NJ, Zuker CS (2015) Sweet and bitter taste in the brain of awake behaving animals. Nature 527(7579):512–515. https://doi.org/10.1038/nature15763

Perusini JN, Cajigas SA, Cohensedgh O, Lim SC, Pavlova IP, Donaldson ZR, Denny CA (2017) Optogenetic stimulation of dentate gyrus engrams restores memory in Alzheimer's disease mice. Hippocampus 27(10):1110–1122. https://doi.org/10.1002/hipo.22756

Poeppel D, Idsardi W (2022) We don't know how the brain stores anything, let alone words. Trends Cogn Sci 26(12):1054–1055. https://doi.org/10.1016/j.tics.2022.08.010

Poll S, Mittag M, Musacchio F, Justus LC, Giovannetti EA, Steffen J, Wagner J, Zohren L, Schoch S, Schmidt B, Jackson WS, Ehninger D, Fuhrmann M (2020) Memory trace interference impairs recall in a mouse model of Alzheimer's disease. Nat Neurosci 23(8):952–958

Power S, Stewart E, Zielke L, Byrne E, Ortega-de San Luis L, Lynch L, Ryan T (2023) Immune activation state modulates infant engram expression across development. Sci Adv 9:eadg9921. https://doi.org/10.1126/sciadv.adg9921

Queenan BN, Ryan TJ, Gazzaniga MS, Gallistel CR (2017) On the research of time past: the hunt for the substrate of memory. Ann N Y Acad Sci 1396(1):108–125. https://doi.org/10.1111/nyas.13348

Ramsey WM (2007) Representation reconsidered. Cambridge University Press

Reijmers LG, Perkins BL, Matsuo N, Mayford M (2007) Localization of a stable neural correlate of associative memory. Science 317(5842):1230–1233. https://doi.org/10.1126/science.1143839

Robins SK (2018) Memory and optogenetic intervention: separating the engram from the ecphory. Philos Sci 85(5):1078–1089

Robins SK (2023) The 21st century engram. Wiley Interdiscip Rev Cogn Sci 14:e1653. https://doi.org/10.1002/wcs.1653

Roediger HL (2007) Encoding. In: Roediger HL, Dudai Y, Fitzpatrick SM (eds) Science of memory: concepts. Oxford University Press, pp 17–21

Ross LN (2021) Causal concepts in biology: how pathways differ from mechanisms and why it matters. Br J Philos Sci 72(1):131–158

Roy DS, Arons A, Mitchell TI, Pignatelli M, Ryan TJ, Tonegawa S (2016) Memory retrieval by activating engram cells in mouse models of early Alzheimer's disease. Nature 531(7595):508–512. https://doi.org/10.1038/nature17172

Roy DS, Park YG, Kim ME, Zhang Y, Ogawa SK, DiNapoli N, Gu X, Cho JH, Choi H, Kamentsky L, Martin J, Mosto O, Aida T, Chung K, Tonegawa S (2022) Brain-wide mapping reveals that engrams for a single memory are distributed across multiple brain regions. Nat Commun 13(1):1799. https://doi.org/10.1038/s41467-022-29384-4

Rule ME, O'Leary T, Harvey CD (2019) Causes and consequences of representational drift. Curr Opin Neurobiol 58:141–147. https://doi.org/10.1016/j.conb.2019.08.005

Ryan TJ, Frankland PW (2022) Forgetting as a form of adaptive engram cell plasticity. Nat Rev Neurosci 23(3):173–186. https://doi.org/10.1038/s41583-021-00548-3

Ryan TJ, Roy DS, Pignatelli M, Arons A, Tonegawa S (2015) Memory. Engram cells retain memory under retrograde amnesia. Science 348(6238):1007–1013. https://doi.org/10.1126/science.aaa5542

Ryan TJ, Ortega-de San Luis C, Pezzoli M, Sen S (2021) Engram cell connectivity: an evolving substrate for information storage. Curr Opin Neurobiol 67:215–225. https://doi.org/10.1016/j.conb.2021.01.006

Ryle G (1949) The concept of mind: 60th anniversary edition. Hutchinson & Co

Sacktor TC (2011) How does PKMζ maintain long-term memory? Nat Rev Neurosci 12(1):9–15. https://doi.org/10.1038/nrn2949

Saigusa T et al (2008) Amoebae anticipate periodic events. Phys Rev Lett 100(1):18101. https://doi.org/10.1103/PhysRevLett.100.018101

Schacter D (2001) Forgotten ideas, neglected pioneers: Richard Semon and the story of memory. Psychology Press

Schacter DL, Addis DR (2007) The cognitive neuroscience of constructive memory: remembering the past and imagining the future. Philos Trans R Soc Lond Ser B Biol Sci 362(1481):773–786. https://doi.org/10.1098/rstb.2007.2087

Semon R (1921) The mneme. George Allen & Unwin. (Original work published 1904)

Semon R (1923) Mnemic psychology (trans: Duffy B). George Allen & Unwin. (Original work published 1909)

Shea N (2018) Representation in cognitive science. Oxford University Press

Shpokayte M, McKissick O, Guan X, Yuan B, Rahsepar B, Fernandez FR, Ruesch E, Grella SL, White JA, Liu XS, Ramirez S (2022) Hippocampal cells segregate positive and negative engrams. Commun Biol 5(1):1009. https://doi.org/10.1038/s42003-022-03906-8

Simon HA (2002) Near decomposability and the speed of evolution. Ind Corp Chang 11(3):587–600

Sterling L, Laughlin S (2015) Principles of neural design. MIT Press

Stone JV (2022) Information theory: a tutorial introduction, 2nd edn. Sebtel Press

Sun X, Bernstein MJ, Meng M, Rao S, Sørensen AT, Yao L, Zhang X, Anikeeva PO, Lin Y (2020) Functionally distinct neuronal ensembles within the memory engram. Cell 181(2):410–423.e17. https://doi.org/10.1016/j.cell.2020.02.055

Tanaka KZ, Pevzner A, Hamidi AB, Nakazawa Y, Graham J, Wiltgen BJ (2014) Cortical representations are reinstated by the hippocampus during memory retrieval. Neuron 84(2):347–354. https://doi.org/10.1016/j.neuron.2014.09.037

Tinbergen N (1963) On aims and methods of ethology. Z Tierpsychol 20:410–433

Tonegawa S, Liu X, Ramirez S, Redondo R (2015a) Memory engram cells have come of age. Neuron 87(5):918–931. https://doi.org/10.1016/j.neuron.2015.08.002

Tonegawa S, Pignatelli M, Roy DS, Ryan TJ (2015b) Memory engram storage and retrieval. Curr Opin Neurobiol 35:101–109. https://doi.org/10.1016/j.conb.2015.07.009

Tsien RY (2013) Very long-term memories may be stored in the pattern of holes in the perineuronal net. Proc Natl Acad Sci USA 110(30):12456–12461. https://doi.org/10.1073/pnas.1310158110

Vetere G, Kenney JW, Tran LM, Xia F, Steadman PE, Parkinson J, Josselyn SA, Frankland PW (2017) Chemogenetic interrogation of a brain-wide fear memory network in mice. Neuron 94(2):363–374.e4. https://doi.org/10.1016/j.neuron.2017.03.037

Wang L, Gillis-Smith S, Peng Y, Zhang J, Chen X, Salzman CD, Ryba NJP, Zuker CS (2018) The coding of valence and identity in the mammalian taste system. Nature 558(7708):127–131. https://doi.org/10.1038/s41586-018-0165-4

Wang C, Yue H, Hu Z, Shen Y, Ma J, Li J, Wang XD, Wang L, Sun B, Shi P, Wang L, Gu Y (2020) Microglia mediate forgetting via complement-dependent synaptic elimination. Science 367(6478):688–694. https://doi.org/10.1126/science.aaz2288

Wheeler AL, Teixeira CM, Wang AH, Xiong X, Kovacevic N, Lerch JP, McIntosh AR, Parkinson J, Frankland PW (2013) Identification of a functional connectome for long-term fear memory in mice. PLoS Comput Biol 9(1):e1002853. https://doi.org/10.1371/journal.pcbi.1002853

# Index




J. Gräff, S. Ramirez (eds.), *Engrams*, Advances in Neurobiology 38,
https://doi.org/10.1007/978-3-031-62983-9

# GPSR Compliance

*The European Union's (EU) General Product Safety Regulation (GPSR) is a set of rules that requires consumer products to be safe and our obligations to ensure this.*

*If you have any concerns about our products, you can contact us on ProductSafety@springernature.com*

In case Publisher is established outside the EU, the EU authorized representative is:

Springer Nature Customer Service Center GmbH
Europaplatz 3
69115 Heidelberg, Germany

The manufacturer's authorised representative in the EU is Springer
Nature Customer Service Centre GmbH, Europaplatz 3, 69115 Heidelberg,
Germany. If you have any concerns regarding our products, please
contact ProductSafety@springernature.com

Printed and bound by CPI Group (UK) Ltd, Croydon, CR0 4YY
30/09/2026
02225598-0007